新工程制图
英汉-汉英
两用词典

NEW ENGLISH-CHINESE & CHINESE-ENGLISH
DUAL PURPOSE DICTIONARY FOR
ENGINEERING DRAWINGS

马中骥　李占利◎主编

中国纺织出版社有限公司

图书在版编目（CIP）数据

新工程制图英汉-汉英两用词典／马中骥，李占利主编．--北京：中国纺织出版社有限公司，2022.9
ISBN 978-7-5180-9680-0

Ⅰ.①新… Ⅱ.①马… ②李… Ⅲ.①工程制图－对照词典－英、汉 Ⅳ.①TB23-61

中国版本图书馆CIP数据核字（2022）第125028号

责任编辑：郭　婷　　责任校对：高　涵　　责任印制：储志伟

中国纺织出版社有限公司出版发行
地址：北京市朝阳区百子湾东里 A407 号楼　邮政编码：100124
销售电话：010—67004422　传真：010—87155801
http://www.c-textilep.com
中国纺织出版社天猫旗舰店
官方微博 http://weibo.com/2119887771
天津千鹤文化传播有限公司印刷　各地新华书店经销
2022年9月第1版第1次印刷
开本：880×1230　1/32　印张：9.75
字数：260千字　定价：55.00元

凡购本书，如有缺页、倒页、脱页，由本社图书营销中心调换

新工程制图英汉-汉英两用词典

编 写 人 员

主 编　马中骥　李占利

编 者　李占利　何 璞　马 艺
　　　　秋兴国　尉朝闻　马中骥

审 定　马中骥

新工程制图英汉-汉英两用词典

内 容 简 介

本词典是一部新编多专业工程制图英、汉互译工具书。它不仅是一部词汇，更是一部对有关重要单词、词语，尤其是标准名词（均标了＊号）、易混淆词、新词、难词，都给了精要注释，可提供读者查阅研究的词典。

词典中收编了工程界各类专业的图示、图解、图算、图表以及计算机辅助设计与绘图和标准化词语，也收录了少量常用的有关词语，英文词语（包括缩写词）约12000条，汉英对照词语近6000条，英－汉例句、汉－英例句近200条；所有例句均选自英、美工程制图书籍原著；附录中的"中国内地与港澳台工程制图名词术语"对照词条，方便于港澳台地区，以及海外读者在工程技术方面的交流，"工程制图英文常用缩写词一览"，方便于工程设计、绘图、审图以及工程制图教学中批改作业的工作。

本词典可供各类工科专业大专院校师生，科研、生产界的广大技术人员、翻译人员使用；特别对高等学校双语教学人员和国内外投入"一带一路"基本建设的广大工程技术人员是一部不可缺少的工具。

前　言

自从中国加入 WTO 以后，我国已成为全球最开放的市场之一、居世界前列的贸易大国，服务贸易开放部门已达 100 多个，与世界多国在经济、科技、工程、文化各个方面的双向交流日益广泛深入。越来越多的中国人走向世界，去学习，去工作，去支援工程建设；越来越多的外国人来华，来学习，来工作，来投入我们的工程建设。双向交流的语言依然以英语为主，在工程界，当然还是以工程界的语言文字——工程图样为唯一媒介。尤其近年来，"一带一路"建设蓬勃发展，我国与世界各地的交流更加广泛深入，工程技术和制图方面的词汇迅猛增加。为了更积极主动地参与到经济与技术的全球化中，教育工作者有必要紧跟形势，广征博引，及时广泛收集工程制图方面的有关资料，编制词典，以满足广大读者在双向交流中的需求。

作为一部工程制图词典，尤其是"双语"词典，重点应以能集中反映工程图学名词术语的国际经典工程图学教材为主要依据去搜集和提炼。本词典编者重点查阅了美国具有一百多年传统的优秀教材：Thomas E. French 所著的大学教本《工程画》（*A Manual of Engineering Drawing: For Students and Draftsmen*）；翻阅了美国具有广泛影响且被工科院校普遍使用的 Paré 教授等所著的《画法几何：公制》（*Descriptive Geometry-Metric*）。近年来，我们又重点研究了在英国具有影响力、普遍使用、多次改版印刷，也是中国现在广泛流行的以 Colin H. Simmons 为主要作者所著的《工程制图教本》（*Manual of Engineering Drawing*），同时参考了多部国内外最新出版的英汉词典、科技专业词汇、工程制图书籍，特别是全国科学技术名词审定委员会公布的标准名词。本词典从这些巨著中吸取了丰富营养。

本书的特色定位于以下四点：

（1）全书分三部分编写：英－汉部分（按字母排序），汉－英部分（按拼音排序），附录（共10个，收录了教学、科研、生产、翻译等工作中常用的重要参考资料）。

（2）在英－汉部分，对英文单词加注了国际音标，以应双语教学之需。

（3）英－汉部分，以工程制图的英文单词为龙头，读者根据工程制图的英文单词，可查到有关该单词的常用词条或在经典著作中的典型应用例句。

（4）在汉－英部分，对重点名词，如标准名词（均标了＊号）、易混淆词、新词、难词，进行了精要注释，使读者有较明确的概念。

本书的学术价值还在于，从其汇集的工程界各行业工程制图的名词、表达方法和内容中，可以看到在各行业中，同一名词所表达的不同对象，或同一对象所用的不同名词，从而启示进一步研究各行业工程制图术语标准化的统一表述方法，以减少人们认识和阅读上的混乱和麻烦。

本书是为多专业工程技术人员、大专院校师生在教学、科研、工程建设与翻译工作中准备的常用工具，也是高等院校双语教学的必备工具。

本书承蒙中国纺织出版社有限公司热情支持出版，在此特表衷心感谢。

本书由西安交通大学工学博士、西安科技大学李占利教授，西安交通大学博士、西安交通大学何璞助理教授，中国科学院大连化学物理研究所博士、陕西师范大学马艺副教授，西安科技大学秋兴国教授、尉朝闻教授、马中骥教授编撰；由全国技术制图标准化技术委员会一届委员、二届顾问，全国图形符号标准化技术委员会一、二、三届委员马中骥教授和西安科技大学计算机科学与技术学院李占利教授主编；马中骥教授审定。

本书虽精工编撰，但由于水平所限，缺点、错误在所难免，敬希读者批评指正。

编者

2022 年 7 月 28 日

目 录

凡例 ··· I

A部 英－汉（ENGLISH-CHINESE）
A部正文 ··· 3

B部 汉－英（CHINESE-ENGLISH）
B部汉语词条音节索引 ··· 167

B部正文 ··· 179

附录

APPENDIX Ⅰ（附录一） 工程制图英文常用缩写词一览 ·· 267

APPENDIX Ⅱ（附录二） 中国内地与港澳台地区工程制图部分术语对照 ······················· 270

APPENDIX Ⅲ（附录三） 常见国际、区域及主要国家标准代号 ·································· 282

APPENDIX Ⅳ（附录四） 国际标准化组织（ISO）成员团体代号 ···························· 284

APPENDIX Ⅴ（附录五） ISO及有关TC、SC简介 ············ 286

APPENDIX Ⅵ（附录六） IEC及有关TC、SC简介 ············ 287

APPENDIX Ⅶ（附录七） 中华人民共和国标准代号 ··········· 288

APPENDIX Ⅷ（附录八）	工程制图 ISO 国际标准号及中国相对应的 GB 国家标准号	289
APPENDIX Ⅸ（附录九）	常见简化汉字、汉字简化偏旁及其繁体字对照	292
APPENDIX Ⅹ（附录十）	希腊字母	296

凡　例

一、本词典英－汉部的单词和缩写词均按英文字母顺序排列，由单词组成的组合词按第一词的字母顺序分条排在该单词后面，并用代字符"～"代替该单词；凡不以该单词开头的组合词，排在以该单词开头的组合词之后。各条组合词用"；"隔开。例：

　　curve 曲线，曲线板；弯曲；弄弯；～ of order 2 二阶曲线，点素二次曲线；～ pen 曲线笔；～ plotter 绘图器；bipartite cubic ～双枝三次曲线；double continuous ～正弦曲线，S 形曲线；faired ～展平曲线；French ～曲线板

这样，读者根据工程制图的英文单词，就可查到有关该单词的常用词条或应用例句。

二、缩写字（或词），其大小写和标点符号的运用，往往因作者的使用习惯而异，如缩写词 AC，就有 AC；Ac；A. C.；A. c.；AC.；ac；a. c.；ac. 等写法，在查阅时，不管其形式，只要字母及顺序与所查找的相同，且其所代表的内容适用于具体的上下文即可采用。

三、与同一个英文词对应的几个中文词，近义的用"，"隔开，远义的用"；"隔开。

四、英文单词的同义词或缩写词与原词用"，"隔开，或用"（ ）"括住。

五、"（ ）"还用于：

1. 英、汉词语中的注释部分。

2. 可省略的字母或汉字，如 concentric（al）同（圆）心的，同轴

的，公共的。

3.词后的（），内写"电"者为电气术语，内写"船"者为船舶术语，以此类推。

六、本词典汉－英部的词语均按词语首字汉语拼音音序排列，当词语首字相同时，按第二字的汉语拼音音序排列，以此类推。

七、中文名词和英文词条左上角带有"*"者，为全国科学技术名词审定委员会审定公布的标准名词。

A 部

英-汉
(ENGLISH-CHINESE)

A

A (academy) 学会, 学院; (addendum) 齿顶高; (area) 面积; (association) 协会, 团体

abac /'æbək/ 诺谟图, 列线图, 线解图, 诺谟曲线; 计算图表; 数字图解; 坐标网

abas /'a:bəs/ 诺谟图, 列线图, 曲线图

abat-jour /əb'ʒu:/ 天窗, 亮窗; 入窗日光反射器

abat-vent 固定百叶窗, 挡风斜板; 烟囱帽; 坡屋顶

abb. (abbr., abbrev., abbreviation) 省略; 缩写; 略语

abbreviate /ə'bri:vieit/ 简化, 省略, 简略; 缩写, 简写; 将……缩短

ABC-Army-STD 美英加陆军规格

ABC-Navy-STD 美英加海军规格

ability /ə'biləti/ 性能; 能力; 技能; interchange $\sim$ 互换性

abort /ə'bɔ:t/ 异常结束

abrase /ə'breiz/ 擦去, 刮掉

abrasion /ə'breiʒn/ 磨光, 磨损, 研磨; 擦掉, 刮掉; surface $\sim$ 表面磨蚀

abscess /'æbses/ 缩孔, (铸造的)砂眼

abscissa /æb'sisə/ (abscissac, abscissas) 横线, 横坐标

abscission /æb'siʒən/ 切断

Abs. E. (absolute error) 绝对误差

abutment /ə'bʌtmənt/ 座, 支承; 邻接; 接合点, 接合器

AC (alternating current) 交流(电); (automatic computer) 自动计算机

access /'ækses/ 访问, 存储

accessory /ək'sesəri/ 辅助; 附件

accompany /ə'kʌmpəni/ 伴同, 伴随

accomplish /ə'kʌmpliʃ/ 完成

accordingly /ə'kɔ:diŋli/ 相应地; 照着办; 因此, 从而

accuracy /'ækjərəsi/ 精度, 精确度; 准确度; 精密性; 精密; 准确; geometric $\sim$ 几何形状准确性; graphical $\sim$ 制图精确度; to (within) the $\sim$ of 精度为……

ACE (ace) (automatic computing equipment) 自动计算设备; (automatic computing engine) 自动计算机

achieve /ə'tʃi:v/ 完成; 达到; 得到

aclinal /ə'klainəl/ 水平的, 无倾角的

aclinic /ə'klinik/ 水平的, 无倾角的

acme /'ækmi/ 顶点, 极点; 极度

acnode /'æknəud/ 顶点; 极点, 孤立点

acorn /'eikɔ:n/ 橡树子; 橡实形管

across /ə'krɔs/ 横过, 穿过; 交叉; 在……的另一边; $\sim$ corners 对角距; 对角; $\sim$ flats 对面距, 对边

actuator /'æktjueitə/ 调节器; 传动机构

acute /ə'kju:t/ 尖, 锐; 尖的, 锐的; $\sim$ angle 锐角

AD (average deviation) 平均偏差

adapt /ə'dæpt/ 修改, 改编; 采用

adaptation /ædæp'teiʃn/ 改进; 适应; 采用

adapter /ə'dæptə/ (adaptor) 改编者, 修改者; 接头, 接合器, 调节板; 座, 衬套; 拾波器

ADCN (advance drawing change notice) (有关) 图纸更改的先期通知

add /æd/ 补充, 增加; 集合; ‖ *Example sen-*

tence: Add a left view of a line AB, given the front and top views. 例句: 给定直线AB的主视图和俯视图, 要求增加左视图。

addendum /əˈdendəm/ * 牙顶高, 齿顶; 齿顶高; 附加物, 附录, 补遗; ～ circle 齿顶圆, 外圆; chordal ～ 弦线齿顶高; nominal ～ 标称齿顶高

additional /əˈdiʃənl/ 另外的, 附加的; ～ view 附加视图

address /əˈdres/ 地址; absolute segment ～ 绝对段地址

ADE (advanced drafting extension) 高级绘图扩展模块

adequate /ˈædikwət/ 足够的, 充分的; 适当的

ADES (automatic digital encoding system) 自动数字编码系统

adjacent /əˈdʒeisnt/ 邻近的, 毗连的; ～ parts 邻近部分

adjust /əˈdʒʌst/ 调节, 校正; ～ the pencil lead 调节铅芯; ～ing the needle point 调节针

adjustable /əˈdʒʌstəbl/ 可调整的, 可校准的, 活动的; ～ head 活动头（丁字尺）

ADMA (automatic drafting machine) 自动制图机

admeasure /ædˈmeʒə/ 量度, 测量; 分配, 配合; ～ ment 测量, 计量; 尺寸

ADP (automatic data processing) 自动数据处理

adumbrate /ˈædʌmˌbreit/ 画轮廓; 暗示, 预示; 遮蔽

adumbration /ˌædʌmˈbreiʃən/ 草图, 素描

advance /ədˈvɑːns/ 前移, 前进

AE (absolute error) 绝对误差

aerial /ˈɛəriəl/ 空中的, 航空的; ～ survey 航空测量

aeroplane /ˈɛərəplein/ (airplane) 飞机; ～

view 空瞰图, 鸟瞰图

aero-projector 航空投影仪; 投影测图

aerovelex /ˈɛərəˈveliks/ 小型投影测图仪

aeroview /ˈɛərəvjuː/ 空中俯瞰图, 鸟瞰图

Af (full annealing) 全退火

A/F (across flats) 对边

affine /əˈfain/ 仿射, 远交; 仿射的, 远交的, 精炼; ～ coordinate 仿射坐标; ～ geometry 仿射几何; ～ group 仿射群; ～ mapping 仿射映射; ～ space 仿射空间; ～ transformation 仿射变换

affinity /əˈfinəti/ 仿射性; 亲合力; 类同, 近似; 爱好

affix /əˈfiks/ 签署, 附件, 附录, 附加物

AFNOR 法国标准协会

ageing /ˈeidʒiŋ/ 老化; 时效; artificial ～ 人工时效; quench ～ 淬火（后自然）时效

aging (ageing) 老化; 时效

aglet /ˈæglit/ 柱螺栓; 金属箍

AGMA (American Gear Manufacturers Association) 美国齿轮制造商协会

aid /eid/ 辅助, 帮助, 辅助设备; ～ section lining 画剖面线的辅助器

aided /ˈeidid/ 辅助的, 半自动的

aileron /ˈeilərɔn/ 副翼（飞机）

air /ɛə/ 空气, 大气; 空中, 天空; ～ map 航空图

aircraft /ˈɛəkrɑːft/ 航空器, 飞机; ～ installation 航空仪表

airfoil /ˈɛəfɔil/ 翼型, 机翼

airphoto /ˈɛəfəutəu/ 航空摄影, 航空照片

airplane /ˈɛəplein/ 飞机, 飞行器

airview /ˈɛəvjuː/ 鸟瞰图

AISI (American Iron and Steel Institute) 美国钢铁学会

Al (aluminum) 铝

alclad /ˈælklæd/ 镀铝的

algebra /ˈældʒibrə/ 代数学; graphic ～ 图

解代数学

ALGOL (algorithmic language) 算法语言
algorithm /ˈælɡəriðəm/ 算法
align /əˈlain/ 对准, 校直; 定位, 定中心; 使成一直线, 列成一行; ～ed cutting 校直剖（旋转剖）; ～ed dimension lines 对齐尺寸线; ～ed system（写尺寸数字的）对齐制; ～ed view 旋转视图
alignment /əˈlainmənt/ 校直, 对准; 准线; 定线, 定心; ～ chart 线规图（包括诺漠图）; highway ～ 公路布置图; negative ～ 负偏差; perpendicular ～ 垂直度; proper ～ 同轴度, 同心性
alike /əˈlaik/ 相似; 相似的, 相同的
aline /əˈlain/ (align) 使成一直线, 列成一行; 定位, 对准
all-around 周围, 环绕, 全周
allocate /ˈæləkeit/ 分配, 分派, 划归
allowance /əˈlauəns/ 允差, 裕度; 修正量, 加工留量; 允许; bend ～ 弯曲裕度; machining ～ 切削裕度; over ～ 上差, 尺寸上偏差; under ～ 下差, 尺寸下偏差; wrenching ～ *旋 紧余量; zero ～ 无公差
alloy /ˈæloi/ 合金; aluminum ～ 铝合金; bearing ～ 轴承合金; hard ～ 硬质合金
alphabet /ˈælfəbet/ 字母, 字母表; 初步, 入门; 规格; cyrillic ～ 希利尔字母; the ～ of line 线之规格, 线型; 格线
alphabetic(al) /ˌælfəˈbetik(l)/ 字母（表）的, 按字母顺序的; ～ly 按字母顺序; ～ order 字母顺序
Alphanumeric Cathode Ray Tube 字母数字阴极管
alt (alternate) 交替（键）; (altitude) 高度
alternate /ˈɔːltəːneit/ 交替, 轮流; 交错的, 邻界的, 另外的; ～ crosshatching 交变剖面线（交替变换间隔的剖面线）; ～ position 第二位置, 交错位置; ～ position line 位置

转变线
alternation /ˌɔːltəːˈneiʃən/ 改变, 变更, 更改, 交错
alternative /ɔːlˈtəːnətiv/ 两者挑一的; 选择对象; 替换物; 改变位置
altitude /ˈæltitjuːd/ 高度, 高线; ～ difference 高（度）差; angular ～ 仰角, 高低角
aluminium /ˌæljəˈminiəm/ 铝; ～ casting 铸铝; ～ plating 镀铝; cast ～ 铸铝
alw (allowance) 公差, 允差, 余量, 间隙
ambit /ˈæmbit/ 轮廓, 外形, 界限, 范围, 周围
amend /əˈmend/ 修正, 修改, 变更
Amer. std (American standard) 美国标准
amount /əˈmaunt/ 量, 多少; ～ of change chart 变量图; extra metal ～ 加工余量
amp (amplification) 放大
amplification /ˌæmplifiˈkeiʃən/ 放大, 扩大; 放大系数, 加强, 详述
amplifier /ˈæmplifaiə/ 放大机, 放大器
amplify /ˈæmplifai/ 放大, 增大, 增强; 详述
Am. std (American standard) 美国标准
amt (amount) 数量, 总计
anaglyph /ˈænəɡlif/ 互补色立体图, 两色体视图, 立体相片, 彩色浮雕（土木）; ～ models 体视模型; the use of ～ models for teaching descriptive geometry 讲授画法几何时体视模型的应用
analog(ue) /ˈænəlɒɡ/ 相似, 类似; 模拟系统, 模拟机; 模拟; ～ machine 模拟机
analogous /əˈnæləɡəs/ 类似的, 相似的, 模拟的
analogy /əˈnælədʒi/ 模拟, 类似, 相似; 类似性
analysis /əˈnæləsis/ 分析, 分解; graphic ～ 图解; graphical ～ 图解, 图解分析; 分析图学
AND/OR 与/或; ～ form 与/或形; ～ graph

与/或图; ～ tree 与/或树

ang (angle)角

angle /'ægl/ 角,角形的;角尺,角钢,挂角,象限; ～ at the center 中心角; ～ between axes 轴间角; ～ between planes 平面间的夹角,两面角; ～ between two lines 二直线间的夹角; ～ between two planes 两面角; ‖ *Example sentence: The angle formed by two intersecting planes is called a dihedral angle.* 例句:两个相交平面所形成的角称为两面角。～ head 弯头; ～ iron 角铁; ～ of ascent 螺旋角,上升角; ～ of chord 弦角; ～ of cirumference 圆周角; ～ of dip 倾角; ～ of helix 螺旋角,螺旋线升角; ～ of inclination 倾角; ～ of obliquity 倾角; ～ of pitch cone 分度圆锥角; ～ of projection 投影角; ～ of pressure 压力角; ～ of slope 倾角; ～ valve 折角阀; ～ of vision 视角; acute ～ 锐角; adjacent ～ 邻角; alternate ～ 对错角; alternate-exterior ～ 外错角; alternate-interior ～ 内错角; apex ～ 顶角; aspect ～ 视线角; at right ～s to 与……成直角; axis ～ 轴间角; azimuth ～ 方位角; blunt ～ 钝角; centric(al) ～ 中心角; central ～ 中心角; complementary ～ 余角; coning ～ 圆锥角; cutting ～ 切削角; dedendum ～ 齿根角; deflection ～ 偏向角度; dihedral ～ 两面角; dip ～ 地层倾角; displacement ～ 位移角; draft ～ 镆斜角; elevation ～ 仰角; face ～ (伞齿轮)面角; exterior ～ 外角; external ～ 外角; first ～ 第一角,第一象限; fourth ～ 第四角,第四象限; half ～ 半角; helix ～ 螺旋角; hexagonal ～ 六角形的角,120°角; intersection ～ (相)交角; lead ～ 螺旋升角; lift ～ 升角; oblique ～ 斜角; obtuse ～ 钝角; offset ～ 偏斜角; pitch ～ (伞齿轮)节面角; point ～ (钻头的)锥尖角; rake ～ 前角,刀面角; relief ～ 后角,后隙角; right ～ 直角; root ～ 伞齿底角; rounded ～ 圆角,修圆角; second ～ 第二角,第二象限; sharp ～ 尖角,锐角; side rake ～ 副前角; side relief ～ 副后角; skew ～ 倾角,斜拱角; spiral ～ 螺旋角; taper ～ 刃角,锥角; third ～ 第三角,第三象限; tilt ～ 倾斜角,摆角; tooth ～ 齿面角; tooth spacing ～ 齿间角; unequal-leg ～ 不等股角钢; vertex ～ 顶角,对顶角; vertical ～ 直角; visual ～ 视角; wedge ～ 楔角,锥角

angle-bar 角铁,角钢

angled /'æŋgld/ 倾斜的; ～ hole 斜孔

angle-shaft 角轴

angular /'ægjələ/ 角的,有角的,角形的,尖锐的; ～ altitude 仰角,高低角; ～ bearing 径向轴承; ～ dimension 角度尺寸; ～ gear 人字齿轮; ～ instrument 量角器; ～ perspective 角透视,斜透视; ～ section 斜剖面; ～ surface 斜面; draft ～ 镆斜角

angularity /ˌæŋgju'læriti/ 倾斜度;棱角,成角度;弯曲度,曲率

animation /ˌæni'meiʃn/ 动画; ～ language 动画制作语言; ～ system 动画制作系统

anisometric /ˌænaisəu'metrik/ 不等轴的

anneal /ə'ni:l/ 退火,韧化,煨,闷火

annex /ə'neks/ (annexe) 附件,附录

annotation /ˌænə'teiʃn/ 注解,注释; bulk ～ 连续注释

annular /'ænjulə/ 环形的; ～ torus 圆环面

annulation /ˌænju'leiʃən/ 环,环形物,环的形成

annuloid /'ænjuləid/ (annulose) 环状的

annulus /'ænjuləs/ 环,圆圈,环形物,圆环域;内齿轮

anomalous /ə'nɔmələs/ 不规则的,异常的

ANS (American National Standard) 美国国家标准

ANSI (American National Standards Institute) 美国国家标准协会；～ X_3H_3 美国国家标准协会计算机图形工作组

anticlockwise /ˌænti'klɔkwaiz/ 逆时针方向的(地)，左旋的(地)

antique /æn'ti:k/ 西文粗体字

aperture /'æpətʃə(r)/ 孔，孔径，缝隙，窗口，靶区；pick ～ 选择窗口

apex /'eipeks/ 顶，顶点；～ angle 顶角；conic ～ 锥顶点

appar (apparatus /æpə'reitəs/) 仪器，仪表，装置，设备，注解，索引；blue print ～ 晒兰图设备；centering ～ 定(圆)心器；plotting ～ 测图仪

apparent /ə'pærənt/ 表面上的

APPD (approved) 批准

appearance /ə'piərəns/ 外貌，外表；表现，出现

append /ə'pend/ 添加，附加，补遗

appendage /ə'pendidʒ/ 备用仪器，附件，附属部分

appendix /ə'pendiks/ 附录

appreciable /ə'pri:ʃəbl/ 看得出的，明显的，感觉得到的

approach /ə'prəutʃ/ 向……靠近，接近，近似；探讨，处理；看待，态度，方法

approval /ə'pru:vl/ 赞成，同意，批准；鉴定，检验，试验

approve /ə'pru:v/ 审定，批准，允许，通过；证明

appr (ox) (approximate) 近似地；～ calculation 近似计算；～ construction 近似画法；～ four-centered ellipse 四圆心近似椭圆；～ method 近似方法；～ value 近似值

approximation /əprɔksi'meiʃn/ 近似；近似法，近似值；～ approximate of an arc 弧线的近似求长法

APRX (approximate) 近似，大约的

APRXLY (approximately) 近似地

arabesque /ˌærə'besk/ 花叶饰，精制的

arbitrary /'a:bitrəri/ 任意，不定的；～ surface of revolution 任意回转面

arbor /'a:bə:/ 轴，主轴，心轴；刀杆柄

arc /a:k/ 弧，圆弧；电弧；～ digraph 弧有向图；～ splines 圆弧样条；～ three point 用三点画圆弧；connecting circular ～ 连接弧；approximate rectification of an ～ 弧线的近似求长法；complementary ～ 余弧

arch /a:tʃ/ 拱；five-centered ～ 五心拱

architect /'a:kitekt/ 建筑家

architectural /a:ki'tektʃərəl/ 建筑上的，建筑学的；关于建筑的；～ composition 建筑布局；～ drawing 建筑制图；～ perspective 建筑透视图

architecture /'a:kitektʃə/ 建筑工程，建筑结构；建筑学；naval ～ 造船学

arcograph /'a:kəgra:f/ 圆弧规

area /'ɛəriə/ 面积，表面，截面，区域，(填充)区；～ diagram 面积图；～ of throat 喉部截面；adjacent ～ 邻近区域；background ～ 字母间的空白；blank ～ 空白区域；fill ～ 填充区；line-closed ～ 封闭线框(区)

arm /a:m/ 臂，幅，杆，柄；～ of wheel 轮辐

arrangement /ə'reindʒmənt/ 排列，布置；设备，装置；～ of views 视图的布置；general ～ 总体布置，总图

array /ə'rei/ 数组，阵列

arrow /'ærəu/ 箭，箭头；指针；～ side 箭边；arc ～ (circular) 环形箭头(旋转符号)；break in the ～ 箭头的转折

arrowhead /'ærəuhed/ 箭头，加注箭头；improve ～s 改善箭头；one-stroke ～ 单笔箭头；two-stroke ～ 二笔箭头

art /a:t/ 技术，技艺

article /'a:tikl/ 论文，文章；项目

artist /'a:tist/ 画家，美术家

artwork /'ɑ:twə:k/ 原图, 布线图, 图 (形), 图模; 工艺 (品); ～ master 照相底图; composite ～ 合成原图; original ～ 原图; photomask ～ 光掩模图

AS (American Standard) 美国标准

ASA (American Standards Association) 美国标准协会

ASAP (Automated Symbolic Artwork Program) 图形符号自动化程序

ASC (American Standard Cord of Information Interchange) 美国信息交换标准码

ASCC (automatic sequence controled calculator) 自动程序控制计算机

ascent /ə'sent/ 斜度, 坡度; 上升

ASCII (American Standard Code for Information Interchange) 美国国家信息交换标准代码

ASE (Amalgamated Society of Engineers) 工程师联合会

ASEE (American Society for Engineering Education) 美国工程教育协会

ASME (American Society of Mechanical Engineers) 美国机械工程师协会; ～ boiler code 美国机械工程学会锅炉法规

aspect /'æspekt/ 缩图; 形态, 样子, 外表, 外貌, 局面; 方向, 方位; 观点

ass. (assembly) 装配, 组件; 集合

ASSEM (assembly) 装配, 组合; 组件, 部件

assembler /ə'semblə/ 汇编程; MACRO ～ 宏汇编

assembly /ə'sembli/ 装配, 组合; 装配单元, 组件, 部件; 装配体, 组合体; 装配图, 总图; 图, 系统, 汇编; ～ drawing 装配图, 组件图, 部件图; ～ drawing number 装配图图号; ～ for installation 安装图; ～ language 汇编语言; ～ working drawing 装配工作图; design ～ 设计装配图; printed board ～ 印制版组装件; pulley bracket ～ 皮带轮托架装配图; sectional ～ 剖视装配图

assign /ə'sain/ 赋值, 赋标号

association /ə,səusi'eiʃn/ 学会, 团体, 社团; 组合; ～ graph 组合图; American Standard ～ 美国标准学会

assume /ə'sju:m/ 假定, 设想, 假设

ASSY (assembly) 装配; 装配体; 装配图

AST (American Standard Thread) 美国标准螺纹

astrolabe /'æstrəleib/ 等高仪

asymmetric(al) /,eisi'metrik(l)/ 不对称的

asymmetry /,ei'simətri/ 不对称 (性), 不平衡 (度)

asymptote /'æsimptəut/ 渐近线

atlas /'ætləs/ 图表集, 地图册, 图集, 图册; star ～ 星图

attribute /ə'tribju:t/ 属性

audio-visual /'ɔ:diəu'viʒuəl/ 直观

audit /'ɔ:dit/ 检查, 审查

auger /'ɔ:gə/ 钻

authorization /,ɔ:θərai'zeiʃn/ 授权

AUTO (automatic) 自动

autocartograph /,ɔ:təu'kɑ:təgrɑ:f/ 自动测图仪, 自动制图

autocontrol /'ɔ:təukən'trəul/ 自控

autodraft /'ɔ:təudrɑ:ft/ 自动绘图

autograph /'ɔ:təgrɑ:f/ 自动绘图仪; 手稿, 亲笔写

autographic /,ɔ:tə'græfik/ 亲笔的; ～ apparatus 自动图示记录器

autographometer /,ɔ:təgrə'fɔmitə/ 自动图示仪

automatic(al) /,ɔ:tə'mætik(l)/ 自动的, 机械的; ～ drafting 自动绘图; ～ drafting machine 自动绘图机; ～ drafting system 自动绘图系统; ～ drawing device 自动绘图设备; ～ drawing digitizing 自动绘图数

字化；~ drawing equipment 自动绘图设备；~ ploter 自动绘图仪；~ ploting 自动绘图

autoploter /ˌɔːtəuˈplɔtə/ 自动绘图仪, 自动编表报机

autoscore 自动画线

Aux (auxiliary) 辅助的, 辅助装置, 附件

auxiliary /ɔːɡˈziliəri/ 辅助的, 辅助设备；~ circle 辅助圆, 参考圆；~ cone 辅助锥（面）；~ cutting plane 辅助截平面；~ cutting surface 辅助截面；~ elevation 辅助正视图；~ figure 辅助图；~ line 辅助线；~-plane method 辅助面法（变换投影面法）；~ plan 辅助俯视图；~ plane 辅助投影面；~ plane of projection 辅助投影面；~ projection 辅助投影；‖ *Example sentence: Auxiliary projection means the application of additional planes and the construction of new views of the object. These additional planes are called auxiliary projection planes.* 例句: 辅助投影意指应用附加的平面作物体的新视图。这些附加平面叫作辅助投影平面。~ section 辅助剖面；~ surface 辅面（垂直面）；~ view 辅视图；~ view method 辅助视图法；front ~ 前辅视图；front-adjacent ~ 前邻辅视图；rear ~ 后辅视图；‖ *Example sentence: An auxiliary view is a view projected on any plane other than one of the three principal planes of projection (frontal, horizontal, or profile).* 例句: 辅助视图是投影在三个主要投影面（正立投影面、水平投影面、侧立投影面）以外的任一个平面上的视图。

Aux-view (auxiliaryview) 辅助视图；double ~ 复辅视图；single ~ 单辅视图

available /əˈveiləbl/ 可用的, 可达到的, 通用的, 有效的

AVG (average) 平均值

avoid /əˈvɔid/ 避免, 躲开

AWG (American Wire Gauge) 美国线规

AX (axis) 轴, 轴线, 轴心

axial /ˈæksiəl/ 轴的, 轴向的；~ angle 轴角；~ plane 轴线的平面

axis /ˈæksis/ 轴, 轴线, 中心线；坐标轴；~ of cone 锥轴；~ of projection 投影轴；~ of revolution 回转轴线；~ of rotation 旋转轴；~ of symmetry 对称轴线；~ of vision 视轴；axonometric ~ 轴测投影轴；body ~ 物体的轴线；central ~ 中心轴线；common ~ 公共轴线；coordinate ~ 坐标轴；fore-and-aft ~ 纵轴（从机头到机尾）；isometric ~ 等角轴；lateral ~ 横轴, 横坐标；longitudinal ~ 纵轴, 长轴；major ~ 长轴, 主轴；minor ~ 短轴；normal ~ 垂直轴；oblique ~ 斜轴；orientation ~ 定位轴线；projecting ~ 投射轴, 垂直轴, 垂直于投影面的轴；revolving ~ 旋转轴；transverse ~ 横轴；vanishing ~ 没影轴；vertical ~ 立轴, 纵坐标

axisymmetric(al) /ˌæksisiˈmetrik(l)/ 轴对称的

axle /ˈæksl/ （轮）轴, 车轴, 轴杆；轴线, 中心线；~ bearing 轴承；~ journal 轴颈；~ sleeve 轴套；bent ~ 曲轴；couple ~ 联动轴；drive ~ 主动轴；slave ~ 从动轴

axonometric(al) /ˌæksənəuˈmetrik(l)/ 正轴测图, 不等角（投影）；轴线测定, 轴测；~ drawing 轴测图, 不等角投影图；~ projection 轴测投影, 不等角投影；三向图

axonometry /ˌæksəˈnɔmitri/ 三向图, 轴测图

AY (assembly) 装配, 组合；总成, 组合件

azimuth /ˈæzimθ/ （测量）方位角, （天文）地平经度

B

bab (babbitt) 巴比合金(铜锑锡合金)，巴氏合金；衬以巴氏合金
back /bæk/ 背面，向后，回；后部的；～ cone 背锥；～ elevation 后视图
backing /'bækiŋ/ 垫片，垫圈，底板；反向，背面的；～ up 后备，备份
backlash /'bæklæʃ/ 间隙，松动；gear ～ 齿轮侧隙
backplan 底视图
backplane 底板
backplate 背板
backsight 后视
backspace /'bæk,speis/ 回退，退格
backstop /'bækstɔp/ 托架，棘爪；挡着，支持
balance /'bæləns/ 平衡
balcony /'bælkəni/ 阳台
ball /bɔ:l/ *球，球形物
ball-bearing 球轴承
balloon /bə'lu:n/ 气球；装配图上标件号的圆圈；～ing 加注件号圆圈
bank /bæŋk/ 堤岸，山坡，沙滩，库；data ～ 数据库
bar /ba:/ 线条，横号，键；杆，棒，棒材，条钢；～ chart 条图，统计表上的标高图；～ diagram 垂直图表；angle ～ 角铁；channel ～ 槽钢；extension ～ 延伸杆；lengthening ～ 延伸杆；rack ～ 齿条；radius ～ 半径杆；spare ～ 空格键；100 percent ～ 百分比条图
barb /ba:b/ 倒钩，羽支，箭头的两翼
barrel /'bærəl/ 圆柱体，圆筒，鼓形体

basal /'beisl/ 基础的，基本的；～ plane 基面，底面
base /beis/ 基点，基线，基面；基础，底部，底座；基准；基数，基地址；～ circle 基圆；～ circular thickness 基圆弧线齿厚；～ frame 底座，支架；～ lacquer 底漆；～ level 基准面；～ map (地质)工作草图；～ of cone 锥底；～ plane 基平面；～ point 基点；～ slope 边坡角，斜面投影；axle ～ 轴间距；data ～ 数据库；engine ～ 机座；wheel ～ 轴距
base-line 基线
basement /'beismənt/ 地下室，底层；～ plan 地下室平面图
baseplane /'beisplein/ 底平面，基面
baseplate /'beispleit/ 底板，底平面
BASIC (Beginner's All-purpose Symbolic Instruction Code) 初学者通用符号指令代码
basic /'beisik/ 基本的，基础的；～ circle 基准圆；～ drawing for engineering technology 工程技术制图基础；～-hole system 基孔制；～-shaft system 基轴制；～ size 基本尺寸，公称尺寸，名义尺寸
basis /'beisis/ 基础，基准，基线；hole ～ 基孔制，孔基准；shaft ～ 基轴制，轴基准
batten /'bætn/ 挂瓦条，板条，压条
batter /'bætə/ 斜坡，内倾，倾度(用于土木工程，为离铅垂线之偏度，相对于斜度)
BB (ball bearing) 滚珠轴承
b. bs. (bus. -bars) 总线，母线；汇流排
B. BZ. (bearing bronze) 轴承青铜
BC (between centres) 中心距；(binary code)

二进制码; (British Columbia)（加拿大的）大不列颠哥伦比亚州;1978年国际画法几何会议在该州的温哥华(Vancouver)举行。

BDY (boundary) 轮廓, 极限, 边界, 界线

beacon /ˈbiːkən/ 标志塔

bead /biːd/ 卷边, 圆缘; 弯头; 目录; 把……像串珠子一样连起来; ～ing 串珠状缘饰; ～ weld 圆缘焊接

beam /biːm/ 横梁, 杆; 光辉, 射照, 射线束; ～ compass 梁规; peripheral ～ 圈梁

bear /bɛə/ 负担, 负荷, 承担, 具有; 产生; 压; 打孔器, 小型冲机

bearing /ˈbɛəriŋ/ 轴承; 方位; 支座, 支承面; 负荷; ～ bras 铜轴衬; ～ circle 方位圆, 方向盘; ～ rest 轴承扶架; ～ ring ＊ 轴承套圈; ～ spacer 隔离圈; ～ surface 支承面; allowable ～ 容许承压; angular ～ 径向轴承; anti-friction ～ 减摩轴承, 滚动轴承; axial ～ 止推轴承, 支撑轴承; azimuth ～ 方位角; ball ～ 球轴承, 滚动轴承; cone ～ 锥形轴承; double-row ～ 双列滚动轴承; globe ～ 球面轴承; inside diameter of ～ 轴承内径; journal ～ 径向轴承, 轴颈轴承; multi-roll ～ 滚针(柱)轴承; needle ～ 滚针轴承; plain ～ 普通(滑动)轴承; radial ～ 向心轴承, 径向轴承; radial thrust ～ 向心推力轴承; roller ～ 滚子轴承; rolling ～ 滚动轴承; self-aligning ～ 自位轴承, 自动调心轴承; sliding ～ 滑动轴承; step ～ 止推轴承; tapered roller ～ 圆锥滚动轴承; thrust ～ 推力轴承, 止推轴承; type of ～s 轴承类型

beat /biːt/ 跳动, 摆动

bed /bed/ 机座, 床身

bee-line /ˈbiːlain/ （两点间的）直线, 最短距离, 捷径, 两地间的直路

before /biˈfɔː/ 在……以前(前面)

behind /biˈhaind/ 在……以后(后面)

bell /bel/ 圆锥体, 钟铃; ～ character 报警符

belong /biˈlɔŋ/ 属于

belt /belt/ 皮带

benched /ˈbentʃid/ （台)阶(形)状的

benchmark /ˌben(t)ʃmaːk/ 水准, 基准

bend /bend/ 弯曲, 弯向; 弯头; 弯管, 曲梁; 圆角, 圆边; ～ line 弯线（展开图上表示棱的线）; square ～ 直角弯头

bending /ˈbendiŋ/ 挠曲, 偏差; ～ strength 抗弯强度

benzin(e) /benˈziːn/ 石油精（可擦图面的铅迹和污垢), 挥发油

BESA (British Engineering Standard Association) 英国工程标准协会

bevel /ˈbevl/ 斜, 斜的; 斜截, 斜角, 斜面, 斜角规; ～ gauge 斜角规, 量角器; ～ gear 伞齿轮, 圆锥齿轮; ～ protractor 角度尺, 万能角尺, 活动量角器; ～ square 斜角规; ～ whel 斜齿轮, 伞齿轮, 斜摩擦轮; combination ～ 组合量角规, 万能量角器

BG (bevel gear) 伞齿轮; (Birmingham gage) 伯明翰规

BH (Brinell hardness) 布氏硬度

BHN (Brinel hardness number) 布氏硬度数

bias /ˈbaiəs/ 偏, 偏差, 偏离, 偏置; 斜线（痕); cut on the ～ 斜截, 斜切

bib(b) /bib/ (bibb. bibcock) 弯管旋塞, 龙头

bibliography /ˌbibliˈɔgrəfi/ 文献(目录), 参考书目

biconcave /baiˈkɔnkeiv/ 双凹的, 两面凹的

biconvex /baiˈkɔnveks/ 双凸的, 两面凸的

big-end 大端(的); ～-down 上小下大(的); ～-up 上大下小(的)

bilateral /baiˈlætərə/ 双向的, 双边的; ～ tolerance 双向公差

bill /bil/ 单(据), 表; 记入; 镰刀; ～ of materials 材料表

binary /ˈbainəri/ -coded 二进制编码的

binding /'baindiŋ/ 边缘, 装订, 联编; language ~ 语言联编

bird's-eye /'bə:dzai/ 俯视的, 鸟瞰的; ~ view 鸟瞰图, 概观

bird's-eye-perspective 鸟瞰图; 大纲

BIRE (British Institution of Radio Engineers) 英国无线电工程师协会

bisect /bai'sekt/ 相交, 交叉; 把……二等分, 对开; ~ing a line 平分线段

bisector /bai'sektə/ 平分, 二等分; 平分点, 平分线, 平分的两部分之一; perpendicular ~ 垂直平分线

bisector /bai'sektə/ 平分线; perpendicular ~ 中垂线

bisectrix /bai'sektriks/ 二等分线

bisymmetric /,baisi'metrik/ 双对称的

bisymmetry /,bai'simitri/ 双对称

bit /bit/ 钻头, 刀头

bitmap /bit'mæp/ 位图; ~ bits 位图字位; ~ dimension 位图大小(尺寸); ~ display format 位图显示格式; ~ font 位图字体; ~ functions 位图函数; ~ graphics 位图图形学; ~ image 位图图像; ~ image in menu 菜单中的位图图像; ~ operation 位图操作; ~ resource 位图资源; ~ structure 位图结构; color ~ 彩色位图

bit-point-forming 钻尖形

BL (baseline) 基(准)线; (blanking) 空白, 间隔, 空位, 留空; (border line) 边界线

black /blæk/ 黑色, 黑墨水; 黑的; ~ ening 涂黑; ~ ening 黑色发蓝; ~ ness 铅笔浓度, 黑度

blade /bleid/ 刀片, 刀口, 叶片, 扁平部; ~ of T-square 丁字尺身

blank /blæk/ 空格, 空白, 空页, 间隔; 毛坯, 半成品; ~ paper 白纸, 无字之纸

blanking /'blækiŋ/ 空位, 留空; 消隐

blinking /'blikiŋ/ 闪烁

block /blɔk/ (标题)栏; 封锁, 块; 部分, 闭塞; 滑轮, 图块(计); ~ data 数据块; ~ letter 方块字; ~ out 布图; change-record ~ 更改栏, 修正栏; drawing ~ 活页画图纸; 拉模板; glass-paper ~ 砂纸板; item ~ * 明细栏; sandpaper ~ 砂纸板; scribing ~ 画线架; text ~ 文本块; title ~ * 标题栏; to ~ out 画草图, 规划, 筹划; v ~ V形块, V 槽块

block-diagram 方块图, 方框图; 简图, 草图

block-system 方框图

blotter /'blɔtə/ 吸墨纸, 吸墨用品

BLOD (block-diagram) 方块图, 方框图

blowback /'bləubæk/ 图像放大

blue /blu:/ 蓝, 蓝色; ~ paper 蓝晒纸; light ~ 浅蓝色

blueprint /'blu:'print/ 蓝印, 蓝图; 晒蓝图; 计划, 订计划; ~ analysis 蓝图分析; ~ reading 识图, 读图

B/M (bill of material) 材料表; 材料单

BN (binary number system) 二进(数)制

BNH (burnish) 抛光, 烧蓝; 精加工

board /bɔ:d/ 木板, 板; chip ~ 刨花板; curve ~ 曲线板; drafting ~ 绘图板; drawing ~ 绘图板; glass drawing ~ 玻璃图板; key ~ 键盘; partition ~ 间壁, 隔断; plastics-faced drawing ~ 塑料面制图板; plotting ~ 图形显示幕(屏); 曲线板; portable drawing ~ 轻便型制图板; printed ~ 印制板; scratch ~ 草图板; shaving ~ 刨花板; trestle ~ 大图板; veneer ~ 胶合板

body /'bɔdi/ 物体, 身体; 主要部分, 多数; 机身; ~ diagonal 体对角线; ~ of curved surface 曲面体; ~ plan 横断面, 横剖型线图, 机身平面图; combined revolving ~ 组合回转; basic ~ 基本形体; curved surface ~ 曲面立体; cutting type combined ~ 切割

式组合体; hollow ～ 空心立体; intersecting ～ 相贯体; penetrated ～ 相贯体; pile up ～ 叠加立体; pile up type combined ～ 叠加式组合体; plane ～ 平面立体; revolving ～ 回转体; revolving combined ～ 组合回转体; rolling ～ 滚动体; simple combined ～ 简单组合体

bold /bəuld/ 粗大的, 醒目的; 大胆的, 勇敢的; ～ face (black face, bold face letters)(印刷上的)粗体字, 黑体字

bolt /bəult/ ＊螺栓, 销子, 插销; 闪, 电光; ～ circle 螺栓分布圆; ～ head 螺栓头; ～ thread 外螺纹; acorn hexagon head ～ ＊六角头盖形螺栓; bay-～ 基础螺栓, 地脚螺栓; belting ～ ＊带用螺栓; binder ～ 连接螺栓; blank ～ 粗制螺栓; bright ～ 精制螺栓, 光螺栓; burnished ～ 精制螺栓; check ～ 防松螺栓; cheese head ～ 圆头螺栓; clip ～ ＊卡箍螺栓; coach ～ 方头螺栓; countersunk (countersink) ～ 埋头螺栓; cup nib ～ ＊圆头带榫螺栓; eye ～ ＊double end ～ 双头螺柱, 柱螺栓; double-screw ～ 双头螺纹螺栓; 活节螺栓, 环首螺栓; expansion ～ 开口螺栓; fillister head ～ 槽头螺栓; flat countersunk nib ～ ＊沉头带榫螺栓; flat head anchor ～ ＊平头固定螺栓; foundation ～ ＊地脚螺栓, 底脚螺栓; half-bright ～ 半光制螺栓; hammer head ～ ＊T形螺栓; hexagon ～ ＊六角头螺栓; hexagon ～ with colla＊ 六角头凸缘螺栓; hexagon ～ with flange＊ 六角头法兰面螺栓; lag ～ 方头螺栓; mushroom head anchor ～ ＊扁圆头固定螺栓; nut ～ 带帽螺栓; octagon ～ ＊八角头螺栓; ring ～ 环端螺栓; round-head short square-neck ～ 圆头短方颈螺栓; round-head square-neck ～ 圆头方颈螺栓; spring ～ ＊弹性螺栓; square-head ～ 方头螺栓; stirrup ～ ＊U形螺栓; stud ～ ＊全螺纹螺柱, 方头螺柱; T ～ T形螺栓; T-head ＊T形螺栓; U ～ ＊U形螺栓; unfaced ～ 不加工螺栓; with feather ～ 带鼻螺栓

bolthead /'bəulthed/ 螺栓头

bolthole /'bəult,həul/ 螺栓孔; ～ circle 螺栓孔分布圆

bolting /'bəultiŋ/ 螺栓连接, 锚杆支护

bonding /'bɔndiŋ/ 连接, 焊接; 搭接, 接地; stitch ～ 自动点焊

border /'bɔ:də/ ＊图框, 边框, 边界, 边缘, 边, 槽; 邻接, 邻近, 接近; ～ line 边界线, 边缘线; ～ pen (画轮廓线用的)绘图笔; 边框笔(专画边框粗线); map ～ 图廓

bore /bɔ:/ 口径, 内径; 孔, 洞, 腔, 钻孔, 镗孔; ～ size 内径; basic ～ 基孔; center ～ 中心孔; counter ～ 柱坑; hub ～ 毂孔

bore-out-of-round (孔)不圆度

boring /'bɔ:riŋ/ 镗; ～ mill 镗床

boss /bɔs/ 凸台(铸件或锻件上); 头; 轮毂, 轴衬, 轴孔, 座

bot (bottom) 底, 地基

bottle /'bɔtl/ 瓶; ～ of drawing ink 绘图墨水瓶

bottom /'bɔtəm/ 底, 底部; 地基; ～ edge of digit 数字底线; ～ edge of letter 字母底线; ～ view 底视图

boundary /'baundəri/ 分界线; 边界, 范围; 极限, 限度

bounded /'baundid/ 有界的; ～ projection 有界投影

bow /bəu/ 弓, 弯曲, 弓形; 圆规; ～ beam 弓形梁; ～ compass 小圆规, 两脚规, 外卡钳; ～ pen 弓形鸦嘴笔; ～ points 弓形小分规, 弓形小圆规(点圆规); center-wheel ～ 中间螺钉弹簧圆规; fast-action ～ 快调弹簧圆规; quick-action ～ 速调弹簧圆规; spring ～ 弹簧圆规

box /bɔks/ 箱, 匣; 方盒法; ～-method 方箱

法; ～ construction 方箱法; ～ of leads 铅笔盒; ～ing method 方箱法(画轴测图的一种方法); core ～ 芯ье箱; gland ～ (gland stuffing ～) 填料函; glass ～ 透明投影箱; junction ～ 接线箱; reducing ～ 减速箱; speed change ～ 变速箱; transmission ～ 传动箱; wheel ～ 齿轮箱

BP (base point) 原点, 基点; (blueprint) 蓝图
Br (brass) 黄铜; (book of reference) 参考书
brace /breis/ 大括号
brachy- /ˈbræki/ 表示"短"; ～ axis 短轴
bracket /ˈbrækit/ 托架, 支架; 括号; 把……装上托架; 把……括在括号内; 把……分类
brain /brein/ 电脑, 电子计算机; 火箭弹上的引导装置; electronic ～ 电脑, 电子计算机
branch /braːntʃ/ 转移, 分支, 支线
branning /ˈbræniŋ/ 抛光
brass /braːs/ 黄铜; 黄铜衬; ～ step 黄铜轴瓦
brazing /ˈbreiziŋ/ 铜焊, 钎焊; ～ seam 焊缝
bread /bred/ 不成熟的, 年轻的; 面包, 食物, 粮食; ～ board 实验性的, 模拟板; ～ board design 模拟设计
breadth /bredθ/ 宽度, 幅度; ～ of tooth 齿宽; extreme ～ 全宽
break /breik/ 截断, 中断; 打破; ～ down 折断; 细分, 分类, 分析; ～ line 折断线, 断裂线, 折线; conventional ～ 习用折断画法; long ～ 长折断面
breakdown /ˈbreikdaun/ 折断, 细分, 分类, 分析; 启开
breaker /ˈbreikə/ 断路器
break-off 断开, 破坏
BRG (bearing) 轴承, 方位
brick /brik/ 砖, 砖状物, 程序块; fire ～ 耐火砖
bridge /bridʒ/ 桥, 跨接; ～ floor 桥面; ～ pier 桥墩; in ～ 并联

brightening /ˈbraitniŋ/ 抛光, 擦光
brightness /ˈbraitnis/ 亮度, 明度, 照度
brilliant /ˈbriljənt/ 明点(物面最亮处); ～ line 明线(物面最亮处为一线者)
Brinell /ˈbrinel/ 布氏(硬度)-布里涅耳; ～ figure (number) 布氏硬度(值); ～ hardness 布氏硬度
Bristol /ˈbristəl/ (纸名)(数层胶合成的极佳纸板)
Brit (Britain, British) 英国, 英国的
bro (bronze) 青铜
broad /brɔːd/ 宽广的, 主要的, 概括的; 粗的, 宽的; ～ line 粗线; in ～ outline 概括地说, 概括地画
broaden /ˈbrɔːdn/ 加宽, 放宽
broadness /ˈbrɔːdnis/ 宽度; 幅面
broadside /ˈbrɔːdsaid/ 宽边
broken /ˈbroukən/ 断开的, 打碎的; ～ circle 虚线圆; ～ line 折断线; 折线虚线; ～ view 局部视图, 断裂视图
broken-out-section 局部剖视图, 断裂剖视
bronze /brɔnz/ 青铜
Brown and Sharpe (Worm Thread) 布朗-沙普(蜗杆螺纹)
BRS (bras)黄铜
BR STD (British Standard) 英国标准
brush /brʌʃ/ 毛刷, 涂漆; air ～ 喷漆; dusting ～ 除尘刷
BRZ (bronze) 青铜
BS (British Standard) 英国标准; (Bureau of Standard)(美国)标准局; (binary scale) 二进位制
B/S (both sides) 双面
B&S (Brown and Sharpe Wire Gage) 布朗沙普线规
BSC (basic size) 基本尺寸; 理想尺寸
BSD (British Standard Dimension) 英国度量

标准

BSF (British Standard Fine Screw Thread) 英国标准细牙螺纹

BSFT (British Standard Fine Thread) 英国标准细螺纹

BSG (British Standard Gauge) 英国标准线规

BSI (British Standards Institution) 英国标准局（学会）

BSP (British Standard Pipe Screw Thread) 英国标准管螺纹

BSW (British Standard Whit-worth Thread) 英国标准惠氏螺纹

BSWG (British Standard Wire Gauge) 英国标准线规

buff /bʌf/ 抛光，擦光，抛光轮

buffer /ˈbʌfə/ 缓冲区

bug /bʌg/ 故障，错误

build /bild/ 造型，构造，建造

building /ˈbildiŋ/ 建筑物，大楼，房屋；～ codes 建筑符号；～ regulations 建筑规则；～ symbols 建筑符号

buildup /ˈbilˈdʌp/ 建立，组成，安装；计算，作图；加强，加厚

builting /ˈbiltiŋ/ 组合；～ model 组合体模型

builtup /ˈbilˈtʌp/ 组合的，装配的；可拆卸的

bul. (buletin) 公报，报告

bulb /bʌlb/ 电灯泡，球状物

bullnose /ˈbul,neuz/ 外圆角

buoy /bɔi/ 浮标

burnish /ˈbə:niʃ/ 磨平，磨光；光泽

bur (burr /bə:/) 毛刺，毛边，毛口；去毛刺；圈；加圈，垫圈，衬片；轴环，套环；磨石

Bur. of Stds. (Bureau of Standards)（美国）标准局

bus /bʌs/ (bus-bar) 母线，汇流排；公共汽车

bush /buʃ/ 衬套，轴瓦

butt /bʌt/ 平接，对接；紧靠；使邻接；伸出；大头（较粗的一端）；桶；～ joint 对接（头）；～ weld 对头焊

butterfly /ˈbʌtəflai/ 蝴蝶，蝶形物；～ nut 蝶形螺母

buttock /ˈbʌtək/ 纵剖线（船）

buttress /ˈbʌtris/ 扶壁，支柱；似扶壁状；～ (screw) thread 锯齿螺纹

butt-welding 对头焊

BW (body weight) 体重；(butt welded) 对接焊

BWG (Birmingham Wire Gauge) 伯明翰线规

Bz (bronze) 青铜

C

C (chipping) 凿平

cabinet /ˈkæbinit/ 橱柜，小房间，盒，壳体；～ drawing 半斜轴测图，斜二测图（q 取 1/2）；～ projection 半斜投影

CAD (Computer Aided Design) 计算机辅助设计，计算机半自动设计

cadastral /kəˈdæstrəl/ 地籍的

CADD (Computer Aided Design and Drawing) 计算机辅助设计与制图

CAD-E (Computer Aided Design and Engineering) 计算机辅助工程设计

CADM (Computer Aided Design and Manu-

facturing) 计算机辅助设计和制造
CAdrawing (Computer Aided Drawing) 计算机辅助绘图
CADS (Conversational Analizer and Drafting System) 综合性自动设计绘图系统
CAE (Computer Aided Engineering) 计算机辅助工程
CAGD (Computer Aided Geometric Design) 计算机辅助几何设计
CAI (Computer Assisted Instruction) 计算机辅助教学
Cal (caliber) 口径, 量规
calcium /ˈkælsiəm/ 钙; ～ hydroxide 熟石灰; ～ lime 生石灰
calculation /kælkjuˈleiʃn/ 计算(法), 计划; graphic ～ 图解计算法, 图算法; mass-properties ～ 物性计算
calculator /ˈkælkjuleitə/ 计算尺, 计算器, 计算图表, 计算装置, 计数器; digital ～ 数字计算器; tabular ～ 表式计算器
calculus /ˈkælkjuləs/ 算法, 微积分学; 演算
caliber /ˈkælibə/ (calibre 美) 直径, 口径; 测径器, 样板, 卡尺, 规
calibrate /ˈkæləbreit/ 核准, 检验; 使标准化; 划分度数
caliper /ˈkælipə/ (calliper) 卡尺, 测径器, 卡钳, 两脚规; beam ～ 大卡尺; combination ～ 内外卡钳; double ～ 内外卡钳; gauge ～ 卡规, 量规; inside ～ 内径规, 内卡; inside micrometer ～ 测微内径规; inside transfer ～ 移动内径规; internal ～ 内卡钳; leg ～ 通用卡钳; micrometer ～ 测微测径器; odd-leg ～ 半内径规; outside ～ 外径规, 外卡; outside micrometer ～ 测微外径规; outside transfer ～ 移动外径规; scribing ～ 内外 (刻线) 卡钳; sliding ～ 卡钳, 卡尺; thread ～ 螺纹卡尺; vernier ～ 游标卡尺

call /kɔ:l/ 调用, 召唤, 命令
calligraphy /kəˈligrəfi/ 书法, 笔迹
cal(l)ipering /ˈkælipəriŋ/ 测量, 量度; 检验
CAM (Computer Aided Manufactory) 计算机辅助制造
cam /kæm/ 凸轮, 偏心轮; 样板, 靠模, 仿形板; elliptical ～ 椭圆凸轮; end ～ 端凸轮; heart ～ 心形凸轮; plate ～ 平板凸轮
camber /ˈkæmbə/ 弯曲, 曲起; 曲率, 弯度, 弧线, 曲面; ～ line 中心线; circular arc ～ 圆弧曲面; variable ～ 变曲面
cambered /ˈkæmbərd/ 曲面的, 弧形的
came /keim/ 顶点, 极度
canonical /kəˈnɔnikəl/ 典型的, 规范的; ～ form 范式
cant /kænt/ 斜面, 倾斜位置
cap /kæp/ 盖, 罩, 帽; 顶; ～ line 顶线; ～ screw 有帽螺钉; bearing ～ 轴承盖
capacity /kəˈpæsiti/ 容量, 能力; 智力
capital /ˈkæpitəl/ 大写字母; 居首的, 首部, 第一等的; 优美的; ～ letter 大写字母; ～s lock 大写字母锁定 (键)
carbon /ˈka:bən/ 炭, 炭质 ～ tetrachloride 四氯化碳 (可擦掉图面的铅迹及污垢)
carburetion /ˌka:bjuˈreʃən/ (carburation) 渗碳
carburisation (carburization /ˈka:bjuraiˈzeiʃən/) 渗碳; local ～ 局部渗碳; surface ～ 表面渗碳
carburize /ˈka:bjuraiz/ 碳化
carburizing /ˈka:bjuraiziŋ/ 渗碳, 碳化硬化处理; cyanide ～ 氰化
card /ka:d/ 图, 表, 卡片, 穿孔卡, 程序, 单, 插件; ～ deck 卡片组; ～ reader 卡片读入机; actual indicator ～ 实际示功图; Chinese character ～ 汉卡; circuit layout ～ 电路布线图; file of ～s 卡片文件; indicator ～ 示功图; instruction ～ 说明片; punched ～

穿孔卡片; test ~ 测试图表; time ~ 时间表

cardboard /'ka:dbɔ:d/ 纸板; ~ roof 纸板屋顶

cardioid /'ka:diɔid/ 心脏线

carriage /'kæridʒ/ 楼梯阁栅, 梯段

carry /'kæri/ 进位

carte /ka:t/ （法语）地图, 海图; 证书, 文件; ~ paper 地图纸

Cartesian /ka:'ti:ziən/ 笛卡儿的, 笛卡儿坐标; ~ coordinate 笛卡儿坐标, 直角坐标; ~ geometry 解析几何

carting /ka:tiŋ/ 制图表, 填表, 编制海图

cartographer /ka:'tɔgrəfə/ 制图员, 制图者

cartographic(al) /ka:tə'græfik(əl)/ 制图的

cartography /ka:'tɔgrəfi/ 绘制图表; 制图学, 绘图法, 地图制图

cartology /ka:'tɔlədʒi/ 地图学, 海图学

carton /'ka:tən/ 厚纸, 纸板; 纸板箱

cartoon /ka:'tu:n/ 卡通片, 动画片, 漫画, 草图

cartouche /ka:'tu:ʃ/ 椭圆形轮廓, 涡形装置

cartridge /'ka:tridʒ/ -paper 图画纸, 厚纸

carve /ka:v/ 切, 切开; 雕刻, 开拓

CAS (Chinese Association of Standardization) 中国标准化协会

CASE (Computer Aided Software Engineering) 计算机辅助软件工程

case /keis/ 外壳, 箱, 盒, 套, 罩, 表面, 情况; 架, 格; ~ of instrument 仪器盒; iron ~ 铁壳; lower ~ 低格; transfer ~ 变速箱; transmission ~ 变速箱, 传动箱; worm gear ~ 蜗轮箱

case-carbonizing 表面渗碳

casehardening /'keis,ha:dəniŋ/ 硬化表面（渗碳, 淬火）, 皮硬

casing /'keisiŋ/ 箱, 盒, 壳, 套, 罩, 套管; 铸件; 铸造

cast /ka:st/ 投射, 铸成, 铸件; 计算; ~ aluminum 铸铝; ~ iron 铸铁; ~ part 铸件; ~ steel 铸钢; D ~ (die-) 压铸件

castellated /'kæstileitid/ 齿形的, 有许多缺口的; 制成蝶状物（如蝶状螺母, 铣有复键的轴）; ~ nut 蝶形螺母

casting /'ka:stiŋ/ 铸造, 铸件; ~ fillets 铸造内圆角; ~ rounds 铸造外圆角; ~ wall thickness 铸件壁厚; malleable ~s 展性铸件; thin ~ 薄壁铸件

cast-iron 铸铁（的）, 硬的; ~ fitting 铸铁接头; ~ pipe 铸铁管

castle /'ka:sl/ 城堡, 城堡形, 皇冠形, 船楼, 大建筑物; ~ nut 皇冠螺母, 槽顶螺母, 蝶形螺母

CAT (catalog, catalogue) 目录, 一览表, 说明书

cavalier /kævə'liə/ 等斜（投影）; 自由自在的, 豪强的; ~ projection 等斜投影, 等斜轴测投影, 斜投影

cbore (c' bore, counterbore) 柱（钻）坑孔, 平底锪钻, 埋头孔

C. Br (cast brass) 黄铜铸件, 铸黄铜

CC (cast copper) 铸黄（紫）铜; (cross correlation) 相关

C/C (between centers) 中心距, 轴间距; (concentric) 同心的, 集中的

CC-DOS (Chinese Character-disk Operating System) 中文磁盘操作系统

C-clamp C形夹

Cd (centre distance) 中心距

CD (cold drawn) 冷拉

CD (committee draft) 委员会草案

CD (compact disk) 光盘

CD-ROM 光盘只读存储器

CDRILL (counterdrill) 带钻柱坑钻头

CEGS (China Engineering Graphics Society) 中国工程图学学会

ceiling /'si:liŋ/ 天花板, 顶棚

cell /sel/ 单元

celluloid /ˈseljulɔid/ 赛璐珞

cement /siˈment/ 水泥，黏合物；～ floor 水泥地板

cementation /siːmenˈteiʃən/ 渗碳

center /ˈsentə/ (centre) 中心，中央；定心，放在中心；顶尖；～ distance 中心距；～ distance gage 中心距量规；～ hole 顶尖孔，中心孔；～ lines 中心线；～ of a conic 二次曲线的中心；～ of a quadric 二次曲面的中心；～ of affinity 仿射中心；～ of connecting circular arc 连接弧圆心；～ of curvature 曲率中心；～ of homology 透射中心；～ of perspectivity 透视中心；～ of projection 射影中心；～ of similarity 位似中心；～ of symmetry 对称中心；between ～ 中心距离；bone ～ 圆心片；coinciding ～ 中心重合；coinciding ～ line 重合中心线；inaccessible ～ 图纸外圆心；revolving ～ 旋转中心；visual ～ 视心点

centered (centred) /ˈsentəd/ 圆心，同心的，同轴的

centerline (centre line) /ˈsentəlain/ 中心线

centimeter (centimetre) /ˈsentimiːtə/ 厘米，公分

central /ˈsentrəl/ 中心的，中央的，主要的；定心；～ hole 中心孔（圆形或矩形）；～ projection 中心射影，中心投影

centre /ˈsentə/ (center) 中心，中央；定心，放在中心；～ around 以……为中心；～ distance 中心距；～ to ～ 中心距；～ to end 中心到端面之距；be placed on 2mm centres 按 2mm 中心距布置

centre-line-average 平均高度，算术平均值

centre-section 中心剖面

centre-to-centre 中心到中心，中心距；～ distance 中心距离；～ method 中心连接法

centric(al) /ˈsentrik(əl)/ 中心的

centring /ˈsentriŋ/ (centering) 定中心，找中心，对准中心

centripetal /senˈtripitl/ 向心的

centrosymmetric (al) 中心对称的

CG (computer graphics) 计算机绘图，计算机图形学

CG (cg, centigram) 公毫（百分之一公克）

CH (case-harden) 表面硬化

Ch (CHK, check) 校对

chain /tʃein/ 链，链接；～ dimension 链状尺寸

CHAM (chamfer) 倒角，倒棱，斜面，切斜面，倒棱面，切角面，槽，刻槽

chamfer /ˈtʃæmfə/ (chamfret) 同 CHAM；～ angle 倒角；～ arc 切弧；～ curve 切弧；～ line 倒角线

chamfered /ˈtʃæmfəd/ 圆角的，倒角的

change /tʃeindʒ/ 变更，改变，更换

change-over 过渡

channel /ˈtʃænl/ 槽，海峡，水道，航道；～(l)ed-iron 槽钢（铁），U 形钢；～ nut 槽形螺母

char 字符，字符型

character /ˈkæriktə/ 特征，性质；字母，符号，字符；～ body 字符框；～ cancel 消去字符串；～ height 字符高度；～ output 字符输出；～ spacing 字符间隔；～ up vector 字符向上向量；～ width 字符宽度；blank ～ 空白字符；block ～s 框形字符；Chinese ～ 汉字；italic ～ 斜体字；pixel ～s 象的素字符；vertical ～ 直体字

characteristic /ˌkæriktəˈristik/ 特征，表示特性的，特有的；～ curve 特性曲线；～ dimension 特性尺寸；～ of projection 投影特性

chart /tʃɑːt/ * 表图，图表，图表图；曲线图，计算图；地图，水路图，海图；图，卡片，

表,一览表;制图;～s and diagrams 图与图表;～ format 图表格式(布局);～ for display 展览图表;～ for production 生产图表;～ of symbols 符号图表,图例;～ rack 图架;～ reader 图表阅读器;～ with contour lines 等高线图; aemilogarithmie line ～ 单对数坐标折线图; aeronautic (al) ～ 航空图; aeronautical planning ～ 航空地形图; alignment ～ 列线图,线规图,诺谟图;列线图解,计算图表;地形图,准线图;图解; amount of change ～ 变量图; analytical ～ 分析图; area ～ 面积图; area bar ～ 面积条图; assembly ～ 装配表; auxiliary ～ 辅助图; bar ～ 条形图,统计图表上的标高图,单式条形图,方框图; bar ～ with plas and minus values 偏差条形图; basis surface ～ 地形图; bathymetric ～ 海图; black-and-white ～ 黑白图; block ～ 方框图,方块图; break-even ～ 损益平衡图; classification ～ 分类图(表); column ～ 柱状图; complex ～ 综合图; compound bar ～ 组合条形图; computation ～ 计算图表; concurrent ～ 共点算图; contour ～ 等高线图,等值线图; control ～ 控制图,质量管理图,质量评估图; conversion ～ 换算图,单折线图; data-flow ～ 数据流程图; Descartes ～ 笛卡儿图; design ～ 设计图表; dimensions ～ 外廓尺寸图; dog ～ 制动爪装配图; dot ～ 点阵图; erection ～ 线路安装图; flow ～ 方框图,流程图,信号流通图,程序方框图,程序操作顺序图表; function ～ 功能表图; Gantt ～ 甘特图,甘特进度表,线条式进度表; graphical ～ 曲线图; indicator ～ 示功图; isentropic ～ 等熵线图; line ～ 线图,折线图; logarithmic ～ 对数坐标图; marine ～ 航海图; multiple line ～ 多折线图,综合折线图; multiple pictorial ～ 复式象形图; multiple-bar ～ 多元条形图,复式条形图; nautical ～ (航)海图; network ～ 辐射算图,网线图; nomographic ～ 列线图表,计算图表; numerical pictorial ～ 数字象形图; one hundred percent bar ～ 百分比条形图; one hundred percent circle ～ 百分比圆形图; operation ～ 运用图,运行图,操作图; operations flow ～ 操作流程图; organization ～ 组织图,分类图; percentage bar ～ 百分比条形图; percentage pie ～ 百分比圆形图,百分比半圆形图; pictorial ～ 象形图,直观图; pie ～ 盘形图,圆形图,单式圆形图,圆瓣(统计)图(或称100 percent circle 百分比图); pipe-organ ～ 管风琴图(铅直式统计图表,像管风琴); polar ～ 极坐标图; principled flow ～ 原则流程图; process ～ 工艺流程图,工艺卡片,流程图表; radial bar ～ 圆式条形图; rate of change ～ 变率图; ratio ～ 比率图; rectilinear ～ 直角坐标图; repair ～ 修理图; roseline ～ 玫瑰线图; route ～ 路线图,流水线工艺卡; sampling and inspection flow ～ 取样和检查流程图; sequence ～ (开关)转接顺序图; service ～ 服务图; sight-reading ～ 模拟系统图; signal interlocking ～ 信号联锁图表; staircase ～ 阶形图(靠拢的铅直式统计图表,像台阶状); statistical ～ 统计图; synoptic (weather) ～ 天气图; system ～ 系统(框)图; technologic flow ～ 工艺流程图; test ～ 试验线图; three-dimensional ～ 三向图表; train operation ～ 列车运行图; trilinear ～ 三角形表图,三线图表,三线坐标图,三角折线图; two-dimensional ～ 两向图表; vector ～ 向量图; weather ～ 天气图; z ～ Z 形算图

charting /tʃɑːtiŋ/ 制图(表),填图
chartography /kaːˈtɔɡrəfi/ 制图法
chartometer /tʃːˈtɔmitə/ 测图器
chart-pattern 图案

chart-projection 地图投影,海图投影
chase /tʃeis/ 车螺纹,在……刻槽
check /tʃek/ 校对,校核,检验,检查,对号,对照,比较,阻止,抑制,妨碍,突停;~ nut 防松螺母;~ing 校对;~ -up 检查,查对;ball ~ valve 球止回阀
checker /'tʃekə/ 校对者,校核员,检查员;测试器
chemical /'kemikl/ 化学;~ engineering drawing 化学工程图
chfr. (chamfer) 斜面,切角面,倒角
chiaroscuro /ki,:rəs'kuərəu/ 图画影光,明暗法
chief /tʃi:f/ 主要的,主任的,首领,主任;~ architect 总建筑师;~ draftsman 制图主任;~ engineer 总工程师;~ resident architect 主任建筑师;~ resident engineer 主任工程师
chill /tʃil/ 冷,冷激,冷淬;~ed 冷硬的,淬火的;~ing 淬火
chimney /'tʃimni/ 烟筒
China-ink 墨(汁)
chip /tʃip/ 削,铲,切屑,碎片,缺口,凹口,芯片
choice /tʃɔis/ 选择,抉择;the right ~ 正确的选择
choose /tʃu:z/ 选择,挑选
chord /kɔ:d/ 弦,协调;~ length 弦长
chordal /'kɔ:dəl/ 弦的;~ addendum (固定)弦(线)齿(顶)高;~ distance 弦长;thickness (固定)弦(线)齿厚
chromium /'krəumiəm/ 铬,镀铬
chromizing /'krəumaiziŋ/ 镀铬处理
chromograph /'krəuməugra:f/ 胶板复制器;用胶板复制器复制
CI (cast iron) 铸铁
CIM (computer intergrated manufacturing) 计算机集成制造
CIP (cast-iron pipe) 铸铁管
CIR (circuit) 线路,电路;(circular) 圆形的;(CIR, circle) 圆
cipher /'saifə/ 零(0)
circinate /'sə:sineit/ 环形的;制图,用圆规画图
circle /'sə:kl/ 圆,一周,环形,圆形物;旋转,做圆周运动;~ arc 圆弧;~ curve 圆弧;~ curve template 圆弧模板;~ dragging 圆的拖动;~ of gorge 喉圆;~ of reference 参考圆;~ runout 圆跳动;absolute ~ 绝对圆;addendum ~ 齿顶圆;auxiliary ~ 辅助圆,参考圆;base ~ 基圆;broken ~ 有缺口的圆;bolt ~ 螺栓圆(排列一组螺栓的点画线圆周);bolthole ~ 螺栓孔分布圆;circumscribed ~ 外接圆;concentric ~s 同心圆;dashed ~ 虚线圆;dedendum ~ 齿根圆;describing ~ 画圆;动圆;eccentric ~ 偏心圆;generating ~ (齿轮的)基圆,母圆;full ~ 全圆;graduated ~ 分度圆,刻度圆,刻度盘;hole ~ 孔圆(排列一组孔的点划线圆周);horizontal ~ 水平圆,水平刻度盘,地平圈;imaginary ~ 虚圆;inner ~ 内圆;inscribed ~ 内切圆,内接圆;isometric ~ 等角圆;locus ~ 轨迹圆;outside ~ 外圆,齿顶圆;parallel ~ 纬圆;pin ~ 引线圆;pitch ~ 节(距)圆;point ~ 齿顶圆;radical ~ 根圆;rolling ~ 滚圆,(齿轮的)基圆;root ~ 根圆;throat ~ 喉圆;top ~ 顶圆,外圆;virtual ~ 假圆,虚圆;whole ~ 根圆,全圆
circuit /'sə:kit/ 电路,电路图;~ breaker panel 线路断流板(电);~ drawing 线路图;~ load 线路负荷(电);control ~ 控制线路(电);main ~ 总电路(电);micro ~ 微电路
circuitry /'sə:kitri/ 电路学,电路图
circular /'sə:kjulə/ 圆的,似圆的;~ arc 圆

弧; ～ arch 圆拱; ～ center line 圆形中心线; ～ cone 圆锥; ～ curve 圆弧曲线; ～ cylinder 圆柱; ～ directrix 圆准线; ～ disc 圆盘; ～ element 圆形素线, 圆形元线; ～ pitch 周节, 弧线节距; ～ thickness 弧线齿厚

circularity /sə:kju'læriti/ 圆, 圆形, 圆度

circular-shaped 圆形的

circumcircle /'sə:kəmsə:kl/ 外（接）圆

circumference /sə'kʌmfərəns/ 圆周

circumferential /sə'kʌmfə'renʃəl/ 圆周的, 周围的

circumgyrate /,səkəmdʒaiə'reit/ 旋转, 回转; 陀螺运动; （使）回转

circumscribe /'sə:kəmskraib/ 圆界线, 限制; 画圆, 确定……界限; ～d circle 外接圆; ～d polygon 外切多边形

cirt. (circuit) 电路, 电路图

civil /'sivl/ 公民的, 民用的; 文明的, 现代的; ～ construction 土木建筑; ～ engineering 土木工程

CK (check) 检验, 检查, 校对

CL (center line, central line) 中心线, 轴线, 中央线路; (class) 分类, 定级; 等级, 种类; (centiliter) 公勺（百分之一公升）

cladding /'klædiŋ/ 表面处理, 镀涂层, （金属）包层

clarity /'klæriti/ 清晰（度）, 透明（度）, 清澈

class /klɑ:s, klæs/ 种, 类; 分类, 等级; ～ of fit 配合类别; ～ of precision 精度等级; American screw thread ～ 美国标准螺纹配合等级; ～ 1 1 级配合; ～ 1A 1 级外螺纹; ～ 1B 1 级内螺纹

classification /klæsifi'keiʃn/ 分类, 分类法, 分等级; ～ chart 分类图（表）; ～ of projections 投影分类; universal decimal ～ (UDC) 国际十进制分类法

clay /klei/ 黏土, 泥土

clean /kli:n/ 清洁的, 整洁的; 匀称的, 整齐的; 美好的; 归零, 清除; ～ing rubber 清洁用橡皮, 软橡皮; ～ing agent 清洁剂; pen～er 笔尖清洁剂

clear /kliə/ 清除, 清晰的; 完整, 整齐; 穿过, 越过; 离开, 不接触, 间隙, 余隙; ～ed land 已开发地

clearance /'kliərəns/ * 间隙, 余隙, 缺口; 清除, 许可; ～ fit 间隙配合, 余隙配合; ～ limit 余隙极限; major ～ * 大径间隙; maximum ～ * 最大间隙; minimum ～ * 最小间隙; minor ～ * 小径间隙; basic ～ 理想余隙; maximum allowable ～ 最大容许余隙; minimum allowable ～ 最小容许余隙

clear-cut 轮廓明显的, 清晰的, 确定的

click /klik/ 爪, 棘爪; 棘轮机构

clinographic /klainə'græfik/ 斜投影画法的, 斜射的, 倾斜的; ～ projection 斜射投影法

clinohedral /,klainəu'hi:drəl/ 斜面体

clinometer /klai'nɔmitə/ 倾斜仪, 量角器

clip /klip/ 截法, 剪法; 裁剪, 剪取, 截短; 夹, 夹住; connecting ～ 接线夹

clm (column) 柱, 圆柱; 栏, 项目

clockwise /'klɔkwaiz/ 顺时针方向

close /kləus/ 关闭, 闭合

closet /'klɔzit/ 壁橱, 套间; 炉室; 盥洗室

cloth /klɔθ/ 布; ～ penwiper 拭笔布; drawing ～ 制图布; dust ～ 拭尘布; pencil ～ 铅画布; tracing ～ 描图布

clout /klaut/ 垫圈, 垫片, 抹布

CLPR (calipers) 圆规, 卡尺, 测径规

CLS (class) 等级, 种类

clutch /klʌtʃ/ 离合器, 联轴器; 联接, 凸轮

cm (centimeter) 厘米

coarse /kɔ:s/ 粗糙的, 未加工的; 不精确的, 近似的, 大型的; ～ pitch 大（粗）螺距; ～ thread 粗（牙）螺纹; ～ thread series 粗螺

纹级

coat /kəut/ 外层,表面;涂镀层,加面层;上涂料,涂镀;armour～ 保护层;first～ 底漆,底涂层;ground～ 底涂层;top～ 外涂层;under～ 内涂层

coating /ˈkəutiŋ/ 镀涂层;black～ 发黑处理,镀黑;protective～ 保护层

coaxal /kəuˈæksəl/ (coaxial)同轴的,共轴的;～ cylinder 同轴圆柱体

coaxality /kəukˈsælɪti/ 同(共)轴性,同轴度

cock /kɔk/ 旋塞,开关

code /kəud/ 代码,编码,代号;规程;～ for materials 材料符号;～ letters for structural elements 构件代号;American～ for pressure piping 美国压力管规号码;conversion～ 转换码;country～ 国家代码

coding /ˈkəudiŋ/ 编制程序;colour～ 彩色编码

coefficient /kəuiˈfiʃnt/ 率,系数,程度;～ of axial deformation 轴向变形率

coffer-wall /ˈkɔfə wɔːl/ 围墙

cog /kɔg/ (齿轮的)齿,轮牙,雄(凸)榫

cogwheel /ˈkɔgwiːl/ 齿轮;～ gearing 齿轮传动装置

coil /kɔil/ 螺形圈,线圈;number of supporting～s 支承圈数;number of working～s 工作圈数;total number of～s(弹簧)总圈数

coil-spring 螺形弹簧

coin /kɔin/ 压印,压花

coincide /kəuinˈsaid/ 重合,与……一致,相符

coincidence /kəuˈinsidəns/ 重合,符合,一致,相等;～ of projections 重影,重影性;use the～ method 使用重合法

coincident /kəuˈinsidənt/ 重合的,一致的,符合的;～ configurations 重合,形状重合;～ line 重叠线

col. (column) 柱

cold-work /ˈkəuldˈwəːk/ 冷加工

collar /ˈkɔlə/ 环,轴环,环状物

collator /kɔˈleitə/ 校对者,校对机

college /ˈkɔlidʒ/ 学院

collinear /kɔˈlinjə/ 同线的,共线的,在同一直线上的;～ planes 共线面

collineation /kəliniˈeiʃən/ 直射(变换),同(素)射(影)变换;共线,共线性;perspective～ 透视共线对应,透视同素对应;singular～ 奇(异)直射(变换)

collocate /ˈkɔləkeit/ 排列,安排,布置

collocation /ˌkɔləˈkeiʃn/ 布置,配置,安排,排列

color /ˈkʌlə/ (colour)颜色,染色,外貌,粉饰;～ circle 色相循环;～ conditioning 色彩调整;～ intensity 色彩浓度;～ matching 配色;～ mixing 混色;～ value 色彩明度;basic～ 原色;cold～ 冷色;contrast of～ 颜色对比;diluting of～ 颜色冲淡;RGB～ scheme 红绿蓝配色法;water～ 水彩

color-harden 着色硬化(发蓝),皮硬至极浅之深度

color-index 颜色指数

column /ˈkɔləm/ 柱,柱状物,列,直行;极,表,栏;座架;polygonal～ 多角柱;remarks～ 备注栏

COM (Computer Output Microfilmer) 计算机缩微输出

combination /ˌkɔmbiˈneiʃn/ 结合,结合体,集团;～ set 组合角尺;～ square 组合角尺

combine /kəmˈbain/ 使结合,使联合,兼有

command /kəˈmɑːnd/ 指令,命令;指挥,操纵

common /ˈkɔmən/ 公共的,公有的,共用的,共同的,普通的;～ difference 公差;～ perpendicular line 公垂线;～ point 公共点,节点;～ tangent 公切的

commutate /ˈkɔmjuteit/ 变换, 转换, 换向, 整流

commutativity /kəmju:təˈtiviti/ 互换性, 可换性

compare /kəmˈpɛə/ 对照, 比较

compartment /kəmˈpɑ:tmənt/ 间隔, 部分, 室

compass /ˈkʌmpəs/ 圆规, 罗盘, 指南针, 指北针, 范围; ~ frame 圆规架; ~es with detachable legs 活脚圆规; ~ attachment 圆规附件; ~ key 圆规用起子; beam ~es（画大圆的）长臂圆规; bisecting ~es 平分两脚规; bow ~es 弹簧圆规,（微调）小圆规, 测径规; caliber ~ 微调小圆规, 弯脚圆规; cardinal points of the ~ 指北标志; drawing ~ 制图圆规; drop ~es 小圆规; friction (hand) ~ 普通圆规; inking point of ~ 圆规用上墨笔头; interchangeable point of ~ 圆规的可换笔头; parallel ~ 平行圆规; proportional ~ 比例分规; tubular beam ~es 管形梁规

compatibility /kəmpætəˈbiliti/ 兼容性, 互换性

compatible /kəmˈpætəbl/ 相似的, 合适的, 协调的

compile /kəmˈpail/ 编译

compiler /kəmˈpailə/ 编译程序

compile-time 编译时间

complementary /kɔmpliˈmentəri/ 补充的, 互补的

complete /kəmˈpli:t/ 完整的, 完全的; 圆满的, 完成的; 完成, 结束; ~ delineation 画出完整外形; ‖ *Example sentence①: Complete the front view of the* two intersecting lines *and add a right-side view.* 例句①: 完成两相交直线的主视图, 并增加右视图。‖ *Example sentence②: Complete the front view of the* plane figure. 例句②: 完成平面图形的主视图。‖ *Example sentence③: Complete the given views of the* truncated pyramid *and add a top view.* 例句③: 完成截头棱锥的给定视图, 并增加俯视图。

complex /ˈkɔmpleks/ 复杂的, 综合的, 多元的, 复合的; 组合体, 全套; ~ chart 综合图; ~ curve 复合曲线; ~ geometrical surface 复杂几何曲面

component /kəmˈpəunənt/ 构件, 元件, 部件, 组件; 成分, 分力; 组成的, 合成的; ~ parts 组成部分; adjacent ~ 邻接组件

compose /kəmˈpəuz/ 构图, 设计, 组成; 著作, 编著

composition /ˌkɔmpəˈziʃn/ 构图, 组成, 出版

compress /kəmˈpres/ 缩短, 压缩

comprise /kəmˈpraiz/ 包含, 包括; 由……组成, 构成

compute /kəmˈpju:t/ 计算, 核算

computer /kəmˈpju:tə/ 计算机, 计算员, 电脑（电子计算机）; ~ aided design (CAD) 计算机辅助设计; ~ aided drafting 计算机辅助绘图; ~ aided graphic design 计算机辅助图形设计; ~ aided layout of masks 计算机辅助掩膜电路图设计; ~ aided manufacturing (CAM) 计算机辅助制造; ~ aided schematic system 计算机辅助简图设计系统; ~ direct drawing 计算机直接制图; ~ directed drawing 计算机控制制图; ~ directed drawing instrument 计算机控制的绘图仪; ~ generated hologram 计算机产生的全息图; ~ generated image 计算机生成的图像; ~ graphic application 计算机制图应用; ~ graphic interface (CGI) 计算机图形接口; ~ graphic system 计算机图形系统; ~ graphics 计算机图学, 计算机制图; ~ graphics aided three-dimensional interactive application 计算机图形辅助三维

交互作用；～ graphics metafile (CGM) 计算机图形元文件；～ graphics system 计算机图示（形）系统；～ image generator 计算机图像发生器；～ image processing 计算机图像处理；～ logic graphics 计算机逻辑图；～ micrographics 计算机微制图学；～ micrographics technology 计算机缩微图像技术；～ output micrographics 计算机输出缩微图形；～ picture 计算机图形；～ plotting system 计算机绘图系统；～ produced drawing 计算机绘图；～ science 计算机科学；decade ～ 十进制计算机；electronic ～ 电子计算机；electronic digital ～ 电子数字计算机；electronic warfare ～ 电子战计算机；general-purpose ～ 通用计算机；graphical ～ 图(解计)算计算机；repetitive (type) 周期运算式计算机；sequence controled ～ 程序控制计算机；special-purpose ～ 专用计算机

computer-aided (tracing) 计算机辅助（描绘）

computer-assisted (mapping system) 电脑（绘图系统）

computerize /kəm'pju:təraiz/ 计算机化

concatenation /kɔnkæti'neiʃən/ 级联，串接

concave /'kɔn'keiv/ 凹的，凹面的；～ pit 凹坑；～ slot 凹槽

conceive /kən'si:v/ 想象，设想；表达，陈述

concentrate /'kɔnsntreit/ 积聚

concentration /,kɔnsn'treiʃn/ 积聚；～ property 积聚性

concentric(al) /kən'sentrik(əl)/ 同（圆）心的，同轴的；公共的；～ circles 同心圆；～ circle method 同心圆法；～ line 公共线

concentricity /kɔnsen'trisiti/ 同轴度，同心度，同心性，同中心，集中

conception /kən'sepʃn/ 构思，想象；概念，想法，看法；fundamental ～ 基本概念

conceptual /kən'septʃuəl, -tjuəl/ 概念上的；方案图；～ phase 草图（初步）设计阶段

concerned /kən'sə:nd/ 有关的

conclude /kən'klu:d/ 结束，断定，决定，终了

concrete /'kɔnkri:t/ 混凝土

concurrent /kən'kʌrənt/ 同时发生的，共点的；～ lines 共点线；～ vectors 共点矢量

condition /kən'diʃn/ 条件，状态；color ～ing 色彩调整；least material ～ 最小实体状态；maximum material ～ 最大实体状态；the geometric ～ 几何条件

cone /kəun/ 圆锥，锥体，锥面，锥形，锥形物；～ center 锥心，锥顶；～ director 准锥面；～ frustum 圆锥台；～ of revolution 直圆锥；～ back 背锥；～ base of 锥底；circular truncated ～ 截圆锥；double ～ 双圆锥，对顶圆锥；elliptical ～ 椭圆锥；female ～ 凹圆锥；frustum of ～ 圆锥台；generating ～ （伞齿轮的）基锥；oblique ～ 斜锥；pitch ～ （伞齿轮的）节锥；right circular ～ 正圆锥体；truncated ～ 截锥体，锥台；warped ～ 翘曲锥面

cone-pointed 锥顶

config /kən'fig/ 配置

configuration /kən,figə'reiʃn/ 外形，形状，轮廓；地形，图形；排列，配置，布置；相对位置

confine /kən'fain/ 边界，边缘；区域，范围；限制

conformal /kən'fɔ:məl/ [数]共形的，保角的；～ mapping 共形映射，保角变换；～ projection 保形射影，共形射影，保角射影，正形射影

conformation /,kɔnfɔ:'meiʃn/ 相应，适应，构造，形态

conge(echinus) /knʒei/ 姆指圆饰

congregate /'kɔgrigeit/ （使）聚集；聚集的

congruent /'kɔgruənt/ 相同的, 相应的, 对应的, 全等的；～ figures 全等图形, 叠合图形；～ forms 左右相反形, 对应形；～ tansformation 全等变换, 相合变换

conic /'kɔnik/ 圆锥体, 圆锥的；圆锥（二次）曲线；～ section 圆锥截面（截交线）, 圆锥曲线, 二次曲线, 割锥线；absolute ～ 绝对锥

conical /'kɔnikl/ 锥形, 锥面, 圆锥形的；～ curves 圆锥曲线；～ helix 圆锥螺线；～ projection 锥面投影

conicity /kəu'nisiti/ 锥形, 锥度, 圆锥度

conicoid /'kɔnikɔid/ 二次曲面

coning /'kəuni/ 圆锥形, 锥度, 锥角；圆锥形的

conjugate /'kɔndʒugit/ 结合, 使成对；共生的, 共轭的；～ axis 共轭轴；～ diameter 共轭直径；harmonic ～ 调和共轭

connecting /kə'nektiŋ/ 连接；expanded-～ * 胀接

connection /kə'nekʃn/ 连接, 关系, 接头；～ to ground 接地；bolted-on ～ 螺栓连接法；fixed ～ * 静联接；flange ～ 法兰连接, 凸缘连接；interference fit ～ * 过盈联接；movable ～ * 动联接；multiple ～ 复接（电路图中的一种简化画法）；profile shaft ～ * 型面联接；rivet ～ 铆钉连接；screw ～ 螺钉连接

connector /kə'nektə/ 连接（符号）, 接插件；claw-plate ～ 爪状板连接器

conoid /'kəunɔid/ 圆锥体, 圆锥形；圆锥体的, 圆锥形的；锥状面, 扭锥面, 劈锥曲面；oblique ～ 斜劈锥体

con. sec (conic section) 圆锥截面

consecutive /kən'sekjutiv/ 连续的, 连贯的, 顺序的, 相邻的；～ elements 相邻素线；～ points 相邻点

consequently /'kɔnsikwəntli/ 因而, 所以

console /'kɔnsəul/ 托架, 控制台；～ printer 控制台打印机；～ typewriter 键盘打字机

conspectus /kən'spektəs/ 路线示意图, 流程示意图；一览表, 提要, 大纲, 简介

constant /'kɔnstənt/ 常数；integer ～ 整型常数；real ～ 实型常数

constitute /'kɔnstitju:t/ 构成, 组成；设立, 制定

construct /kən'strʌkt/ 构造, 建造, 编制, 绘制, 作图；～ devices 作一个样板；～ed profile 示意剖面图

constructing /kən'strʌktiŋ/ 画法

construction /kən'strʌkʃn/ 施工, 建造, 作图, 设计；画法；结构, 构造；图, 解释；～ drawing 构造图, 结构图；～ line 结构线；～ machine 施工机械；～ map 建筑图；～ plan 施工平面布置图；～ survey 施工测量；～s involving the circle 绕圆结构；approximate ～ 近似作图；block ～ 部件结构；four-center ～ 四（圆）心画（椭圆）法；geometrical ～ 几何作图；semi-box ～ 半箱法；show ～ 以图表示；technological ～ 工艺结构

constructor /kən'strʌktə/ 设计者, 设计师

contact /'kɔntækt/ 接触, 啮合, 接触处；～ angle 啮合角；double ～ 二重相切, 复切

contain /kən'tein/ 包含, 包容；等于, 相当于

content /'kɔntent/ 内容, 容量, 目录；满足, 满意的

context /'kɔntekst/ 上下文

continue /kən'tinju:/ 继续, 接续；～ printer 连续复印机

contour /'kɔntuə/ 外貌, 外形, 轮廓；略图；等高线, 等量线, 等深线, 轮廓线, 周线；描画轮廓, 围；大概, 大略；（叶片）型

线,仿形的;异形的;～ character 字体轮廓线;～ gauge 仿形规,板规;～ interval 等高线距;～ line 等高线,轮廓线;～ map 等高线图;～ of any surface 线轮廓度,面轮廓度;～ pen 曲线笔,轮廓笔;～ principle 外形原则;～ view 外形的,轮廓的;～ land 地形等高线; ore body top or bottom ～s 矿体顶底板等高线图

contoured /ˈkɔntʊəd/ 外形的,轮廓的

contourgraph /ˈkɔntuəgraːf/ （三维）轮廓仪,表面轮廓仪

contouring /kənˈtuəriŋ/ 仿形,画的外形,画的轮廓

contracted /kənˈtræktid/ 收缩了的;～ drawing 缩图

contraction /kənˈtrækʃn/ 缩短,缩小,收缩;～ joint(收)缩缝

contraflexure /ˈkɔntrəˈflekʃə/ 反向弯曲,反向曲线变换点,反挠,回折

contrast /kənˈtræst/ 明显的对比

control /kənˈtrəul/ 控制,操纵,调节器,控制系统;automatic ～ 自动控制;carriage ～ 走纸控制;numerical ～ 数字控制;time schedule (variable) ～ 程序控制,按时间控制

convenient /kənˈviːniənt/ 合适的,便利的,附近的,任意的

convention /kənˈvenʃn/ 习惯,惯例;集会,大会;协定,规定;～ for materials 材料的习惯画法

conventional /kənˈvenʃənl/ 习惯的,惯例的;习惯(画法),墨守旧法的;一般的,常规的;～ break 习惯断裂画法;～ construction 习惯结构(画法);～ diagram 示意图;～ intersection 习惯交线画法;～ method 习用方法;～ practice 常用习惯(规定)画法;～ representation 习用表示法;～ section 习用剖面,习惯画剖视法;～ sign 图例,习用符号;～ symbol 习用符号

converge /kənˈvəːdʒ/ 集合,同归一点,收敛,集中于……

converse /kənˈvəːs/ 相反的,颠倒的,逆的;逆,反

conversion /kənˈvəːʃn/ 变换,转换,转化,换算,逆转;coordinate ～ 坐标变换

convert /kənˈvəːt/ 变换,转换,改造,改变,转变

convertibility /kən,vəːtəˈbiliti/ 可逆性,互换性

convex /ˈkɔnˈveks/ 凸的,凸面;～ polygon 凸多边形;～ solid 凸立体

convey /kənˈvei/ 输送,传送;～ing worm 螺旋传送装置

convolute /ˈkɔnvəluːt/ 旋绕,卷绕;旋转的,蟠曲的;复杂的;卷绕面,盘旋面;盘旋形的;～ of two nappes 双叶盘旋面;～ surface 盘旋面;conical ～ 圆锥形盘旋面;helical ～ 螺旋面

convolution /,kɔnvəˈluːʃn/ 回旋,旋绕;折积,结合式

coordinate /kəuˈɔːdinit/ 坐标,同等的,并列的,同位的;～ axis 坐标轴;～ division 坐标网格线;～ method 坐标点法;～ paper 坐标纸,方格纸;～ plane 坐标面;～ plotter 坐标绘图机;～ system 坐标系统;affine ～ 仿射坐标;angular ～ 角坐标;areal ～ 面坐标;Cartesian ～s 直角坐标;circular cylinder ～s 圆柱坐标;device ～s 设备坐标;ellipsoidal ～ 椭圆体坐标;elliptic ～ 椭圆坐标;homogeneous ～ 齐次坐标;isometric ～s 等角坐標;lateral ～ 横坐标;line ～ 线坐标;longitudinal ～ 纵坐标;metric ～ 度量坐标;normalized device ～ 规格(范)化设备坐标;polar ～ 极坐标;projective ～ 射影坐标;rectangular ～ 直角坐标;relative ～ 相对坐标;user ～s 用户坐标;

vertical ～ 垂直(竖)坐标; world ～ 世界坐标(计); zero ～ 零位坐标; z- ～ Z 坐标

coordinatograph /kəuˈɔːdineitəgraːf/ 坐标仪, 坐标制图仪; 等位图, 等位线

coordinatometer /ˌkəuɔːdineiˈtɔmitə/ 坐标尺

cop (copper) 铜, 紫铜

cope /kəup/ 切合, 适应; 顶层盖, 上型箱

copier /ˈkɔpiə/ 复印机, 抄写员, 模仿者

co(-)planar /kəuˈpleinə/ 共(平)面的, 同(一平)面的

coplanarity /ˌkəupləˈnæriti/ 共面性, 同面性

coplane /ˈkəuplein/ 共面

co(-)planer (coplanar) 共(平)面的

copped /kɔpt/ 圆锥形的, 尖头的

copper /ˈkɔpə/ 铜, 工业紫铜, 铜制的, 镀铜; ～ composition 铜合金

COPPL (copper plate) 铜板

copy /ˈkɔpi/ 复制, 模仿, 仿形, 抄录; 副本, 复制品; 拷贝(电影), 样板; 卷, 册, 本, 份; 原稿, 复制图纸; ～ desk 编辑部; ～ paper 复印纸; ～ pattern 复制图; ～ rule 仿形尺; carbon ～ 碳化拷贝; foul (rough) ～ 草图, 草稿; hard ～ 硬拷贝(计); hard ～ unit 硬抄本单元; secondary master ～ 第二原图; soft ～ 软拷贝(计)

copygraph /ˈkɔpigraːf/ 油印图, 油印机

copying /ˈkɔpiŋ/ 复制, 晒印; 仿形切削

copyright /ˈkɔpirait/ 版权

cor. (corner) 角

cord /kɔːd/ 线, 绳; 堆积

core /kɔː/ 芯子, 型芯, 磁心; ～ box 型芯箱, 砂(芯)箱; ～ hole 铸孔; ～ print 型芯座

corner /ˈkɔːnə/ 角, 隅; ～ angle 顶角, 棱角; ～ cube prism 直角棱柱体; ～ dimension 夹角大小, 弯头尺寸; ～ joint 角接; ～ weld 90°(直)角接焊缝; filleted ～ 内圆角;

lower left-hand ～ 左下角; lower right-hand ～ 右下角; rounded ～ (修)圆角; sharp ～ 尖角; square ～ 内尖角

cornice /ˈkɔːnis/ 挑檐, 正檐; ～ & pitch of roof 飞檐与屋面斜度

correction /kəˈrekʃn/ 改正, 纠正; 修改, 核正; 勘误

correlate /ˈkɔrileit/ 相关, 关联, 关系; 相关数

correlation /ˌkɔriˈleiʃn/ 相关, 关联, 相应; 对射, 对射变换; 异射, 异射变换; ～ in space 空间对射变换; ～ of dimension 尺寸关系; angular ～ 角关联; cross ～ (互)相关; involutory ～ 对合对射; vector ～ 矢量相关

correlative /kəˈrelətiv/ 相关的, 相连的

correlatograph /kəˈrelətəˌgraːf/ (correlogram) 相关图, 对比图; ore and rock stratification ～ 矿岩层对比图

correlogram /kəˈreləgræm/ 相关(曲线)图

correspond /ˌkɔrisˈpɔnd/ 对应, 符合; 相当, 同位, 相符; 通讯; ～ to (with) 表示, 相当于, 与……相对应; The broad lines on the map ～s to roads 图中粗线表示公路; to every A there ～ B 对每一个 A 都有 B 与之对应

correspondence /ˌkɔrisˈpɔndəns/ 相当, 相应, 对应, 符合; 一致; 信件; algebraic ～ 代数对应; dualistic ～ 对偶对应; many-one ～ 多一对应; one-to-one ～ 一一对应; perspective-affine ～ of conic 二次曲线间的透视仿射对应; point ～ 点对应; projective ～ 射影对应; singular ～ 奇异对应; symmetrical ～ 对称对应

correspondent /ˌkɔrisˈpɔndənt/ 对应的, 相当的; 通讯者

corresponding /ˌkɔrisˈpɔndiŋ/ 相当的, 对

应的；~ angle 同位角；~ trace 对应迹
coslettise /'kɔzlitaiz/ 磷化处理
cottage /'kɔtidʒ/ 小屋，别墅，小筑
cotter /'kɔtə/ (cotar) 开尾销，销，制销；joint 销连接；split ~ 开口销；cotter-pin 开尾销，扁销；split ~ 开口销
count /kaunt/ 点，数，算数，计数，计算；~ing squares 数格法（用方格纸作比例定尺寸画图法）
counter /'kauntə/ 表示"反""逆""对应""重复"；相反，反击；计数器；~ bore 重复钻孔，柱坑孔，平底扩孔钻；~ parts 相似件，相对件
counterbored (hole) 柱坑（孔）
counterclockwise /'kauntə'klɔkwaiz/ 逆时针方向
counterfeit /'kauntəfit/ 仿造
counterpart /'kauntəpɑ:t/ 复本；配对物；相对应的，相当于
counter-revolution /kauntərevə'lu:ʃn/ 反旋转
countersink /'kauntəsiŋk/ 钻坑，锥坑钻头；钻锥坑；锥口孔，尖底
countersunk /'kauntəsʌnk/ (countersink) 锥口孔，尖底锥坑；埋头孔；锥口钻；~ head 埋头；~ hole 锥坑孔
coupé /'ku:pi/ （法语）（包含其他轮廓的）截面图
coupler /'kʌplə/ 联轴器
coupling /'kʌpliŋ/ 联轴节，连接
course /kɔ:s/ 层，过程，课程，教程；方针，方法，手续；damp-proof ~ 防潮层；heat insulation ~ 隔热层；vapour-proof ~ 隔汽层；waterproof ~ 防水层；watertight ~ 防水层
courtesy /'kə:tisi/ 礼貌，殷勤，谦恭有礼的言辞（行动）；by ~ of author 经作者同意（转载稿件、图片时的作者用语）

cover /'kʌvə/ 盖，罩，壳，套；涂上，镀上；封面；~ plate 盖板；box ~ 箱盖；case ~ 箱盖；end ~ 端盖；sealed ~ 密封盖；wheel hub ~ 轮毂盖
cowl /kaul/ 帽，套，罩，壳，盖；烟囱帽，通风帽；ventilating ~ 通风帽；wheel ~ 轮罩；wind ~ （烟囱）风帽
CP (circular pitch) 周节
CP/M (control program/monitor) 微机监控程序
cpr (copper) 铜，紫铜
CPU (central procesing unit) 中央处理机（单元）
CR (cold rolled) 冷辗（钢）
crack /kræk/ 裂纹，缝
crank /kræŋk/ 曲柄
crankshaft /'krækʃɑ:ft/ 曲轴
crater /'kreitə/ 焊口
craticulation /,kəuɑ:tikju'leiʃ(ə)n/ 分格法，分格转绘法
crayon /'kreiən/ 粉笔，腊笔，彩色铅笔；粉笔画，腊笔画
create /kri(:)'eit/ 生成，创造
crest /krest/ * 牙顶，牙尖，顶部，尖，波峰，峰顶；最大值；~ curve 凸形曲线；~ of screw throad 螺纹牙顶；~ of thread 牙顶，齿顶；tooth ~ 齿顶，齿倒棱（顶部）
crit. (criterion /krai'tiəriən/) 标准，准则
croquis /krəu'ki:/ （法语）尊图，速写，素描
cross /krɔs/ 交叉，相反；十字形，十字相交；~ axis 横轴；~ hatch 用交叉线画成阴影；~-hatched area 阴线区域；ever other ~ hatch line 每隔一个的剖面线；tracking ~ 跟踪十字
crossbreaking /krɔs'breikiŋ/ 横断
crosscorrelation /'krɔskɔri'leiʃn/ 相关
crosshatching /'krɔshætʃiŋ/ (sectionline)

断(剖)面线,断面线法(画剖面线的方法); alternate ～ 交变剖面线; dotted ～ 虚线剖面线

crosshead(ing) /ˈkrɔ(:) shed(iŋ)/ 十字头

crossover /ˈkrɔsəuvə/ 交叉,相交;切断; ～ point 交点

crosspoint /ˈkrɔspɔint/ 交叉点,贯穿点,交点

cross-section /ˈkrɔsekʃən/ 横截面,截面图,横断面图; ～ of stripping work 采剥工程断面图; ～ paper 方格纸,坐标纸; typical road ～ 道路标准横断面图

cross-sectional /ˈkrɔsˈsekʃənl/ 剖面的 cross-sectioned 用交叉线画成阴影线的

crowd /kraud/ 群,一堆;挤,挤满,拥挤;聚集

crown /kraun/ 冠,齿冠;顶,顶部;凸度,轮周; ～ of arch 拱顶; toothed bevel ～ (= ～ gear) 冠齿轮

CRS (cold roled steel) 冷轧钢; (cross section) 横截面; (centres) 中心距

CRT (cathode-ray tube) 阴极射线管

crude /kru:d/ 未加工的,粗糙的,粗制的,原始的

CS (cross section) 横截面; (carbon steel) 碳钢; (cast steel) 铸钢

CSBTS (China State Bureau of Technical Supervision) 中国国家技术监督局

CSG (Constructive Solid Geometry) 实体几何结构表示法

c'sink (countersink) 锥形沉坑孔,锥口孔,埋头孔

CSK (countersink) 锥孔,埋头孔

CSK-O (countersink other side) 对面锥坑

CSTG (castings) 铸件

C to C (centre to centre) 中心距

C to E (centre to end) 中心到端面的距离

CTR (centre) 中心,顶针; (contour) 轮廓,等高线

ctrl (control) 控制(键)

ctrs (centres) 中心

CU (cubic) 立方的

cube /kju:b/ 立方体,正六面体,立方;求……体积; isometric ～ 等角立方体

cubic(al) /ˈkju:bik(əl)/ 立方体的,三次的,立方的;三次曲线,三次方程式; ～ hyperbolic parabola 三次双曲抛物线

cuboid /ˈkju:bɔid/ 长方体;立方体的

CU FT (cubic foot) 立方英尺

CU IN (cubic inch) 立方英寸

culmination /kʌlmiˈneiʃn/ 顶点,极点

cumulate /ˈkju:mjulit/ 累积的

cumulative /ˈkju:mjulətiv/ 累积的,渐增的; ～ errors 累积误差; ～ tolerance 累积公差

cuneate /ˈkju:nit/ 楔形的

current /ˈkʌrənt/ 运转的,流动的;当前

cursor /ˈkə:sə/ 光标,游标; crosshair ～ 十字准线光标; rubber band ～ 皮筋拖动光标

curvature /ˈkə:vətʃə/ 曲率,曲度,弯曲; average ～ 平均曲率; degree of ～ 弯度; double ～ 双曲率

curve /kə:v/ 曲线,弯曲,弄弯;曲线板,曲线尺; ～ length 曲线长; ～ made up of successive arcs 弧成曲线; ～ of order2 二阶曲线,点素二次曲线; ～ pen 曲线笔; ～ plotter 绘图器; ～ ruler 曲线板; ～d face 曲面; ～d surface 曲面; adjustable ～ 自由曲线规; approximate ～ 近似曲线; bipartite cubic ～ 双枝三次曲线; biquadratic ～ 四次曲线,双二次曲线; characteristic ～ 特性曲线; closed ～ 闭合曲线; common ～s 常用曲线; continuous ～ 连续曲线; contour ～ 计曲线; counter ～ 反向曲线; cubic ～ 三次曲线; cycloidal ～s 旋轮类曲线,摆线曲线; development of ～ 曲线的展开; double continuous ～ 正弦曲线, S 形曲线; enveloping ～ 包络曲线; faired ～ 展平曲线; fit ～ 拟合曲线; flexible ～ 挠性曲线尺, 万能曲线尺, 自由曲线尺; flexible French ～ 挠性曲线带; French ～ 曲线板, 曲线尺, 云形板; generating ～ 母曲线, 母线; generating tooth ～ 齿轮曲线; hyperbolic ～ 双曲线; hypergeometric ～ 超几何曲线, 超比曲线; intrinsic ～ 包络线, 禀性曲线; irregular ～s 不规则曲线(板); mathematical ～ 数学曲线; moment ～ 力(弯)矩曲线, 力(弯)矩图; noncircular ～ 非圆曲线; nongeometric ～ 非几何曲线; odd ～ 不规则曲线; ogee ～ 反曲线, S 形曲线, 双弯曲线; open ～ 开口曲线; plane ～s 平面曲线; quartic ～ 四次曲线; quintic ～ 五次曲线; railway ～s 铁路曲线板; regular ～s 规则曲线; reverse ～ 反向曲线; sextic ～ 六次曲线; ship ～ 船体曲线板; sine ～ 正弦曲线; skew ～ 不对称曲线; smooth ～ 圆滑曲线; space ～s 空间曲线; spline ～ 云形曲线; stem ～ 首曲线; tangent ～ 切线曲线; transition ～ 过渡曲线, 介曲线

curved /kə:vd/ 弯曲的, 弧形的; ～ diretrix 曲准线; ～ face 曲面; ～ line 曲线; ～ intersection 曲线相交; ～ rule 曲线规; ～ solid 曲面体; ～ stroke 曲线笔画(字母); ～ surface 曲面表面

curvilineal /kə:viˈliniəl/ (curvilinear / kə:viˈliniə/) 曲线的; 曲线, 曲导线; ～ directrix 曲准线; ～ element 面素线; ～ generatrix 曲动线

curvity /kə:viti/ 曲率

custom /ˈkʌstəm/ 惯例, 习惯

customary /ˈkʌstəməri/ 通常的, 习惯的

cut /kʌt/ 切断, 切去, 割, 截, 加工(锯, 钻, 镗等); 挖, 切口, 掏槽; 割纹; 交叉; 插图, 式样; 图板, 断面图; ～ paper 裁纸; ～ over 开通, 转换; ～ plane 截面; ～ section 截口; ～ centre 中心掏槽, 锥形掏槽; chamfer ～ 斜切面, 楔形切面; cross ～ 横切; finish ～ 精加工; perpendicular ～ 垂直于轴的截面, 垂直切割; rift ～ 破开

cut-away /ˈkʌtəˈwei/ 截去

cutback /ˈkʌtˌbæk/ 截短, 缩减; 急转方向

cutline /ˈkʌtˌlain/ 图例, 插图下的说明文字

cut-off /ˈkʌtɔf/ 切开, 切断, 截止, 关闭, 停车

cut-open /kʌtˈəupən/ 剖开

cut-out /ˈkʌtaut/ 截断, 切口; 切断, 关闭, 中止

cut-over /ˈkʌtəuvə/ 开动, 转换

cutter /ˈkʌtə/ 刀, 刀具; ～ clearance 退刀槽, 刀具余隙

cutting /ˈkʌtiŋ/ 车削, 割切, 切削, 穿插; line 剖切线; ～ plane 剖切面, 切割平面, 截平面; ～ plane method 切平面法; ～ symbol 剖切符号, 剖切(迹线)位置; aligned ～ 旋转剖; a line ～ aplane(直)线穿插平面, 线面相交; by ～ 切割法; compound ～ 复合剖; contiguous ～ planes 转折切割平面; cross ～ 横切, 横割; echelon ～ 阶梯剖; inclined ～ 斜剖; oblique ～ 斜剖; offset ～ 阶梯剖; revolved ～ 旋转剖; stairstep ～ 阶梯剖; turning ～ 旋转剖

CU YD (cubic yard) 立方码

C-washer C 形垫圈

CWG (Chinese Wire Gauge) 中国线规

cyclograph /ˈsaikləugraːf/ 圆弧规

cycloid /ˈsaiklɔid/ 摆(旋轮)线, 摆线形

cycloidal /saiˈklɔidl/ 摆线的; ～ curves 旋轮类曲线; ～ gear 摆线齿轮

cyl (cylinder) 圆柱, 圆筒; 汽缸; 柱面

cylinder /ˈsilində/ 圆柱体, 柱, 柱面; ～ axis 圆柱轴线; ～ bottom 圆柱底; ～ number 圈数; ～ cutting 剖切柱面; elliptic ～ 椭圆

柱; hollow ～ 空心圆柱; hyperbolic ～ 双曲柱面; oblique ～ 斜圆柱; right ～ 直（正）圆柱; ‖ *Example sentence:* There are two ～s, one, being solid and the other hollow. 例句: 有两个圆柱体, 一个是实心的, 另一个是空心的。

cylindric(al) /si'lindrik(l)/ 圆柱形, 柱面

cylindricality /ˌsilindri'kæliti/ 圆柱度, 柱面性; ～ helicoid 柱形螺旋体; ～ helix 柱面螺旋体; ～ oblique helicoid 柱形斜螺旋体; ～ right helicoid 柱形直螺旋体; ～ surface 圆柱面

cylindroid /'silindrɔid/ 柱状面, 扭柱面, 圆柱性面; 椭圆柱, 似圆柱的, 椭圆柱的

cyma /'saimə/ 反曲线

cymatium /si'meiʃiəm/ 反曲线状

D

D (diameter) 直径
DC motor controller 直流电动机控制器
DC (d-c, direct current) 直流电
D cast (die casting) 压铸件
dam /dæm/ 堤坝, 水闸, 挡料圈; 闭合
dark /daːk/ 黑; ～ face 阴暗面; not ～ enough 不够黑
darkening /'daːkn/ 加黑, 加深
dash /dæʃ/ 破折号（——）; 阴影线; ～ (and) dot line (～ed dotted line) 点划线; ～ed double-dotted line 双点划线, 剖面位置线; ～ed line 虚（短划）线; ～ interval 虚线间隔长度; ～ long 虚线短划长度; broken ～ 断裂线; dot-and- ～ed line 一点一划相间的线; long ～es 长划线; long ～ed short ～ed line 长短划线; short ～es 短划线; ‖ *Example sentence:* Run a ～ *dot line* across the sheet. 例句: 在纸上画一条点划线。

dashed(area) /dashd('eəriə)/ 阴影（部分）
dashed-line 虚线; ～ long 虚线长
data /'deitə/ （datum 的复数）数据, 已知数, 资料; 基准, 基准面; 信息; ～-in 输入数据; ～-out 输出数据; ～ bank 数据库集; ～ base 数据库, 信息集; ～ base management system 数据库管理系统; ～ block 数据组; ～ communication 信息传递; ～ medium 数据媒体; ～ processing 数据处理; ～ security 数据安全; ～ structure 数据结构; ～ target 基准目标; ～ transfer 数据传送; digital ～ 数字数据; electronic ～ processing 电子数据处理; engineering ～ 技术数据, 工程数据; experimental ～ 实验数据; graph ～ 图形数据; literal ～ 文字型数据; master ～ 基本数据; operational ～ 运行数据, 运行资料

date /deit/ 日期, 年月日; 注明日期; out of ～ 过时的, 老式的; up to ～ 最新式的

datum /'deitəm/ 资料, 数据; 基准面, 基准, 尺寸基准; ～ level 基准面; ～ line 基准线; ～ mark (point) 基准点; ～ plane 基准面; ～ waterlevel 基准水平; design ～ 设计基准; fixed ～ 已知数, 固定标高, 固定基准; location ～ 定位基准; measurement ～ 测量基准; ordinance ～ 标高; technic ～ 工艺基准; technological ～ 工艺基准

daub /dɔːb/ 涂料, 粗灰泥

daubing /dɔːbiŋ/ 涂抹；～-up 涂上
daylight /ˈdeilait/ 日光，白昼；间隙，间隔
DBMS (Database Management System) 数据库管理系统
DC (digital computer) 数字计算机
D/C (drawing change) 图纸更改
D & C (Drafting and Checking) 制图(设计)与检查
DCD (Design Control Drawing) 设计检验图表
DCL (Drawing Change List) 图纸更改一览表
DCR (Drawing Change Request) 更改图纸的请求
DCS (Drawing Change Summary) 图纸更改一览
DDA (Digital Differential Analyzer) 数字微分分析器
DDC (Direct Drawing Change) 图纸直接更改
DDL (Data Drawing List) 资料图纸清单
dead /ded/ 呆滞的，死的，静寂的；～ center 死顶尖；～ load 静负荷；～ in line 配置于一直线，轴线重合；～ white 洁白
debug /diːˈbʌg/ 调试，排除故障
deburr /diˈbəː/ 去毛刺，倒角，清理毛口，修边
deburring /diˈbəri/ 去除毛刺
dec. (declination) 倾斜，偏差，磁偏角
decade /ˈdekeid/ 十进制的
decagon /ˈdekəgən/ 十边形
decahedron /dekəˈhedrən/ 十面体
decelerator /diːˈseləreitə/ 减速器，减速装置
decide /diˈsaid/ 决定，决心，解决
decimal /ˈdesiml/ 小数的，十进位小数的；～ marker 小数点；～ point 小数点；～ scale 十进位比例尺，小数刻度；～ size 小数尺寸；～ system 小数制
decimeter (decimetre) /ˈdesimitə/ 分米 $(=\frac{1}{10}$ 米)，略作 dm
decision /diˈsiʒən/ 判断
deck /dek/ 甲板，舱面；盖，层面；装饰；(卡片)组；sourse ～ 源程序卡片组
declarator /diˈklærətə/ 说明符
declination /deklineiʃən/ 倾斜，偏差，偏角，磁偏角
declivity /diˈkliviti/ 倾斜，坡面，坡度
dedendum /diˈdendəm/ ＊牙底高，齿根(高)；～ angle 齿根角；～ circle 齿根圆；tooth ～ 齿根高
deep /diːp/ 深的，埋头的，深色的；沉度
deepen /ˈdiːpən/ 加深，延伸
deepening /ˈdiːpəniŋ/ 加深
default /diˈfɔːlt/ 缺省值，缺陷；～ selection 缺省值选择
defect /diˈfekt/ 缺陷，缺点；故障；不足，缺乏；‖*Example sentence:* No ～s did we find in these parts. *例句：* 在这些部件中，我们没有找出任何缺陷。
define /diˈfain/ 限定，规定；解释，明确表示；定义
definite /ˈdefinit/ 确定；确定的，明晰的
definition /defiˈniʃn/ 定义，界说
deflect /diˈflekt/ 偏斜，使偏
deflection /diˈflekʃn/ 偏斜，偏差，偏转；挠曲，挠度；～ angle 偏角
deform /diˈfɔːm/ 使变形
deformation /diːfɔːˈmeiʃn/ 变形，失真，畸变，变态，形变；topological ～ 拓扑形变
DEG (degree) 角度，程度，等第，学位，方次
degree /diˈgriː/ 角度，程度，等第，学位，方次；～ of accuracy 准确度，精度；～ of hardness 硬度；～ of inclination 倾斜度；～ of preciseness 准确度，精确度；～ scale 刻度；circular ～ 圆度
del /del/ 倒三角形

del (delete) /di'li:t/ 消去（键）
delay /di'lei/ 延迟
delineascope /dɪ'lɪnɪəskəʊp/ 幻灯，映画器
delineate /di'linieit/ 描外形，画轮廓，描绘，描写，勾画，刻画
delineation /dilini'eiʃn/ 图解，描绘；轮廓，草图，略图，示意图；叙述
delineative /di'linieitiv/ 描绘的，叙述的
delineator /di'linieitə/ 制图者；图型，描画器
delta /'deltə/ 三角形△；希腊字母 Δ
demagnification /di:mægnifi'keiʃən/ 缩小
demand /di'ma:nd/ 要求，需要，查询
demarcate /'di:ma:keit/ 分开，区分；给……划界
demi- (词头) 半, 部分；～section 半剖面（图）；半节，半段
demicircular /demi'sə:kjulə/ 半圆形的
demilune /'demilu:n/ 半月，新月
demisemi /'demisemi/ 两者各半的，四分之一的
demon. (demonstrative) 指示的，用图表说明的
demonstrate /'demənstreit/ 表明，说明，论证，证明
demonstration /,deməns'treiʃn/ 示明，阐明，证实；挂图，展示图
demonstrative /di'mɔnstrətiv/ 用图表说明的，指示的
demount /di:'maunt/ 拆卸，把……卸下
denominator /di'nɔmineitə/ （分数的）分母
denotation /di:nəu'teiʃən/ 表示，指示；符号，意义
denote /di'nəut/ 表示，指示
dentation /den'teiʃən/ 牙形，齿状；成齿，成牙
dep. (department) 部门，局，部，科，系，室；～ of mechanical engineering 机械工程系；drafting ～ 设计科；engineering ～ 工程系，技术部；process (ing) ～ 工艺科；research ～ 研究室
departure /di:pa:tʃə/ 横距
dependance (dependence /di'pendəns/) 相关性，依赖性；依靠
dependancy (dependency /di'pendənsi/) 相关性，依赖，从属性，从属物
dependant (dependent /di'pendənt/) 相关的，依赖的；～ event 相关事件
dependence /di'pendəns/ 同 dependance
dependency /di'pendənsi/ 同 dependancy
dependent /di'pendənt/ 同 dependant
depict /di'pikt/ 描写，描述，描绘
depicter /di'piktə/ 描绘者
depiction /di'pikʃn/ 描写，描述，描绘，画
depictive /di'piktiv/ 描写，描述，描绘
depictor /di'piktə/ 同 depicter
depicture /di'piktʃə/ 描绘，描述，想象
deploy /di'plɔi/ 展开，部署，调度，开伞
deployment /di'plɔimənt/ 展开，部署
deposit /di'pɔzit/ 复层，镀层，涂，镀
dept. 同 dep., department
depth /depθ/ 深，深度，高度；（颜色的）浓度，厚度，程度；～ axis 深度轴线；～ direction 深度方向；～ of bore 钻孔深度；～ of drilled hole 钻孔深度；～ of thread 螺纹深度；～ of tooth 齿高
depth-gage /dep-geidʒ/ 深度规
depthometer /dep'θɔmitə/ 深度计
depth-width-ratio /depθwid'reiʃiəu/ 高（度）与阔（度）之比
derivation /,deri'veiʃn/ 偏差，偏转，分支，分路；推导，演算；standard ～ 标准偏差
derive /di'raiv/ 引出，引伸出来；推导，从……得出
DES (design) 设计，图纸，设计书；计算

Desargues 代沙格（1593—1662年），里昂的建筑师兼数学家。写有著名的《透视法》一书，建立了几何形体透视绘象的法则——射影法。(代沙格定理是研究射影变换的基础)

descr. (describe) 描写，描述，作图，画图

describe /disˈkraib/ 描写，描述，画图，作图；～ M around N (round N) 绕 N 沿（轨道）M 运行

describer /disˈkraibə/ 制图者，叙述者

description /disˈkripʃn/ 作图，绘图，描述，形容；说明书；种类，形式；～ of a triangle 三角形的绘制

descriptive /diˈskriptiv/ 画法，描写，叙述；图形式的，描述的，画法的，说明的；～ geometry 画法几何；‖ *Example sentence* ①: Descriptive geometry *is the science of graphic representation and solution of space problems.* (引自Paré等著《米制—画法几何》*Descriptive Geometry-Metric.*) 例句①：画法几何是一门图示和图解空间问题的科学。‖ *Example sentence*②: Descriptive geometry *is a versatile tool.* 例句②：画法几何是一种万能工具。

descriptor /disˈkriptə/ 描述符；field ～ 区域描述符（字段说明符）

des. form. (design formula) 设计公式，计算公式

desiccate /ˈdesikeit/ 烘干，晒干

design /diˈzain/ 设计，构思，计划，计算；图样，打样，图纸，图案；雏形，装置；～ activist 设计者；～ assembly 设计图，设计装配图；～ automation 设计自动化；～ chart 设计图表；～ drawing 设计图；～ engineer 设计工程师；～ file 设计文件；～ layout 设计图，设计草图；～ lettering 设计字法（建筑）；～ of stope 采区开采设计图，开采方法设计图；～ of working face 工作面施工（设计）图；～ paper 绘图纸，设计绘图纸；～ phase 设计阶段；～ requirement 设计规则；～ specification 设计规格说明；computer-aided ～ 计算机辅助设计；detailed ～ 详细设计；empirical ～ 经验设计；excavation ～ 采掘设计图；figured ～ 图案画；final pit ～ of opencut 露天矿终了平面图；functional ～ 功能设计图；machine ～ 机器设计图；preliminary ～ 初步设计；schematic ～ 方案设计；standard ～ 标准图；typical ～ 定型图

designate /ˈdezigneit/ 指出，标志

designation /ˌdezigˈneiʃn/ 名称，称号，代号，牌号；指定，标示；～ of structural steel sections 型钢标注；～ strip 名牌；item ～ 项目代号；surface roughness ～ 表面粗糙度符号

designer /diˈzainə/ 制图者，设计员，设计师

designing /diˈzainiŋ/ 设计，计划；～ of construction lines 定出断面，特型设计；～ of section 剖面图，断面图，截面图

designograph /diˈzainəgra:f/ 设计图解（法）

designs /diˈzainz/ 设计，计划，作图；目的，志向

desk /desk/ 台，石板；drawing ～ 绘图桌

dessin /deˈsi:n/ （法语）线画，图案

det (detail) * 详图（放大图，大样图），节点详图；细节，选择，选派；画详图；零件；～ dwg 零件图

detail /ˈdi:teil, diˈteil/ (det)；～ drawing* 零件图，详图；～ed section (removed section) 详细截面（多个截面）；～ element 零件要素；～ er's (est) 详图（用的设备），大样图；～ing 零件设计；～ing pen 特号笔；～ of construction 施工详图，构造详图；～ of design 设计详图；～ of foot 底部详图；～ of joints 节点详图；～ of structure 构造详

图; ～ paper 底图纸,画详图用的描图纸; dominant ～ 主要细节; base ～s 基础图; cast-in sit reinforced concrete element ～s 现场浇制钢筋混凝土构件详图; construction ～s 施工详图; elevation ～ 详细正视图; foundation ～s 基础详图; individual ～ 单个零件图; joint ～ 节点图; measuring and sketching ～ 零件测绘; miscellaneous structural ～s 其他结构详图; pipe piece ～ 管件图; pipe support ～ 管架图; precast reinforced concrete element ～s 预制钢筋混凝土构件详图; prototypes & design ～ing 造型设计; reinforced concrete structure construction ～s 钢筋混凝土结构节点构造详图; spare ～ 备(用零)件; splice ～s 接头详图; standard ～s 标准详图; structural ～ 结构元件; survey and drawing ～ 零件测绘; typical ～ 典型零件图

detail-sketching /diːteilˈsketʃiŋ/ 零件草图

detent /diˈtent/ 棘爪,制动轮,扳手

determine /diˈtəːmin/ 判别,决定,确定,测定,规定; ～ the scale and format 确定比例与图幅; ‖ *Example sentence: Determine the visibility and complete the views. 例句: 判别可见性并完成视图。*

dev. (deviation) 偏转,偏差; (device) 仪器,工具,装置,设备

develop /diˈveləp/ 发展,展开; ～ed pattern 显形图,展开图; ～ed view 展示图

developable /diˈveləpəbl/ 可展的,可展开的,可展曲面; ～ surface 可展曲面

development /diˈveləpmənt/ 展开视图,展开; ～ by triangulation 用三角形法展开; ～ -inner part 内表面部分展开图; ～ of surface 表面展开图; approximate ～ 近似展开; half ～ 半展开图

deviation /ˌdiːviˈeiʃn/ (*尺寸)偏差,偏向,偏移; ～ in lead* 导程偏差; ～ of flank angle* 牙侧角偏差; ～ of stroke* 行程偏差; actual ～ * 实际偏差; angular ～ 角偏差; average ～ 平均偏差; basic ～ 基本偏差; dimension ～ 尺寸偏差; fundamental ～* 基本偏差; limit ～ * 极限偏差; limits of ～ 极限偏差; lower ～ * 下偏差; mean ～ 平均偏差; mean square root ～ 均方根偏差; negative ～ 负偏差; positive ～ 正偏差; standard ～ 标准偏差; tolerance ～ 允许偏差; upper ～ * 上偏差; zero ～ 零偏差

device /diˈvais/ 设计,计划;装置,设备,仪器,仪表,工具,器件,方法,策略; character display ～ 字符显示设备; image input ～ 图像输入设备; image output ～ 图像输出设备; measuring ～ 量具; peripheral ～ 外部设备; plotting ～ 曲线绘制器; stroke ～ 笔划设备(计)

devise /diˈvaiz/ 设计,发明,想出,作出

deviser /diˈvaizə/ 设计者,发明者

dexter /ˈdekstə/ 右手的,右边的

dextr- (词头)右(旋)的

dextral /ˈdekstrəl/ 右的,右旋的,用右手的

dextrogyrate /ˌdekstrəˈdʒaireit/ 右旋的

dextrogyric /dekstrəʊˈdʒairik/ 右旋的

dextrorotary /ˌdekstrəˈrəʊtəri/ 右旋的(物)

dextrorotatory /ˈdekstrəʊˈrəʊtətəri/ 右旋的(物)

dextrorsal /ˈdekstrɔːsl/ 右旋的; ～ curve 右旋曲线

DF (design formula) 设计公式; (drive fit) 密(推入)配合

dft. (draft) 草图,轮廓,草稿,草案;牵引,通风

dftmn (draftsman) 制图员

DH (difference in height) 高度差

DIA (diameter) 直径

diag. (diagonal) 对角线,斜拉条; (diagram)

图表,图解,曲线图;(diagrammatic) 图解的,图表的

diagonal /dia'ægənl/ 对角线,斜拉条,斜剖线(船舶);~ division 对角分割;~ plane 对角面;body ~ 体对角线;principal ~ 主对角线

diagram /'daiəgræm/ * 简图,图形,图示,图表,线图,曲线图,路线图,示意图;平衡图,一览表;图解,制图,作成图形;~ drawing 线图(安装、布线及管线图,常作写生状),图解图;~ flow 操作程序图;~ method 图表法,图解法;~ of curves 曲线图;~ of furnace 炉型示意图;~ of gears 齿轮传动图;~ of hydrostatic curve 静水力曲线图;~ of strains 应变图;~ of work 示功图;accompanying ~ 附图;acess block ~ 存取框图;action ~ 活动流图;action sequence ~ 作用顺序图,动作顺序图;air craft wiring ~ 飞机接线图;alignment ~ 列线图;analytical ~ 分析线图;area ~ 面积图;arrow ~ 矢量图,向量图;asembly ~ 装配图,立体装配示意图;auxiliary piping and instrumentation flow ~ 辅助管道及仪表流程图;base ~ 电子管管底接线图;base line type of connection ~ 基线式接线图;bending ~ 弯折图;beamslab installation ~ 梁板安装图;black-and-white ~ 黑白图;block ~ * 框图;block ~ of optics 光学方框图;business report ~ 业务报告图;cable collocation ~ 电缆配置图;cause and effect ~ 因果(分析)图,鱼刺图,树枝图,特性要因图;chemical technology process ~ 化工工艺图;child ~ 子图;circuit ~ * 电路图,电原理图;circuit-description ~ 接线示意图,电路作用描述图;clock-phase ~ 直角坐标矢量图;complete schematic ~ 总图,总线路图;composition ~ 组合图;connection ~ * 接线图;conventional ~ 示意图;distribution ~ 配线图;draining system ~ 排水系统示意图;draw a ~ 绘图表,作图解;electrical ~ 线路图(电);electrical schematic ~ 电气简图;electronic ~ 电子线路图;electronic schematic ~ 电子系统图;elementary ~ 电气原理图;entity relationship ~ 实体关系图;equipment illustration ~ 设备形象示意图;equivalent-circuit ~ 等效电路图;erection ~ 线路安装图,桥梁、构件等的安装图,建造图,装配图;error check analysis ~ 错误检验分析图;external wirint ~ 外线图(电);flow ~ * 流程图,程序方框图,流量图;flow block ~ 流程框图;force ~ 力图;framing ~ 框架图;free-space ~ 立体图;frequency translation ~ 频率搬移图;functional ~ 方块图,工作原理图,功能图,机能图;general ~ (化工)总工艺流程图;graphical ~ 符号图;hinged connection ~ 铰接图;illustration ~ 说明图;indicator ~ 示功图;installation ~ 立体安装示意图;interconnecting ~ 内接线图,互联接线图;internal wiring ~ 室内配线图(电),安装图;interrupted line type of connection ~ 中断线式接线图;key ~ 索引图,概略原理图,原理草图,解说图;kinematic ~ 机动示意图;level ~ 电平图;line ~ 线路图;location ~ 位置简图;locomotive working ~ 机车周转图;locomotive working ~ with name 记名式机车周转图;logic ~ * 逻辑图;main wiring ~ 主结线图;make a ~ 作图;master ~ 总线图;material balance ~ 物料平衡流程图;mechanical schematic ~ 机构原理图;network ~ 网络图,矢形图,箭头图,统筹图,工序流程图;nozzle orintation ~ 管口方位图;one-line ~ 单线电路图;ore body isoline ~ 矿体等值线图;pareto ~ 排列图;pie ~ 圆形(馅饼状)图

表; pipework ～ 管路图; piping ～ 管路图; piping and instrumentation flow ～ 管道及仪表流程图; point to point type of connection ～ 点对点式接线图; polar ～ 极坐标算图; power system ～ 供电系统示意图; process flow ～ 工艺流程图; process piping and instrumentation flow ～ 工艺管道及仪表流程图; reference ～ 参考图; routing ～ 程序图; scatter ～ 散布图, 相关图; scattering ～ 散布图; schematic ～ *原理图, 简图, 略图; schematic ～ of conductor transposition 导线换位图; schematic ～ of optical system 光学系统图; sector ～ 扇形图; set-up ～ 安装图, (计算系统) 准备 (工作框) 图; single-line ～ 单线电路图; single locomotive working ～ 一元式机车周转图; skeleton ～ 设计轮廓图; space ～ 空间图, 立体图, 示量图, 位置图; stalation ～ for conductor jumper spacer 导线跳线间隔棒安装示意图; state ～ (平衡) 状态图; statistical ～ 统计图表; straight line type of connection ～ 直线式接线图; stress ～ 应力图; stress-deformation ～ (stress-strain ～) 应力应变图; system ～ 系统图; tensile test ～ 拉力试验图; terminal connection ～ 端子接线图; terminal function ～ 端子功能图; test ～ 试验线图; train working ～ 列车运行图; transition ～ 转移图; transportation system ～ 运输系统示意图; triaxial ～ 三轴图; trunk line type of connection ～ 干线式接线图; tube-location ～ 电子管位置图; unit connection ～ 单元接线图; vector ～ 结构线图, 向量图; ventilation system ～ 通风系统示意图; volume ～ 体积图; wall ～ 挂图; winding ～ 绕组图; wiring ～ 接线图, 线路图, 布线图

diagrammatic(al) /daɪəɡrəˈmætɪk(əl)/ 图解的, 图表的; 概略的, 轮廓的; ～ arrangement 构想草图, 图式布置, 原则性布置; ～ chart 示意图, 线图图表; ～ curve 图示曲线; ～ drawing 草图, 略图, 示意图; ～ representation 用图表示; ～ section 图解剖面; ～ view 简图

diagrammatically /ˌdaɪəɡrəˈmætɪkli/ 用图解法, 利用图表

diagrammatize /daɪəˈɡræmətaɪz/ 制成图表, 用图解表示

diagrammed /ˈdaɪəˌɡræmd/ (diagram(m)ing) 用图示出, 用图解法表示

diagraph /ˈdaɪəɡrɑːf/ 绘图器, 放大绘器, 分度画线仪, 分度尺, 仿型仪

dial /ˈdaɪəl/ 表面, 转盘, 标度盘, 拨号; ～ compartor 针盘比测器; ～ gage 千分 (百分) 表, 指示表; ～ indicator 针盘指示器; ～ inside micrometer 内径千分尺

diam. (diameter) 直径

diameter /daɪˈæmɪtə/ 直径; ～ of aperture 孔径; ～ of pitch circle 节圆直径; ～ pitch (齿轮) 径节; ～ run-out 径向跳动; a circle two feet in ～ 直径二尺的圆; angle ～ 中径 (锥形螺纹); apparent ～ 外表直径, 视径; average ～ 平均直径; basic major ～ 螺纹基本大径; basic pitch (of screw) ～ (螺钉的) 基本中径; blank ～ 外径; bottom ～ 螺纹内径, 螺纹底径; conjugate ～ 共轭直径; core (inside) ～ 内径; crest ～ *(螺纹) 顶径; dividing circle ～ 分度圆直径; effective ～ 有效直径, (螺纹) 中径, (齿轮) 节径; form ～ 外形直径; full ～ 最大直径, 主直径; gauge ～ *(螺纹) 基准直径; major ～ *(螺纹) 大径, 外直径; mean ～ 平均直径; minor ～ *(螺纹) 小径; minor ～ of screw thread 螺纹内径; neck ～ 颈部直径; nominal ～ *公称直径, 规定直径; outer throat ～ 最大喉径; outside ～ 外径; overal ～ 总直径, 全径; pitch ～ *(螺纹) 中径,

径节(齿轮),分度圆直径; pore ~ 孔径; root ~ *(螺纹)底径,齿根直径; theoretical pitch circle ~ 分度圆直径; thread ~ 螺纹外径

diametral /daiˈæmitrəl/ 直径的; ~ pitch 齿轮径节; point ~ 前端直径(螺纹的); root ~ 内径,心子直径(铰刀的),齿根直径; sweep ~ 扫描直径; thread ~ 螺纹直径(外径,大径)

diamond /ˈdaiəmənd/ 金钢石,金钢钻; 菱形; ~ penetrator hardness 锥氏金钢石硬度

diapositive /daiəˈpɔzitiv/ 透明照相正片, 幻灯片

diascope /ˈdaiəˌskəup/ 透射映画器

diazo /daiˈæzəu/ 重氮基的; ~ compounds 重氮化合物

dictionary /ˈdikʃənəri/ 词典,字典; mechanical ~ 机械字典,翻译计算机

die /dai/ 搬牙,铸型,螺旋模,胚膜; ~ cast 压铸件; ~ plate 模板; ~ stamp 镁锻件; lower ~ 下模

Diff. E (difference east) 横坐标差

difference /ˈdifrəns/ 差,差额,差别,区别; ~ in elevation 高度差; partial ~ 偏差

differential /difəˈrenʃl/ 差别(的),微分(的); ~ geometries 微分几何; partial ~ 偏微分

Diff. H (difference in height) 高度差

Diff. N (difference north) 纵坐标差

digiplot /ˈdidʒɪplɒt/ 数字作图

digit /ˈdidʒit/ 数(字),位,号; even ~ 偶数元

digital /ˈdidʒitl/ 数字的,(计算机的)数位

digitization /ˈdidʒitaiˈzeiʃən/ 数字化(名词)

digitize /ˈdidʒitaiz/ 数字化(动词)

digitizer /ˈdidʒitaizə/ (digitiser) 数字转换器,数字化仪; rotating drum image scanning ~ 鼓形扫描数字化仪

digraph /ˈdaigrɑ:f/ 有向图

dihedral /daiˈhi:drəl/ 两面角,由两个平面构成的,两面的; ~ angle 两面角; ‖ *Example sentence:* The angle formed by two intersecting planes is called a *dihedral angle.* 例句:两个相交平面所形成的角称为*两面角*。‖ *Example sentence:* The *true size of the dihedral angle* is observed in a view in which each of the given planes appears in edge view. 例句:二面角的真实大小反映在给定两平面都是重影的视图上。

DIL (document issuing list) 文件颁发表

dilatation /ˌdaileiˈteiʃən/ (图像)扩张

Dim (dimension) 尺寸,因次,维,度

dimension /diˈmenʃən/ * 尺寸,尺度,大小,广度,因次,维,度,在……上标出尺寸; ~ analysis 尺寸分析; ~ figure 尺寸数字; ~ forms 尺寸形式; ~ included in frame 加方框的尺寸; ~ inspection 尺寸检验; ~ line 尺寸线; ~ of picture 图形尺寸; altered ~ 更改后尺寸; angular ~ 角度尺寸; assembling ~ 装配尺寸; assembling connecting ~ 装配连接尺寸; assembling position ~ 装配位置尺寸; associative ~ing 相关尺寸标注(计); automatic ~ing 自动标注尺寸; basic ~ 理想尺寸; boundary ~ * 外形尺寸,轮廓尺寸,边界尺寸; chain ~ 链状尺寸; chain ~s 连续尺寸; characteristic ~ 特性尺寸,性能尺寸,基本尺寸,规格尺寸; closed ~ 封闭尺寸; continuous ~s 连续尺寸; coupling ~ 联接尺寸; critical ~ 关键尺寸; datum ~ 理论正确尺寸,基准尺寸; duplicate ~ 重复尺寸; equal ~s 相等尺寸; external ~ 外形尺寸,外部尺寸; extreme ~ 最外尺寸; fit ~ 配合尺寸; fixing ~ 装配尺寸,安装尺寸; form ~ 定形尺寸; free ~ 自由尺寸; full ~ 原尺寸(足

尺）; general ～ 主要尺寸, 概要尺寸; given ～ 已知尺寸; inscription of ～s 尺寸标注; intermediate ～ 分段尺寸; internal ～ 内部尺寸; inside ～ 内尺寸; installation ～ 安装尺寸; largest permissible ～ 最大容许尺寸; leading ～ 主要尺寸, 轮廓尺寸; limiting ～ 极限尺寸; limiting position ～ 极限位置尺寸; linear ～ 线性尺寸; location ～ 定位尺寸, 位置尺寸; longitudinal ～ 纵向尺寸; major ～ 主要尺寸; many ～s 多维; mating ～ 配合尺寸; nominal ～ 公称尺寸; one ～ 一度（的）尺寸, 线性（的）尺寸; non-functional ～ 非功能尺寸; offset ～ 偏位尺寸; omitted ～ 省略尺寸; out-of-scale ～ 不合比例尺寸; outer ～ 外形尺寸, 轮廓尺寸; overall ～ 总体尺寸, 外部尺寸, 最大尺寸, 轮廓尺寸, 全尺寸; parallel ～ 平行尺寸; physical ～ 外形尺寸; photographic reduction ～ 照相缩小尺寸; preliminary ～s 预定尺寸; redundant ～ 多余尺寸; reference ～ 参考尺寸; size ～ 大小尺寸; slot ～ 槽的尺寸; staggered ～s 交错尺寸; superfluous ～ 多余尺寸; three- ～s 三度（的）尺寸; 立体（的）尺寸; three space ～s 三度空间尺寸; total ～ 总体尺寸; two- ～s 二度（的）尺寸, 平面（的）尺寸; underlined ～ 数字下有划线的尺寸; unfinished ～ 非加工面尺寸; unidirectional ～s system 单向注尺寸制; unnecessary ～ 不必要尺寸; untoleranced ～ 不带公差尺寸

dimensional /diˈmenʃənl/ 广度, 大小, 尺寸的, ……维的, ……度空间的; ～ chain 尺寸链; ～ design 设计尺寸; three-～ molding 三维模型

dimensioned /diˈmenʃənd/ 大小（一致的）; 被度量的

dimensioning /diˈmenʃəniŋ/ 寸法, 尺寸注法, 定尺度, 标尺寸; ～ across the chord 尺寸弦长注法; ～ angles 角度标注; ～ around the arc 尺寸弧长注法; ～ by coordinates 坐标标注[曲线]尺寸法; ～ by offsets 支距标注[曲线]尺寸法; ～ by ordinates 坐标标注[曲线]尺寸法; ～ by radii 半径标注[曲线]尺寸法; ～ of circle 圆的尺寸标注法; ～ methods 表示大小的方法; ～ scale 比例尺; ～ standard feature 标准形体注尺寸法; aligned ～ 对齐制注尺寸法; arrowless ～ 不用箭头注尺寸法; base line ～ 基线尺寸（法）; chain ～ 标注连续尺寸; decimal ～ 小尺寸注法; dual ～ 双尺寸注法（公制与英制）; form ～ of tapers 锥体尺寸注法; inch ～ 英寸尺寸标注; mixed fractional and decimal ～ 小数分数并用注尺寸法; order of ～ 注尺寸顺序; ordinate ～ 坐标注尺寸法; over ～ 尺寸过多; progressive ～ 基准式尺寸法; rectangular coordinate ～ 直角坐标注尺寸法; successive ～ 连续注尺寸; tabular ～ 列表注尺寸法

dimetric /daiˈmetrik/ 二等角的, 正方形的, 四角形的; ～ projection 二等角投影, 正二测投影; oblique ～ projection 斜二测投影

DIN (Deutsche Industrie Normen) 德意志联邦共和国工业标准

diorama /daiəˈrɑːmə/ 透视画（面）

dioramic /ˌdaiəˈræmik/ 透视画

dip /dip/ 蘸, 把……放入又取出; 浸; 倾斜, 俯角

diptych /ˈdiptik/ 双幅一联画（雕刻）

dir (direction) 地址; 命令, 指示

direct /diˈrekt, daiˈrekt/ 直接（地）, 笔直（地）; ～ dimensioning size 直接注入尺寸; ～ ratio 正比; ～ tolerancing 直接注入公差

direction /diˈrekʃn, daiˈrekʃn/ 方向, 路线, 方针, 命令; ～ of flow 流向; ～ of projector 投射方向; ～ of turning 旋向; ～ of viewing

观察方向; clock-wise ～ 顺时针方向; projecting ～ 投影方向; reading ～ 识图方向

directly-proportional /di'rektli'prə'pɔːʃənl/ 成正比(例)的

director /di'rektə, dai'rektə/ 导向器, 引向器, 导面; plane ～ 导面; point ～ 导点

directory /di'rektəri/ 目录, 地址录

directrix /di'rektriks/ 准线, 导线; curvilinear ～ 曲导线; rectilinear ～ 直导线

DIS (draft international standard) 国际标准草案

disalignment /disə'lainmənt/ 不同轴, 轴线不重合; 不平行, 偏离中心线, 不同心, 不平行度, 不同轴度

disassemble /disə'sembl/ 拆卸, 分解

disc /disk/ (disk) 圆盘, 圆板, 叶轮, 盘状物,(计算机)磁盘; cam ～ 凸轮盘; crank ～ 曲柄盘, 盘式曲柄; wheel ～ 轮盘

discal /'diskəl/ 平圆盘的, 盘状的

discernible /di'səːnəbl/ 可辨别的, 辨别得出的, 可察觉的, 可看得出的

discoid /'diskɔid/ 平圆形的, 圆盘状的

discontinuity /diskɔnti'njuːiti/ 不连续, 间断; 突变性

discontinuous /diskən'tinjuəs/ 不连续的, 间断的

discrepance /dis'krepəns/ (discrepancy /dis'krepənsi/) 偏差, 误差, 差异; 不符合, 矛盾

disk /disk/ (disc) 盘, 轮, 平圆形物, 圆盘,(计算机)磁盘, 硬盘; fixed ～ 固定式磁盘, 硬盘; floppy ～ 软盘; magnetic ～ 磁盘

diskette /dis'ket/ 软盘; scratch ～ 刻线软盘, 草稿盘; blank ～ 空白软盘; master ～ 主软盘; original ～ 原盘; source ～ 源盘; target ～ 目标软盘, 靶盘

distance /'distəns/ 距离; closely toleranced center ～ 精密公差中心距; edge ～ 边距; linear ～ 直线距离

dismount /dis'maunt/ 拆卸, 卸下

dismountable /dis'mauntəbl/ 可拆卸的

displaced /dis'pleist/ 位移的, 移位的, 代替的

displacement /dis'pleismənt/ 位移, 变位, 置换, 位移度; angular ～ 角位移量, 角偏差

display /dis'plei/ 标记, 表现, 展示, 显示, 陈列; ～ drawing 观览图; ～ unit 显示装置; Chinese character ～ 汉字显示器; colour ～ 彩色显示; graphic ～ 图形显示(器); graphic ～ unit 图形显示装置; monochrome ～ 单色显示; panel ～ 平板显示; raster ～ 光栅显示器; scribing plotting ～ 划线笔绘图显示器; stereo ～ 立体显示(器); three-dimentional ～ 三维显示(器); two-dimentional ～ 二维显示(器)

displayer /dis'pleiə/ 显示器; colour ～ 彩色显示器

disposal /dis'pəuzl/ 配置, 处理, 消除

dispose /dis'pəuz/ 安排, 配置, 处理; 解决; 影响

disposition /dispə'ziʃn/ 配置, 布置, 部署, 安排, 倾向; ～ plan (设备)配置平面图

disproportion /disprə'pɔːʃn/ 不成比例, 不相称

dissymmetry /dis'simitri/ 不对称

dist (distance) 距离; center ～ 中心距; horizontal ～ 平距; ‖ *Example sentence①: The distance between two points is the length of the straight line segment bounded by them.* 例句①: 两点间的距离是两点间直线段的长度。‖ *Example sentence②: The distance of a point from a plane is the length of the shortest line from the point to the plane. This line is perpendicular to the plane.* 例句②: 点到平面的距离是点到平面最短线的长度, 该线垂直于平面。

distinct /dis'tikt/ 明显的, 清楚的; 个别的, 性质不同的; as ~ from 与……不同的

distinction /dis'tikʃn/ 差别, 区别; 特征, 特性

distort /dis'tɔ:t/ 使……变形, 歪曲, 曲解

distortion /dis'tɔ:ʃn/ 变形, 失真; 弄歪

dividers /di'vaidəz/ 分规, 分线规, 脚规; friction (joint) ~ 普通分规; hairspring ~ 徽调分规; proportional ~ 比例分规; spring ~ 弹簧分规

DL (drawing list) 图目; (drill) 钻, 钻孔

dm (decimeter) 分米

DOC (document /'dɔkjumənt/) 文件, 资料, 记录; ~ content security 文件内容安全性; ~ issuing list 文件颁发表; signature ~ 签署的文件

documentation /ˌdɔkjumen'teiʃn/ 文件 (管理); ~ security 文件(系统)安全性

dodecagon /dəu'dekəgən/ 十二角形, 十二边形

dodecahedron /ˌdəudekə'hi:drən/ 十二面体; regular ~ 正十二面体

d. of c (diagram of connection) 连接图, 接线图

dog /dɔg/ 狗; 止挡, 卡箍, 销, 凸轮; ~ hart 制动爪装配图

dog-leg /'dɔg-leg/ 折线形的, <形的

domain /də'mein/ 范围, 区域

dome /dəum/ 圆顶, 圆盖; 成半球形 ~-shaped 圆(屋)顶形的

domic(al) /'dəumik(əl)/ 圆顶的

dominance /'dɔminəns/ 支配性

done /dʌn/ 完成

door /dɔ:/ 门, 闸; double sliding ~ 双扇拉门; hinged ~ 铰链拉门, 门(船); swing ~ 弹簧门; self-closing ~ 自闭门

doorframe /'dɔ:freim/ 门框

dorsal /'dɔ:səl/ 背面的, 背部的; 脊的, 脊部的; ~ view 背视图

DOS (Diskette Operation System) 磁盘操作系统

dot /dɔt/ 小点, 圆点; 打点号, 用点线表示; dot-and-dash line 一划一点相间的线(点划线) ~ dash line (~ted dashed line) 点划线; ~ line (~ted-line) 点线; ~ted curve 点曲线; ‖ *Example sentence:* Run a dashed dotted line across the sheet. 例句: 在纸上画一条点划线; bold ~ 圆点

double /'dʌbl/ 两倍的, 二重的, 双的, 重合的; ~ auxiliary 复辅视图; ~-bevel scale 双斜面比例尺; ~ cone 双圆锥, 对顶圆锥; ~-curved surface 复曲面; ~-dashed dotted line 双点划线; ~-dashed double-dotted line 双划双点线; ~ dots line 双点划线; ~ edge 双边, ~ helical (spur) gear 人字齿轮; ~ pointed nail 双头螺栓, 接合销; ~ size 双倍尺寸; ~ thread 双头螺纹; ~ thread worm 双头蜗杆; ~ T iron 工字钢; ~ vee 双 V 形, X 形的

double-dots-dash-line 双点划线

double-circular-arc 双圆弧

double-track 双线的, 复线的

doublet /'dʌblit/ 双线, 双重线; 一对, 复制品

dove /dʌv/ 鸽子; ~ tailed 鸠尾形

dowel /'dauəl/ 木钉, 暗销, 钉类 ~ pin 定位销, 圆柱销

downspout /'daunspaut/ 雨水管

downward /'daunwəd/ 向下, ……以下, 在下面

DP (diametral pitch) 径节; (deep) 深度, 沉度, 埋头的; (draft proposal) 建议草案; (data processing) 数据处理

DPC (damp proof course) 防潮层

DPM (drafting practice manual) 制图实践手册

DPT. (department) 分类, 部门, 局, 部, 科, 系
Dr (doctor) 博士; (drawer) 制图者, (drawing) 图纸, 草图, 蓝图; 绘图; (drawn) 冷拉的
draft /drɑ:ft/ (draught) 草图, 图样; 起草, 草案; 轮廓, 画……的草图, 制图; 斜度; 出模斜度; 牵引, 通风, 排气; ~ for a machine 机器草图; ~ International Standard (DIS) 国际标准草案; ~ing system 绘图系统; automatic ~ 自动绘图; design ~ 设计草图; forging ~ 模锻斜度; pattern ~ 拔模斜度; rough ~ 草图
drafter /'drɑ:ftə/ (draughtter) 制图机, 制图者
drafting /'drɑ:ftiŋ/ (draughting) 制图, 设计, 图示; 自动绘图机; 起草; ~ board 制图板; ~ machine 制图机; ~ tape 制图带, 胶带; automated ~ 自动绘图机; commercial ~ 商用图; simplified ~ 简化制图; universal ~ machine 万能绘图仪
draftsman /'drɑ:ftsmən/ (draughtsman) 制图员; chief- ~ 制图主任; naval ~ 船舶制图员
draftsman-ship (draughtsman-ship) 制图术, 制图技术, 制图质量
draftsquare /drɑ:ftskweə(r)/ 制图尺, 制图模板
dragging /'drægiŋ/ 拖动, 牵引
draughting /'drɑ:ftiŋ/ (drafting) 制图, 设计; ~ instrument 绘图仪器; ~ room 制图室; ~ scale 绘图比例尺; numerically controlled ~ 数控绘图机
draw /drɔ:/ 画, 划, 作, 绘制; 描写, 起草, 抽出, 引出, 回火; ~ a circle 画一个圆; ~ a line from A to B 从 A 到 B 画一条线; ‖ *Example sentence: Draw* a line *that contains point O, is parallel to plane ABC, and intersects line EF. 例句: 过点O作直线平行于平面ABC并与直线EF相交.* ~ a straight line 画一直线; ~ outlines of 画……的轮廓草图; ~ the temper off 回火, 退火; ~ up 画出, 拟定; ‖ *Example sentence: An essential step in* drawing an orthographic view *is the correct determination of the visibility of the lines that make up the view. 例句: 绘制正投影图, 一个重要步骤是正确判断构成视图的那些直线的可见性.*

drawback /'drɔ:bæk/ 回火; 缺点
drawer /'drɔ(:)ə/ 绘图者, 制图员; 开票者, 抽屉
drawing /'drɔ:iŋ/ * 图, * 图样, 工程图纸, 视图, 附图; 画, 绘画, 制图, 图示; 回火, 退火; 拉, 拔; ~ aid 绘图辅助工具; ~ board 绘图板; ~ by axonometric projection 轴测投影图; ~ callout 图上表示法; ~ change 图纸更改; ~ compasses 制图圆规; ~ control 图样管理; ~ desk 绘图桌; ~ figure 图形; ~ for concrete hull of ships 水泥船船体制图; ~ for hull of ships 船体制图; ~ for metal hull of ships 金属船体制图; ~ free-hand 徒手画; ~ function 作图函数; ~ head 绘图头; ~ ink 上墨图, 绘图墨水; ~ instrument 制图工具; ~ knife 制图用刮刀; ~ machine 绘图机; ~ number 图号; ~ of brick works 砌砖图; ~ of excavation 开挖图; ~ of excavation construction 开挖施工图; ~ of external shape 理论图(飞机); ~ office 制图室, 设计室; ~ of motion trace 运动轨迹图; ~ of partial anlargement * 局部放大; ~ of sample 样板图; ~ of similar parts 列表零件图; ~ paper to suit 尺寸适宜的绘图纸; ~ parallel lines 画平行线; ~ pen 绘图笔, 鸭嘴笔; ~ pencil 绘图铅笔; ~ pin 图钉; ~ record card 图纸记录卡片; ~ room 制图室; ~ scale 绘图比例尺; ~ size 图纸类型; ~ symbol 制图符号; ~ system 绘图系统; ~ table 制图桌,

制图台; ~ with compass 用圆规画图; ~ with two scales 双重比例尺作图; ~ block 活页画图纸, 拉模板; ~ paper 绘图纸; a ~ showing the relation of underground mining method 井下采掘关系图; abbreviated ~ 简图; above and underground comparative ~ 井上下对照图; anisometric ~ 不等角图; architectural ~ 建筑制图, 建筑图; architectural engineering ~ 建筑工程图; architectural free-hand ~ 建筑草图; architectural presentation ~ 建筑示意图; architectural working ~ 建筑施工图; arrangement ~ 配置图; as-built ~ 竣工图; assembly ~ * 装配图, 总图; assembly diagrammatic ~ 装配示意图; audio-visual ~ 直观图; automobile body ~ 汽车车身制图; axonometric ~ * 轴测图, 轴测画法; axonometric exploded ~ 轴测分解图; axonometric ~ from oblique projection 斜轴测投影图; axonometric ~ from orthographic view 正轴测投影图; back elevation ~ 背视图; basic ~ logic 图学逻辑; blank ~ * 空白图, 毛坯图; blueprint ~ 蓝图; boiler ~ 锅炉制图; bright ~ 拉光; building ~ 建筑绘画; cabinet ~ 斜二测图; calculating ~ of reserves 储量计算图; casting ~ 铸件图; cavalier ~ 斜等测图; cave-inaffected ~ 塌陷波及线图; chemical engineering ~ 化学工程图, 化工制图; chemical engineering equipment ~ 化工设备图; civil engineering ~ 土木工程图; cold ~ 冷拉; collective ~ 图集, 装配图; collective assembly ~ 表格图, 表格装配图; collective single-part ~ 表格图, 共用单一图形图; combined ~ 零件装配集合图; combined detail and assembly ~ 零件、装配联合图, 集合图; commercial ~ 商用图; completion ~ 竣工图; conceptual ~ * 方案图; conditional ~ 条件图; construction ~ 结构图, 施工图(水利工程); construction ~ of well and tunnel 井巷施工图, 井巷施工单; contour ~ of an aero-engine 发动机轮廓图; contract ~ 契约图, 发包图; contracted ~ 缩图, 缩绘; control ~ 控制图; controlling system ~ 操纵系统图; conventional ~ 习惯画法; copying ~ 摹图, 抄图; cross-sectional ~ 横断面图; current ~ 适用图; design ~ * 设计图; design assembly ~ 设计装配图; detail ~ * 零件图, 细部图, 细节图, 大样图, 详图; detail ~ of concrete pile 钢筋混凝土桩详图; detail ~ of filter 滤光镜零件图; detail ~ of lens 透镜零件图; detail ~ of optics 光学零件图; detail ~ of prism 棱镜零件图; detail ~ of scale plate 分划板零件图; developing ~ * 展开图; development ~ 展开图; diagram ~ 线图(安装, 布线及管线图, 常作写生状); diagrammatic ~ 草图, 示意图; diagrammatic sectional ~ 剖面草图; die ~ 模具图; digiting ~ 数字化绘图; dimensional ~ 尺度图; dimensioned ~ 有尺度图样; dimetric ~ 二等轴测图; dimetric ~ from oblique projection 斜二等轴测图; dimetric ~ from orthographic view 正二等轴测图; display ~ 展览图, 观览图; disposition ~ 配置图; drive system ~ 传动系统图; earthwork ~ 土方工程图; electrical ~ 电工制图, 电气制图, 电工图; electrical engineering ~ 电气工程图; electronic ~ 电子图; elevational ~ 正面图样; engineering ~ 工程图, 工程画, 工程制图, 工程图样; end-product ~ 成品图; engineering geological ~ 工程地质图; engineering presentation ~ 工程示意图; enlarged ~ 放大图, 局部放大图; equipment placement ~ 设备布置图; equipment platform ~ 设备平台图; equipment support ~ 设备支架图;

erection ～ 安装图, 施工图, 装置图, 架设图; estimate ～ 估价图; exploded assembly ～ 立体分解系统图, 分解装配图; exploded pictorial ～ 立体分解系统图; excavation engineering ～ 采掘工程图; excavation engineering bedding ～ 采掘工程层面图; explanatory ～ 说明图; figuration ～ * 外形图; final ～ 最后图样; first angle ～ 第一角法; first quadrant ～ 第一角(象限)画图法(欧式); fitting arrangement ～ 附件装配图; flipping ～ s 翻动图形; foil ～ 铜片图(电); forging ～ 锻造图, 锻件图; form ～ 木模图, 模板图; formed material ～ 型材图; forming ～ 模板图; formwork ～ 模板图; foundation ～ 基础图; free-hand ～ 徒手画, 徒手图, 草图; full-size ～ 原尺寸图, 实物尺寸图; functional ～ 实用图; furniture ～ 家具制图; gauge ～ 量具图; general ～ 总图; general arrangement ～ 总图; general assembly ～ 总装配图; general ～ of furnace body 炉体总图; generalized ～ primitive 广义绘图图元; geometrical ～ 几何图; gluey unit ～ of optics 光学胶合件图; hand ～ 草图; harbor engineering ～ 港口工程制图; harness ～ 线札图; heating piping system ～ 采暖管道系统图; highway engineering ～ 公路工程图; hydraulic and hydro-electrical engineering ～ 水利电力工程制图; hydrogeological ～ 水文地质图; illustrated ～ 说明图; illustrated working ～ 立体分解系统图; in ～ 画得准确的; individual detail ～ 个别零件图; indoor plumbing system ～ 室内给排水系统图; industrial electronic ～ 工业电子图; initial ～ 初始图; ink ～ 水墨画; inked ～ 上墨图; installation ～ * 安装图, 装置图; installation process ～ 工艺安装图; instrument ～ 工具图; instrumental ～ 用器画, 仪器画; isometric ～ 等角画, 等轴测图; isometric ～ from orthographic view 正等轴测图; jig ～ 夹具图; key ～ 例图, 纲要图, 索引图, 解释图; layout ～ 布置图, 配置图, 外形图; legend ～ 标记符号图; light-and-shade ～ 明暗画, 润饰法; lighting system ～ 照明系统图; line ～ 线图; linear perspective ～ 线透视图; list of ～ 图纸目录; location ～ 定测图, 地盘图, 位置图; machine ～ 机械图; machine movement schematic ～ 机动示意图; manufacturing ～ 机械加工图; map ～ 地形图; masonry structure ～ 圬工结构图; master ～ 样图, 发令图(光学仿型); mechanical ～ 机械制图; Mechanical ～ of national standard 国家标准《机械制图》; mechanical engineering ～ 机械工程图; mine stopping set-on ～ 矿块定型图; mine survey ～ 矿山测量图; miniature ～ 缩小图; model ～ * 毛坯图; model photo ～ 模型照像图; mould ～ 模型图; multi-layer printed board detail ～ 多层印制板零件图; multiview ～ 多面视图; natural-size ～ 实物尺寸, 原尺寸图; nonprojection ～ 非投影图; oblique ～ 斜视图; one-sided printed board detail ～ 单面印制板零件图; optics ～ 光学制图; order of ～ 制图顺序; ore piece geological ～ 矿块地质图; original ～ * 原图; orthographic multiview ～ 正投影的多面视图图样; orthometric ～ 正视画法; out of ～ 画得不准确的, 画错的, 不合画法; outline ～ 外形图, 轮廓图, 草图; outside ～ 外形图; overlapping ～ 重叠图; part ～ 零件图; partial enlarged ～ 局部放大图; parts placement layout ～ 零件布置图(电); patent ～ 专利图; pattern ～ 木模图; pencil ～ 铅笔图; perspective ～ 透视图; phantom ～ 假想图, 幻影图, 幻画, 部分剖视图; photo- ～ 照相图; pictorial ～

直观图, 立体图, 形象图; picture ～ 写生制图法; pipe ～ 管路图; pipe ～ of ore selecting 选矿工艺管路图; pipeline ～ 管系图; pipe-line ～ 管线图; piping ～ 管系图; piping isometric ～ 管道轴测图; piping layout ～ 管道布置图; piping system ～ *管系图; plain view ～ 平面图; plumbing ～ 管路图; pocket ～ 袖珍图; power supply system ～ 电力系统图; preliminary design ～ 初步设计图; preliminary ～ 初步样图; preliminary three view ～ 初步设计三视图(航空); presentation ～ 示意图, 表意图; printed board ～ 印制板制图; printed board detail ～ 印制板零件图; printed board assembly ～ 印制板装配图; procedure ～ 工序图; production ～ *施工图; profile ～ 仿形图; process ～ 程序图; production ～ for casting 铸造图; projection ～ 投影图; proposal ～ 建议图; railway engineering ～ 铁路工程图; reconnaissance ～ 草测图; reduced size ～ 缩尺图; reinforced concrete structure ～ 钢筋混凝土结构图; reinforcement ～ 配筋图, 钢筋布置图; repair ～ 修理图; reproduction of ～ 图之复制; reversed isometric ～ 反向等轴测图; rivet joint ～ 铆接图; road ～ 线路图; rough ～ 草图; scale ～ 缩尺图; schematic ～ 示意图, 略图, 原理图; schematic ～ of an aero-engine 航空发动机简图; schematic design ～ 方案图; screw propeller ～ 螺旋桨图; sectional ～ 剖面图, 断面图, 截面图; sectional detail ～ 剖面详图; setting ～ 安装图; sheer ～ 船型放样图, 船体型线图; shop ～ 工作图, 生产图, 制造图, 装配图; shop detail ～ 装配详图; signal design ～ 信号设计图; simplifying ～ 简化图样; single-part ～ 单图, 单一图形图, 单一零件图; site ～ 区段图, 地区图; site plan working ～ 施工总平面图; skeleton ～ 草图, 原理图, 骨架图; small-scale ～ 缩尺图; spare parts ～ 备件图; special purpose ～ 特种目的图; specification on architectural working ～ 施工建筑图首页; specification on structural working ～ 结构施工图首页; specified ～ 通用图; standard ～ 标准图; steel structure ～ 钢结构图; structural ～ 结构(设计)图, 构造图; structural detail ～ 结构详图; structural working ～ 结构施工图; subassembly ～ 部(组)件装配图, 局部组合图; surveyed ～ 测绘图; surveying coordinate ～ 测量坐标网; symbolic ～ 符号图; tabular ～ *表格图; tabular assembly ～ 列表组合图; tabulated detail ～ 表格零件图; technical ～ 技术制图, 技术绘画; template ～ 样板图; third angle ～ 第三角法; third quadrant ～ 第三角(象限)画法(美式); three-plane projection ～ 三面投影图; three-view ～ 三视图图样, 画三视图, 三面视图; three-view ～ of an airplane 飞机三面图; timber structure ～ 木结构图; tooling ～ 工装图; topographical ～ 地形图; tower position ～ 塔位图; traced ～ *底图; two-sided printed board detail ～ 双面印制板零件图; two-view ～ 两视图图样; 两面图; typical ～ 标准图, 定型图; unfolding ～ 展开图(矿业); unit ～ of glass products 玻璃制件图; unit assembly ～ 部件装配图; ventilation, air-conditioning duct system ～ 通风、空调管道系统图; welding ～ 焊接图; working ～ 工作图, 加工图, 生产图, 施工图, 施工详图; working ～ architectural detail 施工建筑详图; working ～ architectural elevation 施工建筑立面图; working ～ architectural plan 施工建筑平面图; working ～ architectural section 施工建筑剖面图

drawing-back 回火

drawing-board 绘图板
drawing-compasses 绘图圆规
drawing-off 引出的
drawing-paper 绘图纸
drawing-pin 图钉
drawknife /'drɔ:naif/ 刮刀
drawn /drɔ:n/ 画好的, 伸延的, 拉制; cold ～ 冷拉; hard ～ 硬拉
drawshave /'drɔ:ʃeiv/ 刮刀
DRD (Design Requirement Drawing) 设计要求的图纸; (Design Research Division) 设计研究室
DRG (drawing) 图样
DSGN (design) 设计
DWN (drawn by) 制图者
drill /dril/ 钻, 钻头; ～ed hole 钻孔; ～ point 钻尖; center ～ 中心孔钻; depth of ～ tap 预钻孔深（攻螺纹用）
drive /draiv/ 主动, 传动, 驾驶; 驱动器; ～ fit 打入配合; ～ gear 主动齿轮; ～ shaft 主动轴; belt ～ 皮带传动; bevel ～ 伞轮传动; chain ～ 链传动; diskette ～ 磁盘驱动器; gear ～ 齿轮传动; worm-wheel ～ 蜗轮传动
driven /'drivn/ 从动; ～ gear 从动齿轮; ～ motor 电动机; ～ shaft 从动轴; gear ～ 齿轮传动的
driver /'draivə/ 传动轮, 赶锥; screw ～ 螺丝刀, 改锥

driving /'draiviŋ/ 驱动, 传动; 主动的
DRM (Draft Room Manual) 制图室手册
drop /drɔp/ 作, 投; 落下, 下降, 降低; ～ line 底线; ～ pen 坠笔圆规, 骤落笔
drum /drʌm/ 圆筒, 鼓轮, 滚筒; magnetic ～ 磁鼓
drw. (drawing) 图纸, 蓝图, 草图, 绘图
DS (drawing summary) 图纸一览
Dtl. dwg (detail drawing) 详图
duality /dju:'æliti/ 对偶性, 二重性
dual-screen /'dju:əlskri:n/ 双屏幕
duck /dʌk/ 很自然地; 压弯性曲线尺的铅块; draftsmen ～ 压铅
duct /dʌkt/ 管, 槽, 沟
ductile /'dʌktail/ 可展伸的
ductility /dʌk'tiliti/ 延展性
dull /dʌl/ 迟钝的, 呆笨的; ～ pencil 钝头铅笔
duplicate /'dju:plikeit/ * 复制图, 复印图, 复制品; 复制, 重复, 加倍; 双重的
duplicating /'dju:plikeitiŋ/ 复制, 复印; ～ tracing 复制描图, 复制底图
duplicator /'dju:plikeitə/ 复写器, 复印机
dust /dʌst/ 灰尘小尘埃; ～ing brush 清洁剂, 除尘刷; ～ing cloth 清洁用布
DWG (drawing) 图, 图样, 图画, 草图; 制图
dwindle /'dwindl/ 缩小, 变小, 减少; ～ from 6 to 2 从六减少到二
dynamic /dai'næmik/ 动态的

E

E (east) 东; E. &O. E (error and omission excepted) 误差除外

earth /ə:θ/ 泥土, 地, 地面, 地球; 接地
earthwork /'ɜ:θwɜ:k/ 土方工程

easel /ˈiːzl/ 绘图桌, 画架

EASIAC (Easy Instruction Automatic Computer) 教学用自动计算机

east /iːst/ 东, 东方; ～ by north (方向) 东偏北

eavse /iːvz/ 檐口; cantilever ～ 挑檐; projected ～ 挑檐

ebonite /ˈebənait/ 硬橡皮, 胶木

ec (exempli causa) 例如

eccentricity /eksenˈtrisiti/ 离心率; degree of ～ 离心度

echo /ˈekəu/ 模拟, 响应, 回声, 应答, 回显

ECL (Equipment Component List) 设备元件明细表

EDGD (Engineering Design Graphics Division) 工程设计图学分会

edge /edʒ/ 边缘, 边界, 刀口, 边, 棱, 锋; 给……加上边; ～ angle 棱角; ～ distance 边距(板边至铆钉中心之距); ～ iron 角铁; ～ of a surface 面的边; hemmed ～ 折边; ～ round 外圆角; ～ view 边视图(直线或平面在其垂直的投形面上的视图); adjacent ～ 邻棱; base ～ 底面边视图; beveled ～ 斜棱; chisel ～ 楔形笔尖; parallel ～s 平行边; single ～ 单边; straight ～ 直尺, 直边

edge-to-edge 边到边, 边靠边

edge-view 边视, 边视图

edgeways /ˈedʒweiz/ (edgewise) 以边缘向外(或向前), 从旁边, 沿边, 对着边地; ～ view 边视图; ～ weld 沿边焊接

edgewise-view 边视, 边视图

EDGJ (Engineering Design Graphics Journal) 《工程设计制图》期刊

editing /ˈeditiŋ/ 编纂, 编辑, 修改; picture ～ 画面编辑

edition /iˈdiʃn/ 版, 版本

EDP (Electronic Data Processing) 电子数据处理

educe /iˈdjuːs/ 引出, 析出, 推断, 演绎

ee (errors excepted) 允许误差, 误差除外

eg (exempli gratia) (for example) (for instance) 例如, 举例, 实例

egg-shape 卵形

EI (lower deviation of a hole) 孔的下偏差; (lower deviation of a shaft) 轴的下偏差

eidograph /ˈaidəgraːf/ 图画缩放仪, 伸缩画图器

eidography /ˈaidəgrəfai/ 缩放图法

eidophor /ˈaidəˈfɔː/ 大图像投射器

electric /iˈlektrik/ 电的, 用电的; ～ engineer drawing 电气工程图; ～al work 电气工程

electronic /ilekˈtrɔnik/ 电子的, 电子学的; ～ analog(ue) computer 电子模拟计算机; ～ computer 电子计算机; ～ digital computer 电子数字计算机; ～ image storago device 电子录像设备

element /ˈelimənt/ 素线, 母线, 元素, 单元, 零件, 部件, 元件; ～ of a cone 圆锥的素线; ～ of a cylinder 圆柱的素线; ～ of a surface 表面上的素线; ～ of first generation 首次衍生素线; ～ of second generation 再次衍生素线; blade ～ 叶片; consecutive ～ 邻接素线; contour ～ 轮廓素线; corresponding ～s of projective forms 射影形的对应元素; detectable ～ 可检测元素; double ～s 二重元素; geometric ～s 几何元素; machine ～ 机(械零)件; tangent ～ 切于素线

elephant /ˈelifənt/ 象, 象形锅炉, 一种28×23英寸图画纸

elev. (elevation) 立面图, 高度, 高程, 仰角

elevation /eliˈveiʃn/ * 立面图, 正视图, 正面图, 垂直剖面图, 纵剖图, 仰角高度, 标高; ～ angle of view 正视角(最外端之视线

与水平面间之夹角—透视角）；～ auxiliary 立面辅助视图；～ auxiliary view 立面辅助视图；～al drawing 立面图样；～ detail 详细正视图；～ oblique drawing 立面斜视图；～ view 立面图；absolute ～ 绝对标高；architectural ～ 建筑立面图；auxiliary ～ 辅助正视图；back ～ 后视图，背视图，背立面图；complete auxiliary ～ 全辅助视图；design ～ 设计标高；development ～ 展开立面图；down-stream ～ 下游立面图；east ～ 东立面图；end ～ 侧面图，侧视图；excavation engineering ～ 采掘工程立面图；front ～ 正面图，正视图，正立面图；front sectional ～ 前视剖面图；ground ～ 标高图；left ～ 左视图；partial auxiliary ～ 局部辅视图，部分辅视图；north ～ 北立面图；rear ～ 背立面图，后视图；relative ～ 相对标高；right ～ 右视图；sectional ～ 立剖面（图），立面剖视（图），竖直剖面；side ～ 侧视图，侧面图，侧立面图；south ～ 南立面图；tooth-end ～ 锯齿形；up-stream ～ 上游立面图；west ～ 西立面图

ellipse /iˈlips/ 椭圆，椭圆板；～ by trammel method 用椭圆规画椭圆；～-pin-and-string method 椭圆—钉线法；approximate four-centered ～ 四圆心近似椭圆；approximate eight-centered ～ 八圆心近似椭圆；concentric ～s 同心椭圆；Gardener's ～ 钉线椭圆法；eight-centered ～ 八（圆）心椭圆

ellipsograph /iˈlipsəgraːf/ 椭圆仪，椭圆规

ellipsoid /iˈlipsɔid/ 椭圆面，椭球，椭球体；～ of revolution 椭圆回转面；oblate ～ 扁椭球；prolate ～ 扁长椭球

elliptic(al) /iˈliptik(əl)/ 椭圆的，椭性的；～ arch 椭圆拱；～ cone 椭圆锥面；～ cylinder 椭圆柱面；～ paraboloid 椭性抛物面；oblique ～ cylinder 斜椭圆柱

ellipticity /elipˈtisiti/ 椭圆度，椭圆率

elongate /ˈiːlɔŋgeit/ 延长，伸长，拉长

else /els/ 否则

emanate /ˈemaneit/ 发散，放射，发源

empirical /emˈpirkəl/ 经验的，经验主义的；～ equations 经验方程

empty /ˈempti/ 空的

enclosing /inˈkləuziŋ/ 封闭，围合；～ square (method) 方框（法）

enclosure /inˈkləuʒə/ 边框，包围

end /end/ 末端，最后，目的，结果，完结；～ coil（弹簧的）无效圈；～ elevation 侧视图；～ of file 文件结束；～ of thread 螺纹退刀扣；～ plane 端面；～-on 轴向的，纵向的，端对的；～ view 端视图，侧视图；flat ～ 平端；square ～ 方头；～-point 端点，终点；‖ *Example sentence: The projection of a line on a plane is the line connecting the projection on the plane of the endpoints of the line.* 例句：直线在平面上的投影，是直线两端点在平面上投影的连线。

endfile 结束文件

endways /ˈendweiz/ (endwise /ˈendwaiz/) 末端向前地，末端朝上地，竖着；两端相接地，向着两端，在末端

engage /inˈgeidʒ/ 接合，啮合

engineer /ˌendʒiˈniə/ 工程师，技师；chief ～ 总工程师

engineering /ˌendʒiˈniəriŋ/ 工程，技术；～ drawing 工程画，工程制图；～ geometry 工程几何；～ graphics 工程图学；～ parlance 工程俗语；～s scals 工程比例尺；～ technology 工程技术；automatic ～ design program 自动工程设计程序；chemical ～ 化学工程；civil ～ 土木工程；computer-aided ～ 计算机辅助工程；electrical ～ 电气工程；human ～ 人机学，人类工程；hydraulic ～ 水利工程；mechanical ～ 机械工程；structural ～ 建筑工程，结构工程

ENGR (engineer) 工程师

enl. (enlarged) 扩大的, 放大的; (enlargement) 扩大, 放大

enlarge /in'la:dʒ/ 扩大, 放大, 详述; ～d drawing 放大图; ～ in section 截面放大

enlarging /in'la:dʒiŋ/ 放大

entail /in'teil/ 必须, 使承担

enter /'entə/ 打入, 输入

entity /'entiti/ （空间）实体

entry /'entri/ 项目, 入口

envelope /'enviləup/ 包络, 包络线, 包络面; 外壳, 机壳; ～ curve 包络曲线; ～ of curve 曲线的包络线; ～ of surfaces 曲面的包络面; parabolic ～ 抛物线包络

enveloping /'envilәupiŋ/ 球面, 包络面

enneahedron /,eniə'hi:drən/ 九面体

EOB (End of Block) "字组结束" 符, "信息组结束" 符

ep (end point) 端点, 终点

epicycle /'episaikl/ 周转圆

epicycloid /epi'saiklɔid/ 外摆线

epidiascope /epi'daiәskәup/ 实物幻灯机, 透、反射两用幻灯机

epitrochoid /,epi'trәukɔid/ 长短辐圆外旋轮线

epure /e'pju:ə/ 图, 线图, 极图, 原尺寸图案

equal /'i:kwəl/ 相等的; ～ly spaced position 等距分布

equality /(:)'kwɔliti/ 类似

equation /i'kweiʃn/ 公式, 方程（式）; empirical ～s 经验公式

equator /i'kweitə/ 赤道, (球形的) 大圆

equiangular /i:kwi'æŋgjulə/ 等角的

equiaxial /'i:kwi'æksiəl/ 等轴的

equidistance /'i:kwi'distəns/ 等距的

equidistant /'i:kwi'distənt/ 等距离的, 等分; ～ curve 等距曲线; ～ line 等距线

equidistributed /,i:kwidis'tribju(:)tid/ 等分布的

equiform /i:kwi'fɔ:m/ 相似 (的); ～ geometry 相似几何(学); ～ group 相似（变换）群

equilateral /'i:kwi'lætərəl/ 等边, 等边的; 等边形; ～ triangle 等边三角形

equipment /i'kwipmәnt/ 装备, 配备, 设备, 器材, 装置; drawing ～ 绘图工具; measuring ～ 量具; peripheral ～ 外围设备

erase /i'reiz/ 刮去, 擦去, 删除; ～ing shield 擦图片, 擦线板, (擦图)掩蔽盾; ～ing tool 擦图工具

eraser /i'reizә/ 橡皮, 消磁头; motor ～ 电动擦图器

erasing /ɪ'reiziŋ/ ～ knife 刮线小刀; ～ machine 电动橡皮擦

erect /i'rekt/ 直立的, 垂直的; 垂直安装, 装配, 设立, 建立

erection /i'rekʃn/ 竖直, 直立; 建立, 建造, 安装, 装配; ～ diagram (桥梁、构件等的) 安装图, 建造图

ergonomics /ә:gә'nɔmiks/ 人机学, 人类工程学

error /'erә/ 错误, 误差; ～ in encircled area 圆圈内是错误的; ～ of form 形状误差; ～ of location 位置误差; accumulative ～ 累积误差; aggregate ～ 累积误差; allowance ～ 允许误差; average ～ 平均误差; cumulative ～ 累积误差; fatal ～ 致命错误; fractional ～ 相对误差; geometric ～ (图像) 几何误差; permissible ～ 容许误差; projection ～ 投影误差; relative ～ 相对误差

escape /is'keip/ 退刀槽, 空刀槽; 出口, 排气管

esquisse /es'kwi:s/ 草拟图稿, 草稿

establish /is'tæbliʃ/ 建立, 设立; 制定, 规定, 作出, 安置

etalon /'eitələn/ 标准,规格,校准器
E to E (end to end) 末端到末端;全长
E. V. (edge view) 边视图,重影视图
even /'i:vən/ 平的,均匀的;使……平直,使……均匀;～ly spaced 间隔均匀的;～ number 偶数
evenness /'i:vənnis/ 平坦,平面度,平滑度,均匀度
evident /'evidənt/ 明显的,明白的
evolute /'i:vəlu:t/ 渐开线,渐屈线
evolution /i:və'lu:ʃn/ 展开,发展;演变,开方
evolvable /i'vɔlvəbl/ 可展开的
evolve /i'vɔlv/ 展开,开展,发展;引申出
evolvent /i'vɔlvənt/ 渐伸线,渐开线,切展线
exact /ig'zækt/ 精确的,确切的;要求,求出,画出;正好;～ shape 正确轮廓
exaggerate /ig'zædʒəreit/ 夸张,夸大;放大(比例尺);扩大;～ a scale 放大比例
examination /igzæmi'neiʃn/ 考试,检验,检查
example /ig'za:mpl/ 例,例题;榜样
excess /ik'ses/ 超过,过分,过量;～ lines 多余线条
excessive /ik'sesiv/ 过度的,极度的
exchangeability /ikstʃeindʒə'biliti/ 可交换性,互换性
exchangeable /iks'tʃeindʒəbl/ 可互换的
excircle /ek'sə:kl/ 外圆
excursion /iks'kə:ʃn/ 偏差,偏移
exercise /'eksəsaiz/ 练习,习题;训练,运用
exhibit /ig'zibit/ 表示,显示,展出,陈列
exit /'eksit/ 退出;出口
expandable /iks'pændəbl/ 可展开的,可延伸的
expansion /iks'pænʃn/ 展开,扩张,扩大;发展,展开式;shell ～ and framings 外板展开图(船舶)
expedient /iks'pi:djənt/ 方便的,有利的;得计的
explement /'ekspləmənt/ 辅角(360°与该角之差)
explode /iks'pləud/ 分解的,爆炸了的;爆炸;～d pictorial views 分解写生图;～d view(机器的)部件分解图
expression /iks'preʃn/ 表达,表示;表达式;arithmetic ～ 算术表达式;graphic ～ 图示法;integer ～ 整型表达式
ext (external) 外部的
EXT (exterior, external) 外
extend /iks'tend/ 伸长,延长;扩展,连续,伸展;～ through to (一直)延伸到;～ed center line 延伸中心线;～ed surface 展开面
extension /iks'tenʃn/ 伸展,延长,扩充(界限),伸出,外延;扩展名;～ lines 外延线,尺寸界线,延伸线;～ to standard 标准的扩充
extent /iks'tent/ 广度,宽度,程度;长度,尺寸;lateral ～ 横距离;vertical ～ 垂直距离
exterior /eks'tiəriə/ 外部,表面,外形,外观;外表的,外部的,外形的;～ view 外形图,表面图,外观图
exteriority /ekstiəri'ɔriti/ 外表,外形
external /eks'tə:nl/ 外观,外形,外表的,表面的;外部情况;～ diameter 外径;～ dimensions 外形尺寸,外部尺寸;～ fillel 外圆角;～ gear 外啮合齿轮;～ tangent circle 外切圆;～ thread 外螺纹,阳螺纹
extra /'ekstrə/ 非常的,特别的;～ fine-screw 特细牙螺纹;～ heavy series 特重系列;～ light series 特轻系列
extreme /iks'tri:m/ 末端的,最终的,极端

的; 极值, 极端, 末端　　　　　　　　　　　～ bolt 环首螺栓; screw ～ 螺丝眼
extremity /iks'tremiti/ 末端, 极端　　　**eyelet** /'ailit/ 眼圈板, 小孔
extrude /eks'tru:d/ 冲压, 挤压, 锻压　　**eye-measurement** 目测
eye /ai/ 眼, 环, 孔, 入口; 光电管; 信号灯;　**eyeshot** /'aiʃɔt/ 眼界, 视野

f (foot) 呎(英尺)
f & a (fore and after) 前后
FAB (fabricate) 制造
fabric /'fæbrik/ 结构, 构造; 织物, 布; 生产, 装配
facade /fə'sa:d/ 房屋正面; 正面, 外表, 外观
face /feis/ 面, 正面, 表面, 端面; 外观; 刮面; 工作面; ～ angle(圆锥齿轮)齿面角; ～ cone 面锥; ～ hardening 表面硬化; bevel ～ 斜角面; concave ～ 凹面; contact ～ 接触面; convex ～ 凸面; crown ～ 顶面; front ～ 前面; illuminated ～ 受光面; inner ～ 里面; joint ～ 接合面; outer ～ 外面; spot ～ 孔口平面(装螺母或垫圈用); weld ～ 焊道表面
face-off /'feisɔf/ 侧角
faceplate /'feispleit/ 平板; 面板; 荧光屏
facial /'feiʃl/ 正面的, 表面的
facilitate /fə'siliteit/ 简化, 使容易做; 促进, 推进
facing /'feisiŋ/ 饰面, 旋平面, 涂料, 衬片; ～-up 滑配合, 对研, 配研; spot ～ 锪孔
facsimile /fæk'simili/ 摹写, 传真, 影印本, 复制画
factor /'fæktə/ 因素, 要素, 因子, 系数; aesthetic ～ 美学系数; axial deformation ～ 轴向变形系数; human ～ 人工因素, 人机工程学; reduced axial deformation ～ 简化轴向变形系数
factoring /'fæktəriŋ/ 因子分解; scale ～ 比例(尺)选择
factorisation (factorization /'fæktərai'zeiʃən/) 编制计算程序, 因式分解
fake /feik/ 伪造, 捏造; 假货; 假的; ～ perspective 假透视
false /fɔ:ls/ 假的, 伪造的; 临时的, 非基本的; ～ work plan(建筑中)临时支架图
familiar /fə'miljə/ 惯用的, 常见的, 熟悉的
family /'fæmili/ 族, 派, 家; ～ of circles 圆族
fancy /'fænsi/ 幻想的; 想象; 想象力; 精致的
fantail /'fænteil/ 扇状尾, 鸠尾(榫); ～ed 鸠尾的, 燕尾的, 扇形尾的
fan /fæn/ 风扇, 扇子; -type 扇形式
FAO (finish all over) 全部加工
fasten /'fa:sn/ 扣住, 连接
fastener /'fa:snə/ * 紧固件, 扣件, 连接件; paper ～ 图纸扣
fastening /'fa:sniŋ/ 扣紧, 连接; 连接件
fault /fɔ:lt/ 故障
FD (face of drawing) 图纸的正面
FDP (frontal datum plane) 前基准面

fe (for example) 例如

feature /ˈfi:tʃə/ 要素，面貌，特点，特性，特征，定型；另件，部件；细节，画轮廓；～ size 特征尺寸，定型尺寸；actual ～ * 实际要素；asymmetrical ～ 非对称形体；construction ～ 构造特点；external ～ 外部形态；internal ～ 内部形态；real ～ * 实际要素；repetitive ～s 相同结构；uniformly spaced ～s 均布结构

fecit /ˈfi:sit/ ［拉丁语］(某某)画，(某某)作

feedback /ˈfi:dbæk/ 反馈

feeler /ˈfi:lə/ 测隙规，厚薄规

feet /fi:t/ 英尺(英制，foot 的复数)

feint /feint/ 淡的，不明显的；伪装的；～ lines 淡格子线；ruled ～ lines 画有淡格子线的

female /ˈfi:meil/ 阴的，内的，女性的；～ screw 阴螺旋

fettle /ˈfetl/ 修整

FH (flat head) 平头

fibre /ˈfaibə/ (fiber) 纤维板，硬纸板，钢纸，纤维材料

fid /fid/ 钉子，销子，螺钉；双头螺柱，柱螺栓

field /fi:ld/ 字段(计)，半帧，场；～ engineer 工地工程师

Fig. (figure) 图，插图

figurate(d) /ˈfigjurit(d)/ 定形的，表示几何图形的

figuration /figjuˈreiʃən/ 定形，外形，轮廓；图案表现法

figurative /ˈfigjurətiv/ 用图形表现的，象征的

figure /ˈfigə/ 图，插图，图形，图像，形状，外形，轮廓；用图表示；数字，计算，值；用数字表示，有花纹的；～d 图示的，图解的，形象的，带图案的；～ less 无图形的，无数字的；～ signal 图形信号；～-reader 图形阅读器；a geometrieal ～ 几何图形；anaglyph ～ 两色体视图；anyplane ～ 任意平面图形；basic ～ 基本图型；be round in ～ 呈圆形；Brinell ～ 布氏硬度数值；carved ～ 雕刻图样；center of ～ 形状中心；congruent ～s 全等图形，迭合图形；correlative ～ 对射图形，对应图形；curvilinear ～ 曲线图形，曲线形；geometric ～ 几何图形；geometric plane ～ 平面组成之几何形状；homological ～s 同源图形，同素图形；homothetic ～s 同位相似图形，位似图形；perspective ～s 透视图形；plane ～ 平面图形；projecting ～ 投影图形；rectilinear ～ 直线图形；similar ～s 相似图形；solid ～ 立体图形；space ～ 空间图形；strain ～ 应变图；strain ～s 流线，滑移线，吕德斯线；symmetric (al) ～ 对称图形

figuring /ˈfigəriŋ/ 用图形表示；用数字表示，计算

FIL (fillet) 内圆角

file /fail/ 锉刀；排成纵队进行，按次序排列；文件，档案，文件夹；automatic response ～ 自动应答文件；commandgroup ～ 命令组文件；display ～ 显示文件；error ～ 错误文件；lance-tooth steel ～ 油光锉；picture ～ 画面文件；pseudo ～ 伪文件；regular ～ 标准文件；stream ～ 流式文件；temporary ～ 暂时文件

filename /ˈfail‚neim/ 文件名；object ～ 目标文件名

fill /fil/ 填

filler /ˈfilə/ 漏斗；填隙料，填料；(加料)口；垫片

fillet /ˈfilit/ 内圆角，(焊缝)轮廓，痕迹，填角，嵌条，倒圆角；～ed angle 圆角；～ed corner 内圆角；casting ～ 铸造(内、外)圆角；forging ～ and round 锻造(内、外)圆

角; inside ~ 内角焊缝; light ~ 浅角焊缝

fillet-and-round 内外圆角

filleting /ˈfilitiŋ/ 内圆角修整

fillet-weld 圆角焊接

filling /ˈfiliŋ/ 填充区

fillister /ˈfilistə/ 盆状的,填料函,凹槽; filister-head 凹槽头; ~ screw 凹槽头螺丝

fill-up 加注,填上

film /film/ 薄膜照相软片,影片; drafting ~ 制图胶片; transfer ~ 转印膜

FIN (fin, finish) 光制

fin /fin/ 突片,突出之薄肋,完成,结束

final /ˈfainl/ 最后的,最终的

find /faind/ 找出,求,发现,呈现; ‖ *Example sentence: Find the true size of the angle between AB and plane CDE. 例句: 求直线AB与平面CDE之间夹角的真实大小。*

finder /ˈfaində/ 测距器,定向器

finding /ˈfaindiŋ/ 测定,探测,定位,结论,做出

fine /fain/ 细(小)的,薄的,好的,精的;细牙的(螺纹); ~ pitch 小螺距; ~ thread 细牙螺纹

fineness /ˈfainnis/ 精度,光洁度,细度;公差,优良

finish /ˈfiniʃ/ 完成,精加工,抛光,修整,表面光洁度,表面涂层,润饰; ~ all over 全部加工; ~ machining 完工切削; ~ mark 加工符号,光洁度符号; ~ model 最后模型; ~-schedule 加工一览表; ~ surface 加工面; ~ symbol 加工符号; ~ to size 按尺寸加工; black ~ 发黑处理,表面氧化处理; bright ~ 抛光,光面精整,镜面抛光; dead ~ 无光泽抛光; frosted ~ 无光(毛面)光洁度,霜白表面; gloss ~ 抛光; ground ~ 磨光,磨削加工; high ~ 精磨,光制; lapping ~ 研磨; matte ~ 无光(毛面)光洁度; mirror ~ 镜面光洁度,镜面加工; protective ~ 表面处理,防腐处理; rough ~ 低级光洁度,粗加工; sand blast ~ 喷砂(表面)处理; smooth ~ 光面修整; surface ~ 表面光洁度; wrinkle ~ 皱纹漆,波纹面饰

finished /ˈfiniʃt/ 已精加工的,完工的; ~ bolt 精制螺栓; ~ machine drawing 铸件工艺图; ~ surface 光制的表面

finishing /ˈfiniʃiŋ/ 精整,精加工

finite /ˈfainait/ 有限的,限定的; ~ element analysis 有限元分析

FIR (full indicator reading) 量表读数差

fir /fəː/ 冷杉木,枞木

first /fəːst/ 第一,开始,开端;首先;最前的,最先的; ~ auxiliary plane 一次(变换)辅助投影面; ~ quadrant drawing 第一角(象限)画法; ~ story (storey) 第一层(楼); ~-angle projection 第一角(象限)投影法

fit /fit/ * 配合,装配;适合;适当的,相称的; ~ of spline, major diameter (多键)外径配合; ~ of spline, minor diameter (多键)内径配合; ~ of spline, tooth sides (多键)齿侧配合; ~ quality 配合等级; ASA ~s 美国标准的配合种类; clearance ~ * 间隙配合; close ~ 紧配合; close running ~ 紧转配合; close sliding ~ 紧滑配合; close working ~ 紧滑配合; conical ~ 圆锥配合; cylindrical ~ 圆柱配合; drive ~ 打入配合; easy running ~ 轻转配合; force ~ 压入配合; force or shrink ~ 压入收缩配合类; free ~ 自由配合,轻动配合; free running ~ 自由转动配合; general ~ class 一般配合类别; heavy drive ~ 重打入配合; heavy force ~ 重压配合; heavy force and shrink ~ 重压缩配合; hole-basis system of ~s 基孔制配合制度; interference ~ *

过盈配合,干涉(静)配合; light drive ～ 轻打入配合; light press ～ 轻压配合; light running ～ 轻转(动)配合; locational ～ 位置配合; loose ～ 松配合; loose running ～ 松转配合; medium ～ 中级配合; medium drive ～ 中级打入配合; medium force ～ 中级压入配合; medium running ～ 中级转动配合; minor diameter ～ of spline 花键内径配合; normal-running ～ 中转动配合; precision running ～ 精密转动配合; press ～ 压入配合; push ～ 推入配合; running ～ 转动配合; shaft-basis system of ～s 基轴制配合制度; shrink (shrinkage) ～ 冷缩配合,红套配合; slide (sliding, slip) ～ 滑动配合; tight ～ 紧配合; transition ～ * 过渡配合; variation of ～ * 配合公差,配合变化; wringing ～ 轻打配合

fitting /ˈfitiŋ/ 适当的,相称的,装配,修整,装置,设备;接头配件,附件,拟合; arrangement drawing 附件装配图; angle ～ 弯头; duplex ～ 双通接头; female ～ 阴螺纹管接头; reducing ～ 异径接头; 45°-Y ～ 45° Y形接头

fitting-up 装配

fixed /fikst/ 固定的,不变的;～-angle 固定角

fl (full line) 实线; (floor) 地板,地面,楼房的层

flag /flæg/ 标志,标记,特征位

flange /flændʒ/ 法兰盘,凸缘; reducing ～ 异径凸缘; wheel ～ 轮缘

flank /flæŋk/ * 牙侧,齿侧;侧面,位于……的侧面;～ angle*牙侧角;～ of thread 螺纹面;～ of tooth 轮齿侧面;～ profile 齿廓,齿形

flashing /ˈflæʃiŋ/ 泛水;闪光

flat /flæt/ 平,平面,平直的;扁头;(楼房的)一层;扁钢;～ key 平键;～ roof 平顶;～ spring 板弹簧;～ tint 均匀阴影线

flatness /ˈflætnis/ 平面度;平坦度;平直度,平整度;均匀性

flat-plate 平板

flexible /ˈfleksəbl/ 可弯曲的,挠性;～ adjustable curve 挠性曲线板;～ curve 挠性曲线带

flexure /ˈflekʃə/ 弯曲,挠曲,折叠

FLG (flange) 凸缘,法兰

flier /ˈflaiə/ (flyer) 梯级;手轮,飞轮

flight /flait/ (flight of steps)(螺旋推进器的)螺旋片;楼梯的阶段

float /fləut/ 浮点数

flock /flɔk/ 群,毛棉填料;～-paper 毛面纸

floor /flɔː, flɔr/ 地板,地面,(楼房的)层;垫板;～ plan 地板平面图,楼面布置图;～ slab 水泥板; cement ～ 水泥地面; first ～ (=second story)(第)二层楼(英国),第一层楼(面)(美国); first ～ plan 第二层楼平面图(英国),第一层楼平面图(美国); top ～ 顶楼; wood ～ 木地板

floppy /ˈflɔpi/ 软,软磁盘

flow /fləu/ 流,流动,流量;～ chart (或sheet)(工艺)流程图,程序方框图,操作程序图,程序表;～ graph 流向图;～ production 流水作业; operations ～ chart 操作流程图; process ～ 工艺流程(图); technical ～ between work areas 工艺建筑物联系图(选矿厂)

flow-chart (工艺)流程图,操作程序图,程序方框图; operations ～ 操作流程图

flow-sheet (工艺)流程图,程序方框图; basic ～ of coal preparation 选煤工艺原则流程图; equipment ～ of coal preparation 选煤设备流程图; process ～ 工艺(生产)流程图

flush /flʌʃ/ 大量的,齐平的,同平面的,(焊缝的)齐平面

FN (force or shrink fit) 压缩配合类

foci /ˈfəusai/ 焦点,焦距; focus 的复数

focus /ˈfəukəs/ 焦点

fold /fəuld/ 折叠, 折合; ~ line 折线

folding /ˈfəuldiŋ/ 可折叠的; ~ line 投影面与投影面的交线, 折线; ~ rule 折尺

folio /ˈfəuliəu/ 对开纸, 页码; in ~ 对开

following /ˈfɔləuiŋ/ 以下的, 下述的, 后面的

foot /fut/ 英尺; 步, 足; 支点; ~ step 踏步, 梯级; 支座

force /fɔːs/ 力; 加压(力); active ~ 作用力; resultant ~ 合力; thrusting ~ 推力; traction ~ 拉力, 牵引力

forelock /ˈfɔːlɔk/ 开口销, 扁销

foreshorten /fɔːˈʃɔːtn/ (绘画中)按透视法缩短, 缩短; ~ing 缩画, 用透视法缩小画出

foreword /ˈfɔːwəːd/ 序, 前言

forging /ˈfɔːdʒiŋ/ 锻件; ~ rounds and fillets 模锻内外圆角; drop ~ 落锻

fork /fɔːk/ 叉; 叉形物

form /fɔːm/ 形状, 外表, 轮廓, 断面, 造型, 成形, 样式, 模板; ~ advance 换页, 进页; ~ diameter 外形直径; ~ drawing 木模图; ~ feed (打印机) 进页; ~ of thread 螺纹牙型; ~ of title block 标题栏格式; block ~ 方块型; geometric ~s 几何形体; idea geometric ~ 理想几何形状; interior ~ 内部形状

form. (formation) 形式, 构造

format /ˈfɔːmæt/ * 图纸幅面; 格式, 版式, 形式; 尺寸, 大小, 幅度

formatting /ˈfɔːmætiŋ/ 格式化

former /ˈfɔːmə/ 前面的, 以前的; 模型, 量视; wooder ~ 木模

formula /ˈfɔːmjulə/ 公式, 计算, 方案; empirical ~ 经验公式, 试验公式

formwork /ˈfɔːmwəːk/ 模型, 样板; 量规, 定规

forth /fɔːθ/ 向前, 向前方, 向外

FORTRAN (Formula Translator) 公式翻译程序

forward /ˈfɔːwəd/ 在前, 前边

foundation /faunˈdeiʃn/ 基础, 地脚, 地基; ~ bolt 地脚螺栓; ~ plan 基础图

foundry /ˈfaundri/ 铸造车间, 铸件; ~ pig iron 铸件

four /fɔː/ 四, 四个; -center 四圆心; ~ center method 四圆心法(画椭圆); ~ spiral 四心螺线

FR (front) 前面

F & R (fillets and rounds) 内圆角和外圆角

FRP (frontal reference plane) 前辅助投影面

fraction /ˈfrækʃn/ 分数, 小数

fractional /ˈfrækʃənl/ 分数的, 小数的, 部分的, 小于1的; ~ error 相对误差

fracture /ˈfræktʃə/ 折断, 破裂, 破碎; 断口, 断面, 断裂面; ~ plane (或 surface) 破裂面, 断面; ~ section 断裂剖面

frame /freim/ 画面, 框架, 构架; 制订, 发展, 系统, 构造, 结构; ~ of reference 参考系统, 坐标系统; framing 结构, 框架; ~ plan 结构平面图

free /friː/ 自由的, 任意的; ~ fitting 自由配合

freedom /ˈfriːdəm/ 自由(度), 间隙; 摆动, 可能性

freehand /ˈfriːhænd/ 徒手; ~ drawing 徒手画, 徒手图, 草图; ~ line 徒手画的波浪线; ~ sketch 徒手画, 草图; ~ sketching 徒手画, 徒手图, 画草图

French curve 曲线板

fresco /ˈfreskəu/ (fresco painting) 壁画

FRHGT (free height) 自由高度

front /frʌnt/ 前, 前面, 前方, 正面; 外表; 正面的, 前面的; ~ side 正面, 正视图; ~ view (或 elevation)* 主视图, 前视图, 正视图, 正立面图; in ~ of the body 在物体前面 (物体外); in the ~ of the body 在物体前

部(物体上)

frontage /'frʌntidʒ/ 正面

frontal /'frʌntl/ 前面的, 正面的; ～ datum plane 正基准面; ～ line 正线(正平面内之线), 正平线; ～ plane 正面, 正平面; ～ projection 正面投影; ～ projection plane 正面投影面; ～ -profile line 铅垂线

frottage /'frɔtidʒ/ 磨擦拓图法(将图移转另一纸上之法)

frustum /'frʌstəm/ (或 frusta) 锥台, 台, 平头截体; ～ of a cone 平截头圆锥体; ～ of a pyramid 平截头棱锥体

fsd (full size detail) 足尺图, 1:1零件图

ft (foot) 英尺

F to F (face to face) 面到面

full /ful/ 全尺寸, 表示"1:1"或1比1(美国图纸上比例栏内填用全尺寸所画图样); 满, 全, 实, 足; ～ -divided 详细分格; ～ line 实线; ～ scale 足尺(比例), 全尺寸, 原尺寸的; ～ section 全剖视; ～ size 原尺寸, 实物尺寸, 1:1尺寸; ～ view 全视图

full-divided 详细分格

full-scale 原大的, 原尺寸的; 自然的满刻度的

full-size 全尺寸; 原大的

function /'fʌkʃn/ 功能; 函数; locator ～ 定位功能; pick ～ 选择功能; stroke ～ 笔划功能

functional /'fʌkʃənl/ 函数的, 功能的, 实用的; ～ arrangement 函数图; ～ diagram 方块图, 工作原理图; ～ drawing 实用图

fundamental /fʌndə'mentl/ 基本的, 主要的, 基础的

funnel /'fʌnl/ 漏斗, 烟囱

fuselage /'fju:zəla:ʒ/ (飞机)机身; 壳体; 外壳; 弹体

fv (front view) 正面图, 前视图

G

G (grinding) 轮磨

GA (gauge) 量规, 范围; 表, 计; (general arrangement) 总图, 安装图; (general assembly) 装配图

gabarit(e) /'gæbərit/ (法语)外形尺寸, 轮廓, 限界, 净空, 净跨; 样板规, 曲线板

gad /gæd/ 量规, 厚薄规; 销, 键; 车刀, 切刀

gad (general assembly drawing) 总装配图

gage /geidʒ/ (=gauge) 量规, 规号(规格号), 规距(行距), 规格, 标准规, 表, 样板, 计; 测量; 仪器; ～ block 块规; ～ line 量线, 规线(如铆钉孔中心线); ～ of sheet 薄板的厚度; Birmingham wire ～ 伯明翰线号规; British Standard wire ～ 英国标准线规; checking ～ 校对量规; ～ point 标点; caliber ～ 量径规, 厚薄规; feeler ～ 间隙规; fillet-and-round ～ 内外圆角规; finger ～ 厚度规; fixed ～ 固定规; gap ～ 隙规; go ～ 通过规; height ～ 高度规; Imperial Wire ～ 英制线号规; inspection ～ 检验规, 塞规; internal ～ 内齿轮; internal and external ～ 内外径规; international screw pitch ～ 国际螺距规; leaves thickness ～ 分

叶厚度规；level ～ 水准仪；limit ～ 限界量规；limit snap ～ 极限卡规；line ～ 线规；marxing ～ 划线规；metric screw pitch ～ 国际螺距规，米制螺距规；micrometer ～ 测微规；micrometer depth ～ 深度千分尺；no-go (not-go) ～ 不通过规；off ～ 等外品，不合格品；on ～ 合格品；pitch ～ 螺矩规；pitch diameter ～ 节圆直径量规；plug ～ 圆柱塞规；radius ～ 圆角规，半径量规；ring screw ～ 螺纹环规；round and fillet ～ 内外圆角规；screw (pitch) ～ 螺距规；sheet ～ 板规，厚薄规；sheet-metal ～ 金属片规；slide ～ 卡尺，游标规；surface ～ 平面规，粗糙度样板，画针盘；taper ～ 锥度规；thickness ～ 厚度规，测厚仪，塞尺；thread ～ 螺纹规；vernier ～ 游标尺，游标规；wickman ～ 威氏凹口螺纹量规；Washburn & Moen Wire ～ W&M 线号规；wire ～ 金属线规；working ～ 工作量规

gaging /ˈgeidʒiŋ/ 用量规检验；～ dimensional 尺寸检查；～ surface 量度表面

galvanize /ˈgælvanaiz/ 镀锌，电镀

gap /gæp/ 间隔，间隙，缺口，开口，差距，裂口

gap-gauge 间隙（厚度，外径）规

Gardener's-ellipse 钉线椭圆法

garment /ˈgaːmənt/ 外表，外观，外层；～ tag 外表特征

gasket /ˈgæskit/ 垫片，衬；～ material 填料

Gaspard Monge 加斯帕尔·蒙日（1746—1818）被称为"画法几何之父"，诞生于法国波南 (Beaune France)，他发展了投影原理，使之成为现代工程制图的基础

gauche /gəʊʃ/（法语）左边的，非对称的，不可展的，扭的

gauge /geidʒ/ (gage) 量规

GB (GuoBiao)（中国）国家标准，国标

GCS (Graphic Control Software)（计算机）绘图控制软件

ge. (gauge) 量规

gear /giə/ 齿轮，机构，机器，(传动)装置，机关，架；～ addendum angle（圆锥齿轮）齿顶角；～ box (case) 齿轮箱；～ dedendum angle（圆锥齿轮）齿根角；～ in to 啮合；～ pitch angle 节锥角；～ rack 齿条；～ rim 轮缘；～ sector 扇形齿轮；～ tooth 轮齿；angular ～ 人字齿轮；bevel ～ 圆锥齿轮，伞齿轮；cam ～ 凸轮装置；chain ～ 链传动机构；chevron ～ 人字齿轮；common bevel ～ 直齿伞形齿轮；conical ～ 圆锥齿轮，伞齿轮；cylindrical ～ 圆柱齿轮；cylindrical spur ～ 圆柱正齿轮；double helical (spur) ～ 人字齿轮；driven ～ 从动齿轮；driving ～ 主动齿轮；helical ～ 斜齿轮；herringbone ～ 人字齿轮；in ～ 啮合；involute ～ 渐开线齿轮；main drive ～ 主动齿轮；miter ～ 圆锥齿轮，伞齿轮；put into ～ 啮合；ratchet ～ 棘轮；ring ～ 环形齿轮；sector ～ 扇形齿轮；spiral ～ 螺旋齿轮；sprocket ～ 链轮；spur ～ 正齿轮，直齿圆柱齿轮；stepped ～ 塔齿轮；straight bevel ～ 直齿伞齿轮；straight-cut ～ 正（直）齿轮；twin helical ～ 人字齿轮；twisted ～ 斜齿轮；variable ～ 变速齿轮；wheel ～ 齿轮；worm ～ 蜗轮，蜗轮传动装置

gearbox /ˈgiə(r) bɔks/ (gearcase) 齿轮箱，减速机，变速箱

gearbox-case 齿轮箱外壳

gearhousing /giəˈhaʊziŋ/ 齿轮箱；gearing ～ 齿轮传动，齿轮啮合

gear-wheel （大）齿轮

general /ˈdʒenərəl/ 总的，通用的，一般的，普通的，大概的，简略的，全的，综合的；～ drawing 总图，全图，概要图；～ layout 总体布置图；～ location sheet 地盘位置图；～ plan 总（布置）图，总计划图；～ plane 一般

平面；~ section 全剖面图(房屋)；~ view 全视图, 大纲, 概要

generalization /ˌdʒenərəlaiˈzeiʃn/ 通用化

generate /ˈdʒenəreit/ 母点(的), 母线(的), 母面(的)；~ circe 母圆, (齿轮的)基圆；~ cone (伞齿轮的)基锥；~ curve 母曲线；~ line 生成线, 母线

generating /ˈdʒenəreitiŋ/ 产生的, 生成的

generation /dʒenəˈreiʃn/ 形成, 生成

generator /ˈdʒenəreitə/ 母线, 生成元素, 发生器, 编制程序；~ of a ruled surface 直纹曲面的母线；curve ~ 曲线发生器；dot matrix character ~ 点阵字符发生器；stroke character ~ 笔划字符发生器；vector ~ 矢量发生器

generatrix /ˈdʒenəreitriks/ 动线, (产生线、面、体的)母点, 母线, 母面

geodimeter /dʒiːəuˈdimitə/ 光电测距仪

geogram /ˈdʒi(ː)əugræm/ 地学环境制图

geographic(al) /dʒiəˈgræfik(al)/ 地理(学)上的；~ map 地理图, 地形图

geography /dʒiˈɔgrəfi/ 地理, 地形, 地势

geoid /ˈdʒiːɔid/ 大地水准面

geology /dʒiˈɔlədʒi/ 地质学

Geom (geometrical) 几何(学)的；几何形状；(geometry) 几何学

geometric(al) /dʒiəˈmetrik(əl)/ 几何(学)的, 几何形状；~ construction 几何作图, 几何结构；~ figure 几何(图)形；~ letter 几何式字体；~ method 几何方法；~ pattern 几何图样；~ shapes 几何形状；~ shapes combined 几何形的组合；~ solid 几何立体, 几何体

geometrics /dʒiəˈmetriks/ 几何图形

geometrize /dʒiˈɔmitraiz/ 作几何图形, 用几何图形表示；~ ation of ore deposits 矿体几何制图

geometry /dʒiˈɔmitri/ 几何学, 几何形状, 几何图形；~ of machinery 机械几何学；~ of mapping 保形变换的几何性质；~ of ore deposits 矿体几何学；~ of projection 射影几何；~ of projective connection 射影联络几何学；affine ~ 仿射几何(学)；affine differential ~ 仿射微分几何学；analytical ~ 解析几何学；applied ~ 应用几何；constructive solid ~ 结构实体几何表示法；descriptive ~ 画法几何学；differential ~ 微分几何学；graphic ~ 应用几何；infinitesimal ~ 微分几何学；metric ~ 度量几何学；modern ~ 近世几何学, 现代几何学；n-dimensional ~ n 维(度)几何学；n-dimensional descriptive ~ 多维画法几何学；perspective ~ 投影几何学；plane ~ 平面几何学；projective ~ 射影几何学；random-~ 不规则几何形状；solid ~ 立体几何；spherical ~ 球面几何；three dimensional ~ 三维几何；topological ~ 拓扑几何；uniform-~ 规则形状的

Gerard Desargues 代沙格(1593—1662年), 法国的数学家, 有"射影几何奠基人"之称

GI (galvanized iron) 白铁；(gray iron) 灰铸铁

gib /gib, dʒib/ 扁栓, 夹条；具有斜面的立体；~ -head taper key 钩头斜键

gib-head 劈头, 钩头

gibbous /ˈdʒibəs/ (gibbose) 凸圆, 隆起的；~ moon 凸月(大于半月, 小于满月的)

GIN (graphic input) 图形输入

girder /ˈgəːdə/ 梁, 槽钢

girth /gəː/ 周围长度, 曲线长度, 大小, 尺寸, 圈梁；~ line 周围线, 围长线, 展开图基线

given /ˈgivn/ 给定；给予的, 已知的, 假设的；~ line 已知线；~ plane 已知面；‖*Example sentence*①: *The angle between a line* *and a plane lies in a plane that is perpendicular to the* given *plane and contains the given line.* 例句①: 直线与平面之间的夹角位于包含直线并与给定平面垂直的平面内。‖ *Example sentence②: Two skew lines AB and CD, and a point P, are given. It is required to construct a plane parallel to AB and CD and passing through P. (See Fig. 1)* 例句②: 两斜线 AB、CD 和一点 P, 已经给定(见图1), 求作一平面平行于AB和CD, 且通过点P。

GKS (graphical kernel system) 图形核心系统

GL (ground level) 地平线; (ground line) 基准线; (ground location) 地面定位; (use guide lines) 用引线

glacis /ˈglæsis/ 斜坡

glance /gla:ns/ 斜射, 一瞥, 浏览

gland /glænd/ 盖, 压盖, 填料盖; ～ bonnet (轴端)密封盖, 端盖; ～ box 填料函(箱); ～ cover 填料盖; ～ nut 压盖螺母; ～ packing (压盖)填料, 密封垫

glass /gla:s/ 玻璃制品; ground ～ 磨砂玻璃, 毛玻璃, 玻璃粉; magnifying ～ 放大镜; rubbed ～ 磨砂玻璃; sand blasted ～ 磨砂玻璃

glasspaper /ˈgla:sˌpeipər, ˈglæsˌpeipər/ 玻璃纸, 砂纸

globe /gləub/ 球体, 球形物; 灯泡

globular /ˈglɔbjulə/ 球状面的, 圆的; ～ projection 球面投影

gloss /glɔs/ 光泽, 光泽面; 使有光泽, 装饰; 注释; ～ on 润饰

gnomonic /nəuˈmɔnik/ 心射的, (日晷)指时的; ～ projection 心射图法, 球心投影, 心射切面投影; ～ ruler (或 scale) 心射投影尺

gnomonogram /ˈnəuməngrəem/ 心射(切面投影)图

goniometer /ˌgəuniˈɔmitə/ 角度计, 测向器

gorge /gɔ:dʒ/ 咽喉, 峡谷; circle of the ～ 喉圆

Gothic /ˈgɔθik/ 哥德式, 哥德字

GOTO 转向

gr (gear ratio) 传动比

gradation /grəˈdeiʃn/ 分等, 等级, 类别, 层次

grade /greid/ 分等, 等级, 渐变, 坡度(同斜度, 唯以百分比表示), 度; ～ level 坡度线; ～ of fit 配合等级; ～ of slope 坡度; ～ of tolerance 公差等级; ～d tint 渐变阴影线; extra fine ～ 一级精度; fine ～ 二级精度; planin ～ 三级精度, 普通级精度; rough ～ 四级精度

grade-separation 立体交叉

gradient /ˈgreidiənt/ 坡度, 斜度; 坡; ～ of slope 坡度角, 倾斜率

grading /ˈgreidiŋ/ 平坡; 等级; 分类; 校准; 竖向设计图(属建筑总平面设计图)

graduate /ˈgrædjueit/ 分度, 分角, 刻度, 毕业, 得学位

graduation /ˌgrædjuˈeiʃn/ 分度, 分等级; ～ mark 分度符号, 分度线

graph /gra:f/ * 算图, 图形, 图, 图表, 图示, 图解, 曲线图, 标绘图; (作后缀时表示)书写器, 描绘器, 记录器; 用图表表示; 把……绘入图表; ～ coloring 图着色; ～ follower 读图器, 图形跟随器; ～ line 绘折线图; ～ method 图解法; ～ notation for ordinary graph 普通图形符号; ～ of relation 关系图; ～ paper 方格纸; ～ pie 绘圆饼图; ～ plotter 绘图仪, 制图仪; ～ RECDIR (direct inking records 即 graphic recording) 墨水直接记录图, 图示记录; ～ schema 图式; ～ structure 图形结构; ～ theory 图论; ～ type 图类型; analytical ～ 分析图; acyclic ～ 非周期图, 非循环图; AND/OR ～ 与/或图; antisymmetric ～ 反对称图;

area ~ 面积图; arithmetic ~ 算术分度图; association ~ 组合图; asynchronous ~ 异步程序图; bar ~ 直方图, 条形图; bicolourable ~ 双色图; biconnected ~ 双连通图; binary ~ 二叉(树形)图; bipartite ~ 二分图; black-and-white ~ 黑白图; block ~ 块图; characteristic ~ 特征图; characteristic polynomial of a ~ 图的特征多项式; chromatic ~ 色图; complement of ~ 图的补图; complete ~ 完全图; complete n-partite ~ 完全 n 部图; composition of two ~s 二图合成; compositive sectional train ~ 综合分号运行图; connected ~ 连通图; connect-undirected ~ 连通无向图; construct a ~ 绘制曲线图; continuous train ~ 连发运行图; contractible ~ 可收缩图; converse ~ 逆图; coordinate ~ 坐标制图机; covering ~ 覆盖图; cyclic ~ 循环图; cyclic directed ~ 循环有向图; depth in ~ 图的深度; derivation ~ 派生图, 推号图, 引导图; discriptive ~ 描述图; dual ~ 对偶图; empty ~ 空图; engineering ~ 工程图; finite ~ 有限图; flow ~ 流线图; following train ~ 追踪运行图; fundamental train ~ 基本运行图; fuzzy ~ 模糊图; geometric ~ 几何图; independent sectional train ~ 独立分号运行图; line ~ 线图; mathematical ~ 数学图; non-parallel train ~ 非平行运行图; partial ~ 部分图; plane ~ 平面图; sectional train ~ 分号运行图; signle line train ~ 单线运行图; two-line train ~ 双线运行图; train ~ 列车运行图

graphic(al) /'græfik(əl)/ 制图, 图解, 图算, 图示的, 图解的, 绘图的; 图, 图表, 图画, 地图; ~ accuracy 制图精确度; ~ algebra 图解代数; ~ analysis 图解, 图形分析; ~ annotation 图形的注解; ~ application system 图形应用系统; ~ arithmetic 图解算术; ~ arts 图表艺术, 绘画艺术; ~ calculation 图解计算; ~ calculus 图解微积分, 图算; ~ Cathode Ray Tube 图表阴极射线管; ~ character 图形字符; ~ chart 图表, 图解, 曲线图; ~ code 图形码; ~ commands 绘图命令; ~ computation 图解计算; ~ configuration 图形配置; ~ CRT 图表阴极射线管; ~ data 图形数据; ~ design system 图形设计系统; ~ diagram 符号图; ~ differentation 图解微分; ~ display 图形显示(器); ~ documentations of coalmine survey 煤矿测量图; ~ expression 图示、图解; ~ extensions 图形扩展(充); ~ formula 图解式, 立体式; ~ input 图形输入; ~ input language 图形输入语言; ~ instrument 图示器, 自动记录仪; ~ integration 图解积分; ~ interaction 图形的交互作用; ~ interpolation 内插图解法; ~ kernel system 图形核心系统; ~ language 图形语言; ~ layout 图形设计; ~ library 图形库; ~ log 柱状剖面图; ~ master 图形板; ~ method 图解法; ~ model 图模型; ~ monitor 图形监视器; ~ object 图形对象; ~ output 图形输出; ~ package 图形程序包; ~ panel 图形板; ~ presentation of data 数据的图示; ~ primities 图元; ~ processing 图形处理; ~ representation 图示(法); ~ scale 图示比例尺; ~ science 图学; ~ sign 图形标志; ~ solution 图解; ~ statics 图解力学; ~ subroutines 图形子程序; ~ symbol 例图, 图解符号, 图形符号; ~ transmission 图形传输, 图像传输

graphicaly /'græfikli/ 制图地, 用图表示法, 用图解法, 用图……; show ~ 用图表示

graphics /'græfiks/ 图, 图形, 图样; 图示, 图解; 图学, 图形学, 图示学; 制图学, 制法, 图解计算法, 算图; ~ communication 基本图学; ~ library 图库; ~ software 绘

图软件; all points addressable ～ 可定址的全点图形; bit-mapped ～ 位映射的图形; block ～ 框形图形, 块拼图形; block diagram ～ 框图图形学; civil engineering ～ 土木工程制图; coded ～ 编码图形; color ～ 彩色图形(学); computer ～ 计算机图学, 计算机图形, 计算机图形处理; consumer ～ 用户图形; core ～ system 核心图形系统; engineering ～ 工程图学; engineering design ～ 工程设计制图; engineering ～ for mining 矿业图学; generative computer ～ 生成式计算机图形; high-resolution ～ 高分辨率图形; industrial ～ 工业图形显示; initial ～ exchange specification 初始图形交换规范; interactive ～ 交互式绘图, 交互图形显示; interactive ～ system 交互图形系统; kernel ～ system 核心图形系统; line ～ 线连图形, 矢量图形; low-resolution ～ 低分辨率图形; mainframe ～ 大型计算机图形; memory-mapped ～ 存储器映象图形; numerial control ～ 数控制图; pixel ～ 像素图形; raster ～ 光栅(扫描)图形; turtle ～ 龟绘图系统

graphotype /ˈgræfətaip/ 字图电传机
exempligratia /egˈezmplaiˈgreiʃiə/ 例如
graticulation /grətikjuˈleiʃən/ 方格画法 (在设计图上画上方格, 在方格纸上作图, 以便缩放)
graticule /ˈgrætikjuːl/ (reticule) 方格画法; 分度线, 标线, 十字线, 网格, 方格图; map ～ 制图网
gray (grey /grei/) 灰色; 本色的, 半透明的
graze /greiz/ 相切, 接触; 抛光
GRD (grind) 研磨
grease /griːs/ 油脂; ～ cup 滑脂杯, 油杯
greaten /ˈgreitən/ 放大, 增加

Greek /griːk/ 希腊, 希腊人, 希腊语; 希腊的
grid /grid/ 栅档, 格子, 栅格, 网, 网状物, 坐标网, 方格网, 网点; drawing ～ 绘图网格; map ～ 地图方格网; normal ～ 直角坐标网; perspective ～ 透视格网, 透视栅档纸
grinding /ˈgraindiŋ/ 磨削, 磨光
groove /gruːv/ 槽谷, 沟, 开槽; ～ pulley 槽轮, 三角皮带轮; double bevel ～ K 型坡口; dovetail (ed) ～ 燕尾槽; key ～ 键槽; oil ～ 油槽; single bevel ～ 半 V 型坡口
gross /grəus/ 总的, 全部的; 总重, 毛重; 显著的, 肉眼能看到的
grotesque /grəuˈtesk/ 奇形怪状的图形 (东西)
ground /graund/ 地面, 广场; 研磨, 光的; 基础, 底子, 底; ～ level 地平(面); ～ line 地平线, 境界线; 基线(投影面与投影面的交线)投影轴; ～ neutral 地线(电); ～ plane 地平面
group /gruːp/ 群, 组; affine ～ 仿射群
grub /grʌb/ 钻研, 掘除; ～ screw (平头) 螺丝, 木螺丝
gr. wt. (gross weight) 总重, 全重, 毛重
guide /gaid/ 指导, 引导, 引导者, 导轨; 手册; ～ book 参考手册; ～ing the needle point 导引针头(找圆心); ～ line 导线, 格线; lettering ～ 书写模板
gunmetal /ˈgʌnˌmetəl/ 炮铜, 锡锌青铜
GW (gross weight) 总重, 毛重, 全重
gyrate /dʒaiəˈreit/ 回转, 旋转
gyration /dʒaiəˈreiʃn/ 回转, 旋转; cetre of ～ 回转中心; radius of ～ 回转半径
gyro /ˈdʒaiərəu/ 陀螺仪; ～-axle 回转轴
gyro-rotor 回转体

H

H (hammering) 锤击

H. (too heavy) 太粗;(height) 高,高度; (hour) 时,小时

habit /'hæbit/ 习惯,习性

habitual /hə'bitjuəl/ 习惯的,日常的

hachure /hæ'ʃuə/ (法语)山坡线,(表示地形、断面等的)影线

hairline /'hɛəlain/ 瞄准线,十字线

hairspring /'hɛəspriŋ/ 细弹簧,游丝;～ divider 细弹簧分规

half /ha:f/ 半,半尺寸,表示"1比2"或1/2（美国图纸标题比例栏内填写用半尺寸所画图样的比例）；～ as much (many, large, fast) again as……比……多（多,大,快）一半,或"……是……的一倍半"；～ as much (many, large, fast) as……比……少（少,小,慢）一半,或"……是……的一半"；～-bearing 轴瓦;～-bush 轴瓦;～ center 半缺顶尖;～ line 半(直)线,射线;～ moon 半月(形);～ section 半剖面;～ sectional side elevation 半剖侧视图;～ sectional view 半剖视(图);～ size 半尺寸(1:2); A is ～ as long as B. A 比 B 短一半, A 是 B 的一半长; unsectioned ～ 未剖切的一半; wheel A turns ～ as fast again as wheel B. A 轮的转动比 B 轮快出半倍, A 轮转动的速度是 B 轮的一倍半

half-bright 半光制的

half-center 半缺顶尖

half-moon 半月形

half-plane 半面; front ～ 前半面; lower ～ 下半面; rear ～ 后半面; upper ～ 上半面

half-round 半圆形;半圆的,半圆形的

half-size 半尺寸,缩小一半的

half-thread 半螺纹

half-width 半宽度

halt /hɔ:lt/ 停机

halve /ha:v/ 对分,平分,二等分

handbook /'hænd,buk/ 手册

handle /'hændl/ 手柄

handling /'hændliŋ/ 处理

hardcopy /'ha:dkɔpi/ 硬抄本

harden /'ha:dn/ 硬化,淬火;～ed case 表面渗碳硬化;～ed and tempered 调质的; color-～ 着色硬化; face ～ing 表面硬化

hardening /'ha:dəniŋ/ 硬化,淬火; case ～ (或 ～ case)表面硬化; flame ～ 火焰淬火; full ～ 全硬化,淬透; local ～ 局部淬火; oil ～ 油淬火; shallow ～ 表面硬化; shell ～ 渗碳,表面淬火; torch ～ 火焰淬火; water ～ 水淬

hardness /'ha:dnis/ 硬度,硬性,刚度,强度;难解; ball ～ 布氏(球测)硬度; Brinell ～ 布氏硬度; degree of ～ 硬度(铅笔); diamond penetrator ～ (或 Vickers diamond ～)维氏硬度; Herbert pendulum ～ 赫伯特摆式硬度,赫氏硬度; Moh's ～ 莫氏硬度; Rock'wel ～ 洛氏硬度; Shore ～ 肖氏硬度; skin ～ 表面硬度

hardometer /ha:də'mıtə/ 硬度计

hardware /'ha:dwɛə/ 硬件,实物,附件;计算机部件

hardwood /'ha:dwud/ 硬木

harmony /'ha:məni/ 谐调,和谐

hatch /hætʃ/ 舱口，开口，人孔；影线，阴影；画阴影线

hatchet /ˈhætʃit/ 刮刀，小斧

hatching /ˈhætʃiŋ/ 影线，阴影线，剖面线，斜的断面线；画阴影线；～ convention 习用剖面线；～ lines 剖面线，阴影线；cross ～ 断面线，剖面线，双向影线；general ～ 通用剖面线

hatchures /ˈhætʃəz/ 阴影线，短线

H-beam 工字钢，H 形梁

HB (Brinell Hardness Number) 布氏硬度数

H-Br (Brinell Hardness) 布氏（球测）硬度

HCBS (Host Computer Basic Software) 主计算机基础软件

HD (hand drawn) 手工绘制的；(head)（螺栓）头；(hard drawn) 硬拉

hdbk (handbook) 手册

HDN (harden) 使硬化，淬火

H. D. P (horizontal datum plane) 水平基准面

HE (height of the eye) 视线高度

head /hed/ 头，首长，领导，主任，标题；顶部，箭头；（漏）斗；～ of window 窗楣；angle ～ 弯头；arrow ～ 箭头；bolt ～ 螺栓头；bottom ～ 底盖；closing ～ 铆钉头；counter-sunk ～ 埋头；cup ～ 圆头，半圆头；end ～ 末端；flat ～ 平头；full bearing ～ 全承面螺钉头；hexagon socket ～ 内六角头；index ～ 分度头；key ～ 键头；leader ～ 水斗；open ～ 开尾式(箭头)；rain-water ～ 雨水斗；read/write ～ 读写头；recessed ～ 凹槽头（用改锥的螺钉头）；rivet ～ 铆钉头；round ～ 圆头；screw ～ 螺钉头；snap ～ 半球头；solid ～ 填实式箭头；tooth ～ 齿顶（高）；water pipe ～ 水管接头

heading /ˈhediŋ/ 方向，标题，题目

headless /ˈhedlis/ 无头的；～ set screw 无头止动螺钉

headroom /ˈhedˌruːm, ˈhedˌrʊm/ 头顶，空间，楼梯上的空间

heater /ˈhiːtə/ 加热器

heating /ˈhiːtiŋ/ 采暖，镦粗（锻）；～ ventilation & air conditioning 采暖通风空调

heat-treat 热处理，对……进行热处理；～ment 热处理

heavy /ˈhevi/ 重的，重型的，粗大的；～-line 重型线，粗线；not ～ enough 不够粗；too ～ 太粗

heel /hiːl/ 鞋后跟，梯子的底脚，（圆锥齿轮轮齿的）大端

height /hait/ 高，高度，顶点，标高；～ board 高度测绘板，高度绘图仪；～ of addendum 齿顶高；～ of arc 弧高，拱高；～ of center 中心高度；～ of dedendum 齿根高；～ of tooth 齿高；extreme ～ 全高；free ～* 自由高度，有效高度；over-all ～ 全高；total ～ 总高度；unsupported ～ 自由高度；vertical ～ 垂直高度；working ～* 工作高度

helical /ˈhelikəl/ 螺旋，螺线，螺旋形的；斜；～ bevel gear 螺旋圆锥齿轮；～ convolute 螺旋形的蟠曲面（盘旋面）；～ curve 螺旋曲线；～ gear 螺旋齿轮，圆柱斜齿轮；～ line 螺旋线；～ spring 螺旋弹簧

helices /ˈhelisiːz/ （helix 的复数）螺旋线

helicograph /ˈhelikəgrɑːf/ 螺旋规

helicoid /ˈhelikɔid/ 螺旋面，螺旋状的；Archimedes spirel ～ 阿基米德螺旋面；conical oblique ～ 圆锥形斜螺旋体；conical right ～ 圆锥形直螺旋体；curvilinear generatrix ～ 曲线螺旋面；involute ～ 渐开线螺旋面；oblique ～ 斜螺旋面；rectilinear generatrix ～ 直线螺旋面；right ～ 正螺旋面；screw ～ 轴向直螺旋面（螺旋杆轴向截面内具有直线齿廓的螺旋面）

helix /ˈhiːliks/ * 螺旋线；～ angle 螺旋角；～ of cone 圆锥螺旋线；～ of cylinder 圆

柱螺线; conic ~ 圆锥螺线; conical ~ 圆锥螺线

hem /hem/ 滚边, 卷边, 折边

hemisphere /'hemisfiə/ 半球面, 半球体; 范围

hemispheric(al) /hem:'sferik(əl)/ 半球形的

hendeca- 表示"十一"

hendecagon /hen'dekəgən/ 十一角形, 十一边形

hendecahedron /ˌhendekə'hedrən/ 十一面体

heptagon /'heptəgən/ 七角形, 七边形; ~al 七角形的

heptahedron /'heptə'hedrən/ 七面体

herringbone /'heribəun/ 人字齿轮, 人字形; 人字形的; 鲱鱼骨; ~ tooth 人字齿, 双螺旋齿

herring-gear 人字形齿轮

hex. (hexagon) 六边形, 六角形; (hex-agonal) 六角形的

hexagon /'heksəgən/ 六角形(平面); 六角体; ~ socket head 内六角头; ~ socket screw 六角承窝螺钉, 内六角螺钉

hexagonal /hek'sægənl/ 六角形的; ~ head 六角头; ~ prism 六角柱; ~ pyramid 六角锥

hexagram /'heksəgræm/ 等边六角形, 六星形, 六线形

hexahedral /'heksə'hedrəl/ 六面体的, 有六面体的

hexahedron /ˌheksə'hedrən/ 六面体, 立方体; regular ~ 正六面体, 六方体

hexbolt 六角头螺栓

HEX (hexagon) 六角形

hex. hd (hexagon head) 六角头

hf (half) 一半; (hard surface) 硬面

hf-h (half-hard) 中等硬度

hg (helical gear) 斜齿轮; (harden and grind) 硬化与磨削

HGT (height) 高度

HI (horizontal interval) 水平间距

hidden /'hidn/ 虚的, 隐匿的; ~ detail 隐蔽细部; ~ feature 隐蔽形体; ~ line 隐线, 看不见的棱线, 虚线; ~ outline 看不见的轮廓线, 隐藏轮廓线; ~ point 隐蔽点; ~ surface 隐藏面

high /hai/ 高, 高度; 强的; 显著地

highlighted 醒目

high-lustre 镜面光亮

highway /'haiwei/ 公路, 大路

hill /hil/ 山, 丘, 岗

hinder /'hində/ 后面的, 后部的; 阻碍, 阻止, 挡住

hint /hint/ 提示, 暗示; 点点滴滴

hip /hip/ 屋脊, 节点, 屋顶之尖角; ~ roof 四坡屋顶

histogram /'histəgræm/ 直方图

history /'histəri/ 历史(学); 经历, 过程, 变化规律, 图形关系曲线, 时间关系的图示法; temperature ~ 温度与时间的关系曲线; time ~ 时间关系曲线图

hl. (hole) 孔, 洞

hls (holes) 孔, 洞

hob /hɔb/ 毂, 蜗杆, 螺旋杆, 齿轮滚刀

hobber /'hɔbə/ 滚铣刀, 螺杆, 蜗杆; 滚齿机

holder /'həuldə/ 柄把, 支架, 座; 套, 圈, 夹; drawing pen ~ 绘图笔杆; fine line lead ~ 细笔心自动铅笔

hole /həul/ 孔, 眼, 槽, 穿孔; ~ circle 孔位圆(即 bolt hole circle), 螺栓孔分布圆; ~-basis system of fits 基孔制; ~ center 孔之中心; ~ method 刺孔法(画图); ~ pattern 孔排列模式; ~ table 空白表; ~ tolerance 孔的公差; air (blow) ~ 气孔, 砂眼; basic ~ * 基准孔; blind ~ 盲孔, 未穿孔; blind tapped ~ 盲螺孔, 未贯穿的螺孔; bolt ~

螺栓孔; bore ～ 钻孔; central ～ 中心孔, 顶针孔; circular ～ 圆孔; countersunk ～ 锥坑孔; dead ～ 盲孔, 未穿孔; depth of tapped ～ 螺孔深度; inclined ～ 斜孔; key ～ 键槽; locating ～ 定位孔; mounting ～ 机械安装孔; oil ～ 油孔; oil drain ～ 放油孔; pin ～ 销钉孔, 小孔; plated through ～ 金属化孔; rectangular ～ 方形孔; round ～ 圆孔; slotted ～ 槽孔, 长圆孔; smooth ～ 光滑孔; splined ～ 花键孔; sprocket ～ 定位孔; taper ～ 锥形孔; terminal ～ 引线孔; threaded ～ 螺纹孔; through ～ 贯通孔; top ～ 顶孔

hollow /ˈhɔləu/ 空心, 杯形, 中空; 空心的

homologize /hɔˈmɔlədʒaiz/ (homologise) 使相应, 使类似, 使同系, 表示与……同系, 同系, 同源

homologous /hɔˈmɔləgəs/ 相应的, 相似的, 对应的, 同调的, 同系的, 同调于, 对应; ～ pair 对应线对; ～ series 同系列

homology /hɔˈmɔlədʒi/ 相同, 相似, 相当, 相应; 相互射影; 同调, 透射对应, 异体同形; 同系; 透射, 透射变换; ～ of conic 二次曲线间的透射对应; axial ～ 轴性透射

homothetic /ˌhəuməˈθetik/ 位似对应

hook /huk/ 钩; grab ～ 起重钩

hopper /ˈhɔpə/ 漏斗; coal ～ 煤漏斗; conveyer ～ 输送器漏斗

HOR (horizon) 地平线, 限界; 水平线, 水平仪; (horizontal) 地平的, 水平的, 水平线; 卧式的

horizon /həˈraizn/ 水平线; 水平仪; 地平（线）; ～ line 视平线; ～ plane 视平面

horizontal /ˌhɔriˈzɔntl/ 水平, 水平的, 水平线, 水平面; ～ axis 水平轴; ～ datum plane (H. D. P) 水平基准面; ～-frontal line 侧垂线; ～ line 水平线; ～ ordinate 横坐标; ～ plan 水平投影, 平面图; ～ plane 水平面; ～

profile 水平剖面, 水平断面; ～ profile line 正垂线; ～ projection 水平投影; ～ projection plane 水平投影面; ～ reference plane 水平辅助投影面; ～ system 水平制（写尺寸数字的单向制）; ～ trace 水平迹点（线）

horocycle /ˌhɔrəˈsaikl/ 极限圆

horosphere /ˈhɔrəsfiə/ 极限球面

house /haus/ 房屋, 建筑物; 室, 箱

housing /ˈhauziŋ/ 套, 壳, 罩, 机架; ～ cover 套盖; axle ～ 轴套

HP (horizontal parallax) 水平视差; (horizontal plane) 水平面; (hp, horsepower) 马力

H-parallel 平行于 H 面的（水平面）

H-projection 水平投影

H-projector H 面投射线

HR (Rockwell hardness) 洛氏硬度

HRB (Rockwell Hardness Number, scale B) B 标尺洛氏硬度数

HRC (Rockwell Hardness Number, scale C) C 标尺洛氏硬度数

HRP (horizontal reference plane) 水平面

HRS (hrs, hot-rolled steel) 热轧钢

HS (shore scleroscope hardness) 肖氏（回跳）硬度

H. T. (heat treatment) 热处理; (height) 高度

H. tr (heat treat) 热处理

H-trace 水平迹

HT TR (heat treatment) 热处理

hub /hʌb/ 轮毂, 中心, 衬套

hull /hʌl/ 机身, 船体; 壳, 皮

HV (Vickers hardness) 维氏硬度

HVAC (heating ventilation & air conditioning) 采暖通风空调

hydrograph /ˈhaidrəugræf/ 水流测量图, 水位图, 水文图

hydryzing /ˈhaidraiziŋ/ （防表面氧化的）氢气热处理

hyperbola /haiˈpəːbələ/ 双曲线, 双曲线

形；confocal ～ 共焦双曲线；equilateral ～ 等轴双曲线，等边双曲线；rectangular ～ 直角双曲线

hyperbolae /haiˈpəːbəliː/ （复数）双曲线

hyperbolic(al) /haipəːˈbɒlik(əl)/ 双曲(线)的；太大的；～ lines 双曲线性直线；～ paraboloid 双曲抛物面

hyperbolograph /haiˈpəːˈbəuləgraːf/ 双曲线规

hyperboloid /haiˈpəːbɒlɔid/ 双曲面，双曲线体；～ of one sheet 单叶双曲面；～ of revolution 回转双曲面；～ of revolution of one sheet 单叶回转双曲面；circular ～ 圆形双曲面体；elliptical ～ 椭性双曲面

hypergeometry /ˌhaipəˌdʒiˈɒmitri/ 多维几何(学)，多度几何(学)

hypergraph /ˈhaipəgraːf/ 超图；balanced ～ 平衡超图

hyperplane /ˈhaipəplein/ 超平面

hyperspace /ˈhaipəːˈspeis/ 多维空间，多度空间，超空间

hypersurface /ˌhaipəˈsəːfis/ 超曲面；～ curved 弯曲的超曲面

hyphen /ˈhaifn/ 连字符；连字号

hypocycloid /ˈhaipəuˈsaiklɔid/ 内摆线，次摆线，圆内旋轮线

hypodispersion /ˌhaipəudisˈpəːʃən/ 平均分布

hypoid /ˈhaiˌpɔid/ 准双曲面的；～ gear 准双曲面齿轮，螺旋圆锥齿轮

hypotenuse /haiˈpɒtinjuːz/ 斜边，弦

hypoth. (hypothesis) 假设，前提；(hypothetical) 假设的

hypsography /hipˈsɒgrəfi/ 地形测绘学，等高线法，测高法；表示不同高度的地形图

hypsometric(al) /hipsəˈmetrik(əl)/ 测高学的；～ curve 等高线

I

I-bar 工字钢

I-beam 工字梁

IC (integrated circuit) 集成电路

ICDG (International Conference on Descriptive Geometry) 国际画法几何会议

ichnograph /ˈiknəugraːf/ 平面图

ichnography /ikˈnɒgrəfi/ 平面图(法)

ichthyoid /ˈikθiɔid/ 流线形的，鱼(状)的

icon /ˈaikɒn/ 像，图像，插画，肖像；图标，图符；～ system 图符系统；application ～ 应用程序图标

iconic /aiˈkɒnik/ 图像的；～ representation 图像表示

iconography /ˌaikəˈnɒgrəfi/ 插图，图解

icosahedron /ˌaikəsəˈhedrən/ (icosahedra) 二十面体；regular ～ 正二十面体

ID (inside diameter, internal diameter) 内径；(inside dimensions) 内部尺寸；(installation drawing) 安装图；(interconnection diagram) 相互联系图；(idem) 同上，同前，同在一处

idea /aiˈdiə/ 概念，思想，计划

ident (identification /aidentifiˈkeiʃn/) 标志，标识，打印；确定，辨认；注法

identical /aiˈdentikəl/ 同一的，相同的，完全

相似的, 全等的; ~ map 等角投影地图; ~ relation 恒等式; ~ly equal 恒等, 全等

identified /aiˈdenti,faid/ 表示

identifier /aiˈdentifaiə/ 标识符

identify /aiˈdentifai/ 识别, 鉴定; 确定在……上的位置; 使等同于, 视为同一

identometer /aɪˈdentəˈmi:tə/ 材料鉴定仪

ideogram /ˈidiəgræm/ (ideograph)(表意)文字(符号); Chinese ~ 汉字

ideograph /ˈidiəugra:f/ 示意图

idiomatic(al) /idiəˈmætik(əl)/ 成语的, 惯用的; ~ ally 按照习惯用法

i. e. (idest 拉丁字)那就是; 即(=that is), 换言之

IEC (International Electrotechnical Commission) 国际电工技术学会(委员会)

ig (involute gear) 渐开线齿轮

IGES (initial graphics exchange specifiction) 初始图形交换规范

I-girder 工字大梁; 大工字钢

ih (inverted hour) 逆时针的

ill. (illustrated) 举例说明的, 图解说明的; (illustration) 实例, 图解

illuminate /iˈl(j)u:mineit/ 阐明, 启发; 照射, 照明

illumination /il(j)u:miˈneiʃən/ 照射

illust. (illustrated, illustration)(用图或例子)说明的, 插图的; 说明, 图解, 例证, 插图

illustrate /ˈiləstreit/ 解释, 例解, 图解, 插图, 用图说明, 证明; ‖ *Example sentence:* The method of orthogonal projection *will be illustrated* in the next chapter. 例句: 正射影法将在下一章中用图说明。

illustration /iləsˈtreiʃən/ 插图, 注解, 实例, 例证, 说明, 润饰, 图解; cite Instances in ~ of 举例说明; manufacturing ~ 制造说明书; technical ~ 工程说明图

illustrative /ˈiləstreitiv/ 直观的, 说明性的

im. (image) 图像, 影图

image /ˈimidʒ/ 图像, 影像; 镜像, 相似物, 作……的像; 画面, 反射; ~ data 图像数据; ~ element 像元, 像素, 图素; ~ encoding 图像编码; ~ graphics 图像显示; ~ space 图像空间; ~ transformation 图像变换; active ~ 活动图像; background ~ 背景图像; background display ~ 背景显示图像; bi-level ~ 双值图像; binary ~ 二进制图像; binary ~ analysis 二进制图像分析; binary ~ processing technique 二值图像处理技术; bit ~ 位图像; block ~ transfers (BLIT) 块图传输; coded ~ 编码图像; composite ~ 合成图像; foreground ~ 前景图像; run ~ 运行图像; static ~ 静止图像; ‖ *Example sentence:* The object *throws an image* on the plane. 例句: 物体映影于平面上。

imaginary /iˈmædʒinəri/ 假想的, 虚构的, 虚的; ~ axis 虚轴

imaginary-line 假想线, 虚线

imagination /imædʒiˈneiʃn/ 想象力, 创造力, 假想

imaginative /iˈmædʒinətiv/ (富于)想象(力)的; ~ power 想象力

imagine /iˈmædʒin/ 想象, 设想, 料想

imitate /ˈimiteit/ 模仿, 仿造, 临摹, 模拟

impeller(impelor) /imˈpelə/ 叶轮, 转子的叶片, 压缩器, 推进器, 刀盘; water pump ~ 水泵叶轮

imperceptibility /impəseptəˈbiliti/ 看不见, 极微, 极细

imperceptible /impəˈseptəbl/ 看不见的, 难以察觉的, 细微的

implement /ˈimplimənt/ 仪器, 工具, 器具; 执行, 实现

implementation /implimenˈteiʃən/ 工具,

器具;完成,补充

implicit /im'plisit/ 隐含;不明显的,含蓄的,绝对的

imply /im'plai/ 隐含

impress /im'pres/ 痕迹,印象;特征;to ~ upon 加到……上

impression /im'preʃn/ 印模,印象,印刷;版,印次

improve /im'pru:v/ 改进,改善;~ arrowheads 改善箭头;~ form or spacing 改善形式或间隔

improvement /im'pru:vmənt/ 调质

in /in/ 在……里,在……方向,朝……方向,按,以,用;~ bridge 并联,跨接;~ extenso(拉丁语)全部;~ place 在适当位置;~ series 串联的,顺序的,串联式;~ some cases 在一些情况下;~ tandem 串联的,前后排列的;~ the main 基本上,主要部分;~ -line 串联的,轴向的,一列式的

in. (inch) 英寸

inaccuracy /in'ækjurəsi/ 偏差,误差,不精确,不准确;不精确性,不准确度;~ of dimensions 尺寸不合格;distortion ~ 变形误差

inboard /'inbɔ:d/ 内部

incenter /'insentə/ 内(切圆)心

inch /intʃ/ 英寸,少量,少许;~ based 英制

incircle /in'sə:kl/ 内切圆

incise /in'saiz/ 切割,切开,切入,雕刻

incision /in'siʒən/ 切割,切开,切口,刀痕

inclination /inkli'neiʃn/ 倾斜,倾度,倾角,斜角,斜度,坡度;downward ~ 下倾角

incline /in'klain/ 倾斜,斜度,斜坡,斜线,斜面;an ~ of 1 in 5 坡度为 1:5(竖:斜)的一个斜面;an ~ of 1 on 5 坡度为 1:5(竖:横)的一个斜面;an ~ of 1 to 5 坡度为 1:5(横:竖)的一个斜面

inclined /in'klaind/ 倾斜;倾斜的;be ~ to 倾斜于

inclinometer /inkli'nɔmitə/ 测斜仪,量坡仪

include /in'klu:d/ 包括,包含;~d angle 夹角,焊接角

inclusion /in'klu:ʒən/ 包括,包含

inclusive /in'klu:siv/ 总括,在内的

incognizable /in'kɔgnizəbl/ 不可辨别的,不能认识的,不可知的

incommutable /inkə'mju:təbl/ 不能变换的,不能交换的

incomplete /inkəm'pli:t/ 不完全的,未完的

inconvertibility /inkənvə:ti'biliti/ 不能交换性,不可逆性

inconvertible /inkən'və:təbl/ 不能变换的,不能转换的

incorrect /inkə'rekt/ 错误的,不正确的;~ shape 不正确的形状

increaser /in'kri:sə/ 异径接头(管)

incremental /inkri'məntəl/ 增量;~ coor-dinate 增量坐标;~ vector 增量向量

incurvation /inkə:'veiʃən/ 向内弯曲

IND (index) 索引,指数

indefinite /in'definit/ 模糊的,不明确的,不定的;一般式

indent /in'dent/ 锯齿形,刻成锯齿形;合同,契约

indentation /inden'teiʃn/ 呈锯齿形,缺口,缩进,空格

indention /in'denʃn/ (indentation) 缩进,空格,凹入,锯齿形

independent /indi'pendənt/ 独立的;~ of 与……无关的

index /'indeks/ 标高,指数,指标,分度头,索引,目录,变址,下标;~ed plane 标高平面;~ed projection 标高投影;chart ~ 图幅

拼接，索引；colour ~ 颜色索引；to ~ out 指出

India /'indjə/ 印度（亚洲）；~ ink 墨（汁）；~ paper 薄而坚韧的印刷纸

indic. (indicative) 表示，指示；(indicator) 指示器，指针

indicate /'indikeit/ 表示，表明，指出，指示

indication /indi'keiʃn/ 标注，注写，显示，表示，指示，迹象，征候

indicator /'indikeitə/ 指示器，示功器，百分表；~ chard 示功图；~ diagram 示功图；etching ~ 蚀刻指示图；inside dial ~ 内径千分表

indiscernible /indi'sə:nəbl/ 难辨别的，分辨不出的

indistinct /indis'tikt/ 不明显的，模糊的

indistinguishable /indis'tigwiʃəbl/ 不能辨别的，难区分的，无特征的

industrial /in'dʌstriəl/ 工业的，产业的；~ drawing 工业制图

inequality /ini:'kwɔliti/ 不平均，不等量，不相同，不平坦

in. ex (in extenso) 全部

inexact /inig'zækt/ 不准确的，不正确的，不仔细的

infinite /'infinit/ 无限的；极小量；~ point 无限远点

infinity /in'finiti/ 无穷（大），无限（性），无限（远），无穷不连续点

inflexion /in'flekʃn/ 弯曲，挠曲，拐折，回折点；~ point 转折弯，拐点

info /'infəu/ 信息

information /infə'meiʃn/ 资料，数据，信息，报告；~ processing 信息处理；~ retrieval 信息检索

inherent /in'hiərənt/ 原来的，固有的

initial /i'niʃəl/ 最初的，开始的，初期的

initializing /i'niʃə,laiziŋ/ 初始化

ink /iŋk/ 墨水，油墨，用墨水写（画），上墨水线；~ed drawing 上墨图；~ holder 钢笔杆；~ in（或 over）上墨，上墨水线，再用墨水笔加描；~ in a drawing 在铅笔底线上用墨水笔加描；~ out 用墨涂去；black Indian ~ 黑墨水；China ~ 中国墨；colored ~ 色彩墨水；drawing ~ 绘图墨水；Indian ~ 黑墨水

inkbottle /'ik'bɔt/ 墨水瓶

inking /'iŋkiŋ/ (ink-in) 上墨水线，涂油墨，上墨；order of ~ 上墨顺序

inkless /'iŋklis/ 无墨水（汁）的

ink-pencil （复写用）颜色铅笔

ink-pot 墨水瓶

inkspot /'iŋkspɔt/ 墨水点

inky /'iki/ 有墨迹的，墨黑的

inlet /'inlet/ 注入，引入，入口，进口；海湾，小湾

in-line 在线的，串联式的，轴向（式）的

inner /'inə/ 内部的，里面的

input /'input/ 输入；~ device 输入设备；~ /output 输入/输出；manual ~ 人工输入

inquiry /in'kwaiəri/ 查询，询问

ins (insert) 插入（键）

ins-and-outs 来龙去脉

inscribed /in'skraibed/ 内接的，内切的 ~ circle 内切圆；~ cone 内切锥；~ sphere 内切球；~ triangle 内接三角形

inscription /in'skripʃn/ 标题，题词，编入名单

insection /in'sekʃən/ 切开，切断，切口；齿纹，锉纹

inset /'in'set/ 插图，插画，插页，插入，嵌入

inside /'in'said/ 在内部；~ diameter 内径

inspect /in'spekt/ 检查，检验，观察，审查

inspection /in'spekʃn/ 检查，检验，验收，调查，观看，视察；dimension ~ 尺寸检查；hand ~ 目测检查，手检；surveyer of the technical ～ 技术鉴定者

inspector /in'spektə/ 检查员

inst. (institute) 学会, 协会, 学院; (instrument) 仪器, 工具, 器械

instal(l) /in'stɔ:l/ 安装, 装配, 拧入, 装入

installation /instə:'leiʃn/ 装置, 设备, 装配, 安装; ～ diagram 装置图, 安装图; ～ drawing（机器等的）装配图

instance /'instəns/ 例子, 实例, 举例; 情况; 步骤; 要求; for ～ 例如, 举例来说; inthis ～ 在这种情况下

instead /in'sted/ 代替, 顶替; 不是……而是……

institute /'institju:t/ 学会, 协会, 学院; 研究所; 设置; 实行; American National Standard ～ (ANSI) 美国国家标准协会

institution /insti'tju:ʃn/ 建立, 设立, ‖*Example sentence:* The institution of system of three projection planes. *例句:* 三投影面体系的建立。

instl. (installation) 装置, 设备, 安装

instr. (instructor) 教师, 讲师; (instrument) 仪器, 工具, 器械

instruct /in'strʌkt/ 指示, 通知, 教（授）

instruction /in'strʌkʃn/ 指令, 指南, 说明, 说明书, 指示书; 细则, 规程; 教学的, 讲授, 教导, 指导, 指示; ～ book 说明书; ～al modules 指令模式; ～al film 教学影片; computer assisted ～ 计算机辅助教学

instructor /in'strʌktə/ 教师, 讲师, 讲授者, 指导者

instrument /'instrumənt/ 仪器, 器械, 工具; 文件; ～ drawing 用器画, 仪器图; ～ joints 仪器的关节; Ames ～ 字格板; Ames lettering ～ 字格板; angular ～ 量角器; case ～s 成套绘图仪器; drafting ～ 绘图仪器; drawing ～ 绘图仪器; special ～ 特制（绘图）仪器; springbow ～ 弹簧弓形仪器; surveying ～ 测量仪器; tracing ～ 描图仪器; use of ～ 仪器使用

instrumental /instru'mentl/ 助成的, 仪器的, 器械的; ～ drawing 仪器制图, 机械制图, 仪器图

int. (intermediate) 中间的, 中等的; (internal) 内面的; (interval) 间隔

intact /in'tækt/ 完整的, 原封未动的, 未触动的

intarometer /intə'rɔmitə/ 盲孔千分尺

integer /'intidʒə/ 总体, 整体; 整数, 定点数

integrable /'intigrəbl/ 可积分的

integral /'intigrəl/ 组成的, 完整的, 积分的, 总体的

integraph /'intigra:f/ 积分仪, 积分描图仪, 积分曲线仪; graphic ～ 图解积分法

integrate /'intigreit/ 积分, 积累, 汇集, 集成, 使结合, 使完整; ～d circuit 集成电路（积体电路）; ～ing circuit 积分电路

integration /inti'greiʃn/ 完整的, 完全的, 主要的, 整数的; 集成, 总合, 积分（法）; ～ technology 集成技术（积体技术, 积体工艺）; graphical ～ 图解积分法

integrator /'intigreitə/ 积分仪

inter. (intermediate) 中间的, 中等的

interactive /intər'æktiv/ 相互作用的; ～ modes 交互方式

intercept /intə'sept/ 截取, 相交, 交叉, 贯穿; 截距

interchangeability /intə,tʃeindʒə'biliti/ 互换性; 可交换性; 替代

interchangeable /intə'tʃeindʒəbl/ 可互换的, 可拆卸的, 通用的

interconversion /intəkən'və:ʃən/ 变换, 互换

intercostal /intə'kɔstəl/ 加强肋, 肋间的

interdependence /intədi'pendəns/ 互相关联, 相关

interface /ˈintəfeis/ （交）界面，分界面；面线；对接，接合，接口；~ (M) with N 使 (M) 与 N 面接合；~ requirement 界面需求；parallel ~ 并行接口；serial ~ 串行接口

interference /intəˈfiərəns/ * 过盈，干涉；~ checking 干涉检验；~ fit* 过盈配合，干涉配合；maximum ~ * 最大过盈；minimum ~ * 最小过盈

interfix /ˈintəˌfiks/ 相关，相互确定

intergrind /ɪntəgˈraɪnd/ 互相研磨

interior /inˈtiəriə/ 内部

intermediate /intəˈmiːdjət/ 中间的；中间体；~ radius 中间半径

intermittent /intəˈmitənt/ 间断的，间歇的，断断续续的

internal /inˈtəːnl/ 内部的；~ fillet 内圆角；~ thread 内（阴）螺纹；~ly tangent 内切，内切线

international /intəˈnæʃnəl/ 国际的

interpenetrate /intəˈpenitreit/ （互相）贯穿，穿插

interpenetrating /ˌintə(ː)ˈpenitreitiŋ/ 穿插的，互相贯穿的

interpolation /intəːpəuˈleiʃn/ 插补；arithmetical ~ 算术内插法；circular ~ 圆弧插补；linear ~ 直线插补

interpolate /inˈtəːpəuleit/ 插入，内插；~d section 内插剖面（即旋转剖面，重合剖面）

interpret /inˈtəːprit/ 翻译，译码，说明，解释；graphic ~ation 图解法

intersect /intəˈsekt/ 相交，交叉；横断，横切，贯穿；~ing axes gears 二轴线相交的齿轮；~ing body 相贯体；~ing point 交会点，转角点，贯穿点；‖ *Example sentence①: Unless a line is in or parallel to a plane, it must intersect the plane.* 例句①：如果直线与平面不重合不平行则必定相交。‖ *Example sentence②: Any two planes either must be parallel or they must intersect.* 例句②：任意两平面不平行必相交。‖ *Example sentence③: Any two lines in a plane must either intersect or be parallel.* 例句③：平面内的任何两直线不是相交就是平行。

intersection /intə(ː)ˈsekʃn/ 相交，相贯，交线，交点，交集；交切，交切点，横断，截断；~ line 交线（面与面相交的交线）；~ line of two surfaces 二面交线；~ of line and plane with body 平面、直线与立体相交；~ of more bodies 多个立体相交；~ of plane body with curved surface body 平面主体与曲面主体相交；~ of plane and right circular cone 平面和正圆锥体相交；~ of surfaces 面的交线；~ point 交点；apparent ~ 视图上的交点；theoretical line of ~ 理论交线；‖ *Example sentence①: The intersection of two planes is a straight line common to the planes.* 例句①：两平面的交线是两平面共有的一条直线。*Example sentence②: Intersection of plane and right circular cone.* 例句②：平面和正圆锥体相交。

interspace /ˈintəˌspeis/ 中间，空间，空隙，留空隙，星际

interval /ˈintəvəl/ 间隔，距离，区间，范围，空隙；at ~s of 每隔，相隔；at regular ~s 每隔一定间隔；contour ~ 等高线间隔；level ~ 水平间距

intricate /ˈintrikit/ （形状）复杂的

intrinsic(al) /inˈtrinsik(əl)/ 内部的，固有的，原设计的；~ curve 包络线，裹性曲线

introduce /intrəˈdjuːs/ 引进，设立；输入，传入；插入；介绍

introduction /intrəˈdʌkʃn/ 引言，前言，绪言；引进，引入，介绍；brief ~ 简介

INTSTD THD (International Standard Thread)

国际标准螺纹
intuition /intjuˈiʃn/ 直觉，直观；直觉知识
intuitional /intjuˈiʃənəl/ 直觉的，直觉的
intuitive /inˈtjuitiv/ 直观的，直觉的；intuitiveness 直观性
invalid /ˈinvəlid/ 无效的
invariant /inˈvɛəriənt/ 不变式，不变的
inverse /ˈinvəːs/ 相反的
invert /inˈvəːt/ 倒置，转换
invisibility /invizəˈbiliti/ 不可见(性)，看不见(的东西)
invisible /inˈvizəbl/ 看不见的，不可见的，无形的；～ line of an object (投影中)物体不可见的线；～ -line technique 画虚线的技巧；～ shadow 看不见的影区；be ～ to the naked eye 是肉眼看不见的
involute /ˈinvəluːt/ 渐伸线，渐开线，切展线；渐开，渐伸；～ gear 渐开线齿轮；～ of circle 圆的渐开线；～ of pentagon 五角形的渐开线；～ rack 渐伸线齿条；～ teeth 渐开线齿形
involution /invəˈluːʃən/ 乘方，对合，对合对应，对合变换，回旋，内卷；absolute ～ 绝对对合；hyperbolic ～ 双曲型对合
involve /inˈvɔlv/ 包括，包含；包围，环绕；涉及；促成；乘方
I/O (input/output) 输入/输出
I&R (interchangeability and replacement) 可互换性与置换
iron /ˈaiən/ 铁；angle ～ 角铁；cast ～ 铸铁，生铁；malleable ～ 展性铸铁；pig ～ 铸铁；round ～ 圆铁，圆钢；section ～ 型钢
irreg. (irregular) 不规则的，不整齐的
irregular /iˈregjulə/ 不规则的，不平坦的，不均匀的，有凹凸的；～ curves 不规则曲线板，万能曲线板，不规则曲线
irregularity /iregjuˈlæriti/ 不规则性，不均匀性，不对称度；surface ～ 表面不平度
IS (internal surface) 内表面
ISA (International Federation of the National Standardizing Association) 国际标准化协会
I-section 工字形剖面
ISO (International Organization for Standardization) 国际标准化组织；～ (recommendation) 国际标准组织规范
isocline /ˌaisəʊˈklain/ 等倾线，等向线；等倾，等斜；～ planes 等斜平面
ISO/DAD (ISO/Draft Addendum to an International Standard) 国际标准的补充草案
ISO/DAM (ISO/Draft Amendment to an International Standard) 国际标准的修正草案
ISO/DIS (ISO/Draft International Standard) 国际标准草案
ISO/DP (ISO/Draft Proposal) 国际标准建议草案
isogonal /aiˈsɔgənl/ 等角的；～ affine transformation 等角仿射变换
isogonality /aiˌsɔgəˈnæliti/ 等角变换
isogram(s) /ˈaisəugræm(z)/ 等(值)线图；～ of ash content 灰分等值线图
isograph /ˈaisəugraːf/ 等(值)线图，求根仪
isohypse /ˈaisəˌhaips/ 等高(度)线
isometric(al) /aisəuˈmetrikəl/ 等距的，等大的，等角的，等比例的，等体积的，等轴的，等径的；～ correspondence 等距对应；～ diagram 等轴测图；～ drawing 等角图，等距画法，等轴测图；～ line 等角线；～ paper 等轴测格纸；～ projection 等角投影，正等测投影，三轴等比正投影；～ scale 等轴测比例尺；～ sketch 等轴测草图；～ surface 等距曲面
isometrography /ˌaisəməˈtrɔgrəfi/ 等角线规
isoplanar /aisəuˈpleinə/ 同平面的

ISO/R (ISO/Reference) 国际推荐标准
isosceles /ai'sɔsili:z/ 等腰的; ～ triangle 等腰三角形
ISO/TC97/SC5/WG2-Graphics 国际标准化组织第97技术委员会第5分技术委员会第2工作组—计算机图形工作组
ISO/TR (ISO/Technical Report) 国际标准技术报告

issue /'iʃ(j)u:/ 出版, 发行; ～ and distribution 出版和发行阶段（计）
IT (ISO Tolerance) 国际公差, 标准公差
italic /i'tælik/ 斜体的; 斜体字
item /'aitəm/ 项目; ～ bolck* 明细栏
IWG (Imperial Standard Wire Gauge) 皇家标准线规

J

jack /dʒæk/ 千斤顶, 起重器; 弹簧开关; 插孔, 插座; ～ shaft 中间轴
jalousie /'ʒælu:zi:/ 百叶窗
JAP (Japan) 日本
japanning /dʒə'pæniŋ/ 涂漆, 涂黑
jargon /'dʒa:gən/ 术语, 行话
jaw /dʒɔ:/ 爪, 虎钳牙, 夹片; 销, 键
JES (Japan Engineering Standard) 日本工业标准（旧名）
jig /dʒig/ 钻模
JIS (Japan Industrial Standard) 日本工业标准, 日本工业规格
jitter /'dʒitə/ 跳动, 抖动, 起伏
job /dʒɔb/ 零件, 加工件; 工作, 任务, 部件, 成果, 作业; ～ sheet 说明图, 零件图纸; ～ shop 加工车间; ～ bing shop 修理车间
jog /dʒɔg/ （面或线上的）凹进, 凸出; 突然的转向
joggled /'dʒɔgld/ 啮合的, 接合的
joggling /'dʒɔgliŋ/ 弯接, 弯接头; 折曲, 偏斜
join /dʒɔin/ 连接, 接合; 结合处, 接缝; 会合

joint /dʒɔint/ 接头, 接口; 接合; 枢扭, 关节, 连接处; 共同的; ～ of framework 节点; angle ～ 角接; bolt ～ 螺栓连接; butt ～ 对接; corner ～ 角接, 弯头连接; double V butt ～ 双面 V 形对焊接; double-riveted butt ～ 双列铆钉对接; double-riveted lap ～ 双列铆钉搭接; double strap butt ～ 双列对接; double welded butt ～ 双面对焊接; elbow slip ～ 曲折滑动接缝, 肘管接缝; end to end ～ 对（平）接; expansion ～ 伸缩缝, 伸缩接头; flare ～ 喇叭口接合; jump ～ 对接; key ～ * 键联接; lap ～ 搭接; miter ～ 斜接合; mortise and tenon ～ 榫头榫孔接合; pinned ～ *销联接; plain corner ～ 平头角焊接; plumb ～ 锡焊接; riveted ～ * 铆钉联接, * 铆接; screwed ～ * 螺纹联接; tee ～ T 形接; welded ～ 焊接
jointing /'dʒɔintiŋ/ 填料, 接合, 连接, 焊接; asbestos ～ 石棉填料; hemp ～ 麻填料
joist /'dʒɔist/ 阁栅; 工字钢
jour (journ, journal) 会刊, 杂志, 日报
journal /'dʒə:nl/ 轴颈; 杂志, 日报, 日记
joystick /'dʒɔi,stik/ 操纵杆

JSME (Japanese Society of Mechanical Engineers) 日本机械工程师学会
jt. (joint) 结点, 接头, 接缝
judge /'dʒʌdʒ/ 审查员; 鉴定, 判断, 下结论
jumping /'dʒʌmpiŋ/ 跳动（的）, 跃变, 跳跃（的）, 转移; ～ correspondence 跳动对应
junction /'dʒʌkʃn/ 连接, 接头, 接合处, 项缘过渡处; 腊焊; 熔焊; 联络; make a ～ 连接起来, 连接点, 结合处, 接头, 汇结点; 交叉; 取得联络
juncture /'dʒʌktʃə/ 接合, 接合处, 交界处, 连接点
JWG (joint work group) 联合工作组

K (constant) 常数
Kanji （日语）汉字
Karnaugh (map) 卡诺图（一种简化开关函数的方格图）
Karton 厚纸
KB (key board) 键盘
keep /ki:p/ 保持, 处于; -ring 保持环
keeper /'ki:pə/ 夹子, 锁紧螺母; 管理人
kent /kent/ 制图纸, 绘图纸
kerf /kə:f/ 截口, 切口, 锯口, 槽; 截断, 切开, 切槽, 开缝
key /ki:/ * 键; 扳手, 开关, 钥匙, 关键, 要害; 主要的, 基础的; 销, 双头螺栓; 楔, 按钮, 线索, 答案, 解题; ～ aspect 主要趋向; ～ board 键盘; ～ diagram 概略原理图, 解说图; ～ drawing 解释图, 索引图, ～ way* 键槽, 销槽; axle ～ 轴键; Barth ～ 巴斯键; dive ～ * 导向平键; feather ～ * 滑键, * 导向平键; flat ～ * 平键; forelock ～ 开口键; function ～ board 功能键盘; general flat ～ * 普通平键; general taper ～ * 普通楔键; gib-head taper ～ * 钩头楔键; gib headed ～ 钩头键; half round ～ 半圆键; Lewis ～ 切线键, 吊楔键; logitudinal ～ 长方键; Pratt-Whitney ～ 圆头平键; saddle ～ * 鞍形键; square ～ 方键; stock ～ 普通（平）键; sunk ～ 嵌入键; tangential ～ * 切向键; taper ～ * 楔键; thin (flat) ～ * 薄型平键; wedge ～ 楔形键; whitney ～ 半月键, 月牙键, 半圆键, 月弧销; woodruff (woodroof 或 woodrow) ～ * 半圆键
keyseat /'ki:si:t/ 键座（键在其中不能滑动）
keyslot /'ki:slɔt/ 键槽; 销槽
keyway /'ki:wei/ 键槽; 销槽
keyword /'ki:wəd/ 保留字
kg. (kilogramme) 千克, 公斤
kickpoint /'kikpɔint/ 转折点
kinematics /kaini'mætiks/ 动态仿真
kip /kip/ 千磅
kit /kit/ 仪器, 成套工具, 配套元件
km. (kilometre) 千米, 公里
knee /ni:/ 弯头, 曲线的弯曲处; 膝形杆, 膝; 直角的; ～ bend 弯管; ～ point（曲线）弯曲点, 拐点
knife /naif/ 刀片, 刀, 小刀; drawing ～ 制图刮刀; erasing ～ 刮刀
knob /nɔb/ 捏手, 旋钮
knoop number 努普硬度值

knoop scale 努氏(硬度)标度
knurl /nə:l/ 滚花,压纹;节,瘤,小的隆起物;～ for press fit 压花; hatching ～ing 网纹滚花; rhombic ～ing 菱纹滚花; straight ～ing 直纹滚花
KST (key seat) 键座
kW (kilowatt) 千瓦
KWY (key way) 键槽

L

label /ˈleibl/ 标号,标记,标识,标签,标牌;标明,把……列为,把……称为; statement ～ 语句标号
labyrinth /ˈlæbərin/ 曲折,曲径;迷宫;～ seal 迷宫式密封
lack /læk/ 缺少,不足,没有;需要
lacquer /ˈlækə/ 漆,涂漆,喷漆; antirust ～ 防锈漆; priming ～ 底漆; spray ～ing 喷漆
ladder /ˈlædə/ 梯子,梯形物,阶梯,梯形裂缝;～ (type) network 梯形网络;～ step 扶梯;阶级
ladderlike 梯(子)状的
ladder-type 梯形
laevo- 左(旋)的,在左方,向左方
laevorotation /ˌli:vəurəuˈteiʃən/ 左旋
laevorotatory /ˌli:vəuˈrəutətɔri/ 左旋的
lam (laminate) 迭片
lamp /læmp/ 灯泡,光源;电子管
lanai /ˈla:na:i/ 外廊,(上有顶棚的)门廊,门庭
land /lænd/ 地面,陆地,土地;连接盘(导电图形的一部分);棱,齿刃;～ map 区域图;～ mark 陆地标志
landscape /ˈlændskeip/ 风景,山水,风景画;眺望;～ architect 美景师;～ map 风景地图;～ painting 风景画
lane /lein/ 狭路,小巷,通道;～ of traffic 航线
language /ˈlæŋgwidʒ/ 语言,术语,(机器)代码,文字; algorithmic ～ 算法语言; architectural blockdiagram ～ 结构框图语言; assembly ～ 汇编语言; command ～ 命令语言; high level ～ 高水平代码; machine ～ 机器语言; object ～ 目的语言; sign ～ 符号语言学; visual ～ 视觉语言
lap /læp/ 盖板,搭接,余面,周围,折,怀抱,互搭;研磨;～ angle 余面角;～ circle 余面圆
lapping /ˈlæpiŋ/ 研磨,抛光
lapped /ˈlæpt/ 重叠,极光的,研磨光的
large /la:dʒ/ 大的,粗的,多的;～ compasses 大圆规;～ end 大端;～ -scale 大比例尺的;～ -size(d)大尺寸的,大型的
laser /ˈleizə/ 激光,激光器,光量子放大器
lat (lateral) 横的
latch /lætʃ/ 封闭,闩,凸轮
late /leit/ 后
lateral /ˈlætərəl/ 横的,横向的,侧面的,侧向的;～ angle of view 侧视角(透视图中最外端两视线之夹角);～ edge of a prism 棱柱体的侧棱线;～ face 侧面
latest /ˈleitist/ 最后的,最迟的,最近的,最新的; the ～ thing 最新的发明,新产品
lathe /leið/ 车床;～ centre 车床顶尖; chas-

ing ～ 螺纹车床; metal turning ～ 金工车床

latitude /'lætitju:d/ 范围, 幅度, 宽度, 地区, 地方, 纵距, 纬圆, 纬度

latter /'lætə/ 后面的, 后半的, 末了的, 后者, 近来的

latus /'leitəs/ （拉丁语）边, 弦

lavatory /'lævətəri/ 盥洗室

law /lɔ:/ 定律, 规律, 规程; associated ～ 结合律; associative ～ 结合律; combination ～ 并合律; communicated ～ 交换律; graphical ～ 制图规则; the ～ of projection 投影规律

lay /lei/ 位置, 方向, 层; 放, 摆, 安排, 铺形势; ～ off 划分, 画出, 量出; ～ on 放样; ～ out 摊开, 展开, 放样, 计划, 绘样, 设计, 布置; 区划图; ～ out machine 测绘缩放仪; the ～ of the land 地势, 地形; to ～ on 涂（颜料等）, 放置, 安排

layer /'leiə/ 层; ～ discrimination 层辨别; ～ freeze 层冻结; ～ thaw 层解冻; representation in ～s 分层揭示

laying /'leiŋ/ 布置, 铺设, 衬垫, 安装; ～ off 作出

layout /'lei,aut/ 设计, 布局, 布置, 安排, 规划, 编排; 划线, 划分; 划定; 方案, 图, 草图, 略图, 示意图, 外形图, 计划图, 线路图, 平面图, 格式外形, 轮廓, 区划; ～ design 图纸设计, 草图设计; ～ drawing 外形图, 方案图; ～ machine 测绘缩放仪; ～ of equipment 设备布置图; ～ of railway junctions 铁路枢纽布置图; ～ of semiconductor IC chip 半导体集成电路芯片图; ～ of station 车站（线路、房屋）布置图; ～ plan 规划, 设计, 平面图; ～ work 起稿图; air-conditioning ～ 空调平面图; air-conditioning machine room ～ 空调机房平面图; cable ～ 电缆配线图, 电缆敷设图; design ～ 设计草图; electrical ～ 电路布置图; general ～ 总平面图, 总体布置, 综合图; general ～ of an aeroplane assembly 飞机总体布局图; general ～ of electrical supply 供电总平面图; general ～ of outdoor water supply and drainage system 室外给排水总平面图; general ～ of pipe systems 管道综合图; general ～ of signal 信号总布置图; heating ～ 采暖平面图; indoor ～ of plumbing system 室内给排水平面图; lighting ～ 照明平面图; lighting protection grounding ～ 建筑物防雷接地平面图; plant ～ 设备（车间）布置图; power supply ～ 电力平面图; principle ～ 总布置图; problem ～s 题目的图面布置; skeleton ～ 草图, 初步布置; torminal ～ 枢纽布置图; ventilation ～ 通风平面图

LBP (length between perpendiculars) 垂直线间的距离

LC (location clearance fit) 定位间隙配合

lead /li:d/ *（螺纹）导程, 导引, 导致; 居首的, 上面的; 铅, 镀铅; ～ angle* 螺纹升角, 导前角; ～ line 指引线; ～ screw 丝杠, 导螺杆

leader /'li:də/ 领导者, 指引线, 引出线, 导线, 导杆, 引导, 指导者; 社论, 重要文章; ～ line 带箭头的指引线; ～ record 标题记录; ～ bevel 铅笔尖斜面; fine line ～ 细铅笔芯; sharpen ～ 削尖铅芯

lead-out 引出线, 引出（端）

leaf /li:f/ 叶, 叶片, 薄片, 箔; ～ spring 板弹簧; spring ～ 弹簧片

learn /lə:n/ 学习; ～ing to read by models 借模型学习读（图）; computer-asisted ～ing 计算机辅助学习

leather /'leðə/ 皮革, 革制物

leatheroid /'leðə,rɔid/ 人造革, 纸皮, 薄钢纸

leave /li:v/ 出行, 脱离; ～ out 遗漏

ledge /ledʒ/ 凸出部分，凸耳，凸缘

leeway /ˈliːwei/ （活动）余地，可允许的误差；leave some ~ 留有余地

left /left/ 左；~ auxiliary 左辅视图；~ elevation 左立面图；~ side view 左侧视图；~-hand part 左手件；~-hand view 左视图

left-handed 左旋的，用左手的，笨拙的

leftmost /ˈleftˌməust/ 最左的

leg /leg/ 角尺，脚，腿，支柱，床脚，角边；~ of a right triangle 直角边；~ of angle 角边；~ of compass 圆规的腿，dog ~ 犬蹄弯，死弯；perpendicular ~ 直立边（焊接）

legal /ˈliːgəl/ 法律上的，法定的，合法的，正当的；~ -size 规定尺寸的

legend /ˈledʒənd/ 图例，图表，符号，符号表，代号，说明书，插图的说明，地图的图例；传说，铭文

lemma /ˈlemə/ 标题，主题，题词，命题

LENG (length) 长度，距离

length /leŋθ/ 长度，距离；~ gauge 长度规；~ of development 展开长度；~ of engagement 旋入长度（螺纹）；~ of fit 配合长度；~ of increment 焊段长度；~ of side 边长；~ of thread 螺纹长度；arc ~ 弧长；approximate ~ 近似长度；assessment ~ 鉴定长度；取样长度；chord ~ 弦长；curve ~ 线长；cut off to ~ 截成一定长度；draw a bar to ~ 按一定长度画一线段；effective ~ 有效长度；focal ~ 焦距；entrance ~ 旋入深度（螺纹）；extreme ~ 全长；inside pitch line ~ 齿根高度；indefinite ~ 无限长；isometric ~ 等角轴上线长；outside pitch line ~ 齿顶高度；over ~ 过长；overall ~ 全长；serew-in ~ 旋入长度；tail ~ 螺尾长度；to ~ 按一定长度；total ~ 总长；true ~ 实长，真长

lengthen /ˈleŋθən/ 加长，延长，放长，拉长，变长；~ing bar 延伸杆

lengthwise /ˈleŋθwaiz/ 纵向的；~ section 纵断面，纵截面

lessen /ˈlesn/ 减少，使……变小，使……变少

let /let/ 让，赋值

letter /ˈletə/ 字母，文字；信(件)；~ steel 钢字；~s and lettering 字母与写法；black ~ 黑体字，粗体字；bold face ~ 粗体字，黑体字，印刷体；capital ~ 大写字母；commercial Gothic ~ 商用哥德体字母；compressed ~ 狭体字，长体字，密集字；dotted ~ 点线字母；ellipse ~s 椭圆形字母；extended ~ 宽体字，横体字；hook ~ 曲钩形字母（如 m, n, r 等字母）；inclined ~ 斜体字；inclined capital ~ 斜体大楷字母；italic ~ 斜体字；light face ~ 细体字；loop ~ 环形字母（如 a, b, d, e 等字母）；lower-case ~ 小写字母；modern Roman ~ 近体罗马字；old Roman ~ 古体罗马字；one-stroke ~ 单笔字；phantom ~ 点线字母；Roman ~ 罗马字；small ~ 小写字母；straight-line ~ 直线型字母（如 i, k, A 等字母）；vertical ~ (~ vertical) 直体字，垂直体字

lettering /ˈletəriŋ/ * 字体，字型，字法，书法；aids 写字辅助器；architect ~ 建筑图字体；composition in ~ 字母组合；~ device 写字用具；~ diverse form pens 各种书写笔；~ guide 书写模板；~ joint 针笔写字接头；~ machine 字法机；~ pen 书写笔；~ plate 书写模板；~ stencil 书写模板；~ template 书写模板；~ triangle 字法三角板；office ~ 常用字法（建筑）；proportional ~ 比例字体；single stroke ~ 单笔字法；tabular ~ 表格状字体

level /ˈlevl/ 高度，标高，水平，水平面，水平尺，水平仪，水准仪，找平，平衡，水准测量，等级；~ line 等高线；~ plate 水平板；~ surface 水(准)平面；~ing pole (leveling rod) 水平尺; datum ～ 基准面; engineering ～ 技术水平

levelness /ˈlev(ə)lnəs/ 水平度

LG (length) 长度; (long) 长的

lgth (length) 长度

LH (left hand) 左旋; 左面, 左边; (left head) 左首

LHS (left-hand side) 左边, (等式) 左边

LHTH (left hand thread) 左旋螺纹

L. I. (line of intersection) 交线

library /ˈlaibrəri/ 图书馆, 库 (数据的); graphics ～ 图库

lid /lid/ 盖, 罩, 帽

lie /lai/ 躺, 平躺, 平放, 伸展, 位于, 位置

life /laif/ 生命, 生活, 寿命; 实物, 原形; 使用期限

light /lait/ 光明, 灯; 轻的, 细的; ～ comes from the left 光线由左上方来; ～ pen 光笔; ～ ray 光线; ～ ray direction 光线方向; ～ series 轻系列; ～ source 光源

lighting /ˈlaitiŋ/ 照, 照明; 发亮的, 画面的明亮部分; conventional ～ 习用光线; electric ～ 电器照明

light-pen 光笔

lim. (limit) 极限, 限度, 界限

limb /lim/ 分度弧; 针盘; 零件; 电磁铁心

limes /ˈlaimi:z/ 边界; ～ inferiores 下极限; ～ superiores 上极限

liminal /ˈliminəl/ 最初的, 开端的; 最低量; ～ value 极限值

limit /ˈlimit/ 极限, 范围, 界限; 限度, 限制; 公差; ～ high 上限; ～ of size 极限尺寸; ～ system 极限制(度), 公差制; ～ing point 极限点; bilateral system of ～s 双向公差制; commercial ～s 商品界限(尺寸); go ～ 通过极限; interior ～ 下限, 最小尺寸; lower ～ 下限; manufacturing ～s 制造公差; maximum ～ of size 极限尺寸;

minimum ～ of size 最小极限尺寸; not go ～ 不通过极限; superior ～ 上限; surface ～ 面的极限, 转向素线; tolerance ～ 公差, 界限

limn /lim/ 素描, 画; ～er 绘画者, 描画者

lin. (linear) 直线的, 一次的, 线性的, 纵的

linable /ˈlainəbl/ 排成一条直线的

linchpin /ˈlintʃpin/ 开口销, 保险销, 关键

line /lain/ * 图线, 直线, 路线; 作业线; 行; 管路; 布成一条线; 界线; 专业; 轮廓, 草图; 画线于; ‖ *Example sentence: The term* line *is generally used to designate a straight line unless otherwise specified.* 例句: 除非另有规定, 线这个名词通常用来表示直的线。～ cut 截交线; ～d in 连成一线, 上墨; ～ dimensions 图线尺寸; ～ drawing 线画(线条画, 铅笔画); ～ element 线素; ～ extension 线延伸; ～ feed 进行(打印机的按钮, 按下一次可使纸向前走一行); ～ font 线型; ～ kind 线型; ～ number 行号, 语句标号; ～ of action 施力线, 压力线, 作用线; ～ of centers 连心线; ～ of connection 连接线; ～ of contact 接触线; ～ of deflection 挠曲线; ～ of excavation construction 基坑开挖线; ～ of greatest inclination 最大倾斜线; ～ of intersection 交线, 相贯线; ～ of intersection given 已知交线; ～ of maximum inclination 最大倾斜线; ～ of parallel 平行线; ～ of projection 投影线; ～ of recall 忆线, 联线; ～ of reference 参考线, 基准线; ～ of section 截交线; ～ of shaft 轴线; ～ of sight 视线; ～ of symmetry 对称线; ～ of transition 过渡线; ～ of vision 视线; ～ on general position 一般位置直线; ～ out the route on the map 在地图上把该路线表示出来; ～ printer 行式打字机; ～ segment 线段; ～ smoothing 线光顺; ～ spacing 线间距离; ～ thickness

线型的粗度; ~ tone 线条色调; ~ type 线型; ~ up the board 竖线; ~ weight(粗)实线; ~ width 图线宽度; a ~ of 一行, 一系列; aclinal ~ 无倾线; actual ~ 实线; addendum ~ 齿顶线; adjacent ~ 邻线; adjacent parts ~ 邻接件线; along the ~s 根据……方法; alphabet of ~s 图线; alternate position ~ 位置转变线; appellative ~ 名称线; arbitrary ~ 任意直线; arc ~ 弧线; ‖*Example sentence:* Arbitrary line: a reference line, the direction of which does not necessarily coincide with cardinal direction. *例句: 任意直线:* 是一条参考线, 它的方向不一定符合基本方位。*(自BS 3618-1:1969 Glossary of Mining terms—Term Definition, Confirmed January 2011)* assembly ~ 装配线; auxiliary ~ 辅助线; auxiliary straight ~ 辅助直线; base ~ 基线, 准线; basic conventions for ~s 图线的基本规定; bee- ~ 最短距离, (两点间的)直线; bend ~ 弯线(展开图上表示棱之线), 转折线; bold ~ 粗线; body ~s 横剖线(船体); border ~s 边缘线, 边界线, 图框线; boundary ~ 边界线; bounding ~ 外缘线, 边线; break ~ 破断线, 断裂线, 折线; brilliant ~ 亮线; broad ~ 粗线; broken ~ 虚线(即……), 断裂线, 折线; buttock ~s 纵剖线(船), 船股线, 船尾线; camber ~ 脊线, 上弯线, 弧线, 梁拱线(船); cap ~ (capital line) ~ 大楷字母高度线, 顶线; center ~ 中心线, 中线; central ~ 中心线; chain-dotted ~ 点划线, 链线; chain ~ 点划线, 链线; chain thick ~ 粗点划线; chain thin ~ 细点划线; character of ~ 线的属性; closed ~ 闭合线; coincident ~ 重叠线; command ~ 命令总线; 指令传送线; comment ~ 注解行; common ~ 共有线; common perpendicular ~ 公垂线; concentric (al) ~ 公共线; concurrent ~s 共点线; configuration of ~s 图线构形; connecting ~ 连接线; connecting ~s for floor end on top 肋板边线; connecting ~ of projection 投影连线; construction ~ 作图线; continuation ~ 延续行, 继续行(计); continuous ~ 实线; continuous thick ~ 粗实线; continuous thin ~ 细实线; contour ~ 等高线, 等雨量线, 等深线, 轮廓线; contour rain ~ 等雨量线; crest ~ 螺纹峰线; curved ~ 曲线; cutting ~ 切面线; cutting plane ~s 切面线, 剖切位置线; dark ~ 阴影(暗)线; dashed ~ 虚线; dashed dotted ~ 点划线; dashed double-dotted ~ 双点划线; dashed triplicate-dotted ~ 三点划线; dashed spaced ~ 间隔划线; datum ~ 基准线; deck ~ at center 甲板中线; deck ~ at side 甲板边线; dedendum ~ 齿根线; demarcation ~ (尺寸)界线, 分界线; deviation in ~ width 图线宽度的偏差; diagonal ~ 对角线; dimension ~ 尺度线, 尺寸线; ditto ~ (repeat line) 细节省略线; divide ~ 分割线段; dot-and-dash ~ 链线; dot-dash- ~ 点划线; dotted ~ 点线, 虚线; double-dashed dotted ~ 双划点线; double-dashed double-dotted ~ 双划双点线; double-dashed triplicate-dotted ~ 三点双划线; dranghting of ~s 图线画法; drop ~ 底线; dump ~ 倾卸线(填方线), 洗管道的汽管线; edite ~ 编辑行; element ~ 素线; equilibrium ~ 等高线; every ~ 每一直线; excess ~ 多余线条; extension ~ 延伸线, 尺寸界线, 引出线; faulty ~ 错误线; feint ~ 淡色线, 不明显的线; fine ~ 细线; flow ~ 流程线; fold ~ 折线; frame ~s 肋骨型线; freehand ~ (徒手画的)波浪线; freehand continuous ~ 波浪线; frontal ~ 正平线, 平行于正面的线; frontal-horizontal ~ 侧垂线; frontal-profile ~ 铅垂线; full ~ 粗实线; faint pencil ~ 轻细铅笔线; fathom- ~ 英寻线; front profile ~ 铅垂线; gage ~ 量线, 规线 (如铆钉孔之中心线); general ~ 任意直线; general position ~ 一般位置直线; generating ~ 母线; girth ~ 周围线; given ~ 已知线; gravity ~ 重力线; ground ~ 地平线, 基线, 投影轴; guide ~ 引线, 导线; half ~ 半 (直) 线, 射线; heavy ~ 粗线; height ~ 重型线, 高度线; helical ~ 螺旋线; hidden ~ 虚线, 不见线; 隐 (藏) 线 (例: removal of hidden line 消除隐藏线); hinge ~ 投影轴, 枢纽线, 铰合线; horizon ~ 视平线; horizontal ~ 水平线; horizontal-profile ~ 正垂线; heavy broken ~ 粗折断线; imaginary ~ 虚线, 假想线; in ~ 成一直线; in ~ with 和……成一直线; inclined ~ 斜线; initial ~ 初始行; ink ~ 墨线; intersecting ~s 交线; ‖ *Example sentence: A plane is perpendicular to a line if the plane contains* two intersecting lines *each of which is perpendicular to the given line*. 例句: 如果平面内相交两直线都垂直于给定直线, 则平面垂直于该直线。invisible ~ 不可见线; irregular boundary ~ 徒手线; isometric ~ 等角线; knuckle ~s 折角线; lead ~ 指引线; leader ~ 引出线; level ~ 水平线; light pencil ~ 轻铅笔线; long-chain thin ~ 点划线; long dashed dotted ~ 长划点线; long dashed double-dotted ~ 长划双点线; long dashed double-short dashed ~ 长划短划线; long dashed short dashed ~ 长划双短划线; long dashed triplicate-dotted ~ 长划三点线; longitudinal center ~ 纵向中心线; long break ~ 长折断线; main ~ 干线; mean ~ 中线, 二等分线; meander ~ 曲折线, 蜿蜒; meaning of ~ 图线的意义; measuring ~ 量线, 测量线; meridian ~ 子午线; mesh ~s 网格线; mold ~ 模线 (即圆角的理论尖角线); molded base ~ 基线; moving ~ 动线; multiparty ~ 合用线, 同线; margin ~ 边界线; medium ~ 中线 (线条粗细); nominal pitch ~ 标称节线; non-coplanar ~s 非共面直线; nonisometric ~ 非等角线; non-symmetrical reference ~ 非对称参考线; normal ~ 法线; nonperpendicular ~s 非垂直线; nonplanar ~s 非同平面线; object ~ 物线, 物体轮廓线; oblique ~ 斜线, 一般位置直线; ‖ *Example sentence*①: An oblique line *is one that is not parallel to any of the principal planes: frontal, horizontal, or profile.* 例句①: 倾斜直线是与任何主要投影面 (正立投影面、水平投影面、侧立投影面) 都不平行的直线。‖ *Example sentence*②: Oblique lines *that appear parallel in two or more principal views are parallel in space.* 例句②: 倾斜直线在两个或两个以上的主要视图中平行, 它们在空间也平行。off ~ 脱机; on ~ 连机; on the ~s 按照……方向; orientation ~ 定位线; out of ~ 不在一条直线上; parallel oblique ~s 平行的倾斜线; parting ~ 分离线, 铸件上的分箱线; pencil ~ 铅笔线; porpendicular ~ 垂直线; perpendicular ~ of the projection plane 投影面垂直线; pipe ~ 导管, 管线; pitch ~ 节线; plumb ~ 铅垂线; polar ~ 极线; pressure ~ 压力线; profile ~ 侧平线, 平行于侧立投影面的线; projcting ~ 投影线, 投射线; projection-coincided ~ 重影线; phantomline ~ 假想线; real ~ 实线; reference ~ 参考线, 基准线; required ~ 求作线; rhumb ~ 方位线, 恒向线, 等角线; right ~ 直线; root ~ 齿根线; rounded ~ 转圆线 (船); ruled ~ 尺子线 (用尺子画的线); run-out ~ 渐灭线; scanning ~ 扫

描线; secondary object ~ 物体表面线, 面内线, 辅助(作图)线; section ~ 剖面线; shade ~ 阴影线; shaft ~ 轴线; shore ~ 水果; short break ~ 短断裂线(波浪线); short chain ~ 短链线; short-dash thin ~ 细虚线; sighting ~ 视线; simplified ~ 简化线; single ~ 单线, 单行; skeleton ~ 轮廓线, 骨架线, 概略; skew ~ 斜(直)线; skew ~s 交叉线, 相错线; slope of a ~ 线的斜率; solid ~ 实线; special-position ~s 特殊位置直线; stop ~ of thread 螺纹终止线; straight ~ 直线; ‖ *Example sentence*①: *A straight-line segment is established by locating its endpoints.* 例句①: 直线段的位置由它两端点的位置确定。‖ *Example sentence*②: *An infinite number of lines that contain the point may be drawn in the plane.* 例句②: 在平面内过一点可作无数条直线。‖ *Example sentence*③: *Any two lines in a plane must either intersect or be parallel.* 例句③: 平面内的两直线不是相交就是平行。stretch-out ~ 直伸线, 展开线, 周围线; strike ~ 走向线; stub ~ 短线, 线段, 短截线; supply ~ 供电线路, 供水管线, 供气管线; symmetrical reference ~ 对称参考线; tangent ~ 切线; the ~ maximum inclination 最大斜度线; theoretical ~ of intersection 理论交线; thick dash and dot ~ 粗点划线; thickened ~ 加粗之线; thickness of ~ 线之粗细; thin ~ 细线; thin solid ~ 细实线; thin-light ~ 轻细线条; tie ~ 连接线; transition ~ 过渡线; trim ~s 外形线; types of ~s 线型; uniform ~ 等粗线; uniform spiral continuous ~ 规则螺旋连续线; uniform wavy continuous ~ 规则波浪连续线; uniform zigzag continuous ~ 规则锯齿连续线; up-hill ~ 上行管道; use guide ~s 用引线; vanishing ~ 没影线, 灭线; vertical ~ 垂直线, 铅垂线; viewing plane ~ 视向平面线; visible ~ 可见线, 粗实线, 外形线; visual ~ 视线; warp ~ 翘曲线; water ~ 水管线路, (船面的)水线, 水平面线; water supply ~ 给水管线, 上水道; wavy ~ 波浪线; weight of ~ 线的粗细; widest ~ 最粗线; width of ~ 线的粗细; witness ~ 目测线, 作图线, 引证线(细实线); word ~ 字线; working ~ 工作线 (即物件的重力线, 用于计算构件的应力与大小), 工作线路; zero ~ * 零线, 零位线; zigzag ~ 曲折线

lineal /ˈliniəl/ (直)线性的, 线状的, 纵的
linear /ˈliniə/ 线的, 直线的, 长度的, 线性的, 一次的; ~ construction 用直尺作图; ~ equation 一次方程; ~ measure 长度测量; ~ perspective 直线透视(作图)法, 直线透视图; ~ pitch 直节(直线节距长, 等于其配合齿轮之周节); ~ programming 线性规划; ~ shrinkage 线收缩; ~ size 线性尺寸
linearity /liniˈæriti/ 直线性, 线性度; ~ error 线性误差
lineation /liniˈeiʃən/ 轮廓, 画线
lined /laind/ 格线, 格子线; 带格子的; blue ~ paper 蓝色的格子纸, 蓝格子的纸
linen /ˈlinin/ 麻布; tracing ~ 描图布
lineograph /ˈlainəˌɡrɑːf/ 描线规
lineoid /ˈlainɔid/ 超平面
liner /ˈlainə/ 衬垫, 画线的人, 画线的工具; geo- ~ 可调三角板; lay-out ~ 划线(工件上的); section ~ 剖面线器
lines /lainz/ 线型, 线条; 型线图(船舶); alphabet of ~ 线型; hatched ~ 剖面线; moulded ~ 型线; nonintersecting ~ 非相交线; nonparallel ~ 非平行线
line-segment 线段
linetype /laiˈnetaip/ 线型
linewidth /ˈlainwidθ/ 线宽; ~ scale factor

线宽缩放因子

lining /'lainiŋ/ 衬,画线;section ～ 剖(断)面线法,画剖面线的方法;symbolic section ～ 不同材料剖面线画法

link /liŋk/ 环节,链环;连杆;线路,网络节;连接,装配;链接,链路

lintel /'lintl/ 门窗楣,过梁

liny /'laini/ 画线的,似线的,细的,有皱纹的

lip /lip/ 底,边,缘,凸缘,法兰盘,切削刀

liquidation /likwi'deiʃn/ 清理

L-iron 角铁(钢)

list /list/ 表式,列表,清单,表格,表,一览表,目录,编目;～ of drawing 图目;～ of symbols 符号表,图例;comprehensive material ～ 综合材料表;door ～ 门户一览表;equipment ～ 设备一览表;input ～ 输入表列;item ～ 明细表(栏);material ～ 材料表;output ～ 输出表列;parts ～ 零件表,明细表;piping support ～ 管架表;source ～ing 源列表

lithoprinting /'liθəuprintiŋ/ 石印,(照相)胶印

LL (lower limit) 下界,下限

LM (list of materials) 材料单

LN (locational interference fit) 定位过盈配合

LOA (length over all) 总长,全长

load /ləud/ 装配,装载,负担

loader /'ləudə/ 装配器

loam /ləum/ 黏土

lobe /ləub/ 凸起部,凸角;瓣,波瓣

LOC (location) 位置,配置,定位,布置

loc (local) 局部的,地方的

local /'ləukəl/ 局部的,地方的,部分的;市内的;～ contraction 局部收缩;～ unconformity 局部不整合;～ view 局部视图

localizable /'ləukəlaizəbl/ 可定位的

localization /ˌləukəlai'zeiʃən/ 定位,定域,局部化;局限(性);～ of disturbance 确定扰动的位置

localize /'ləukəlaiz/ 定位,确定位置

localized /'ləukəˌlaiz/ 局部的,固定的;～ high temperature 局部高温

locally /'ləukəli/ 局部地,在本地

locate /ləu'keit/ 确定位置,定位,定线,放样,位于,判明;位置;把……设置在;be ～d at (或 in) 位于,放在;to ～ points or lines on the plane 在平面内取点或直线

location /ləu'keiʃn/ 地点,位置,定位配置,探测(计算)单元;～ by angle 角度定位;～ by offsets 支距定位;～ dimension 定位尺寸;～ layout 定位布置;～ pin 定位销;～ plan 位置平面图;～al interference fit 定位过盈配合;～al tolerance 定位公差;memory ～ 存储单元

locator /ləu'keitə/ 定位物,定位器;～ center 中心定位物

lock /lɔk/ 封闭,锁,锁紧,缺乏,不足,防松;Pittsburgh corner ～ 匹兹堡扣缝(板金)

locking /'lɔkiŋ/ 关闭,锁紧,联锁,制动,防松;～ device 防松装置;～ equipment (device) 锁紧装置;cotter pin ～ 开口销防松;double nuts ～ 双螺母防松;lug washer ～ 带耳止退垫圈防松;spring washer ～ 弹簧垫圈防松;steel wire ～ 钢丝防松

lockout /'lɔkaut/ 切断,分离,闭厂

locus /'ləukəs/ 轨迹,地方,位置;‖*Example sentence: A locus is the path of a point, line, or curve moving in some specified manner.* 例句:轨迹是点、直线或曲线按某一规定方式运动的路径。

lodge /lɔdʒ/ 舍

lofted /'lɔftid/ 翻样,翻画样子

lofting /'lɔftiŋ/ 模线;放大样,理论模线的绘制;翻样

log /lɔg/ 原(圆)木，大木料；符号，对数，记录；～ on 进驻；graphic (strip) ～ 柱状录井图，柱状剖面图

logarithm /'lɔgəriθəm/ 对数

logarithmic /lɔgə'riθmik/ 对数的；～ ruling 对数格线；～ scale 计算尺

logical /'lɔdʒikəl/ 逻辑型

logistic /ləu'dʒistik/ 逻辑的，对数的，计算的，比例的

logout /'lɔgaut/ 注销

long /lɔŋ/ 长，长的，纵的；经度；～ break 长断裂线；‖ *Example sentence:* This line is one-third as ～ as that line. *例句:* 这条线只有那条线的三分之一长。

longitude /'lɔndʒitju:d/ 长，长度；经度，经线

longitudinal /lɔndʒi'tju:dinl/ 纵的，纵向的，经度的；～ axis 纵轴；～ grain 纵纹；～ key 长方键；～ section 纵向截面；～ view 纵视图

longl. (longitudinal) 纵的

long-radius 大半径(的)，长半径的，大范围的

long-term 长期的，长远的

look /luk/ 外观，看来，面貌，样子；～ into 查找；～ round 环顾；～ sharp 注意准确；～ up 查，检查

loop /lu:p/ 圈，环，环形天线，回路，回线，相贯线；循环

oose /lu:s/ 松的，宽的，自由的；～ fit 松配合

lop (line of position) 定位线，目标线；(list of parts) 零件目录

los (line of sight) 视线，瞄准线

lotus /'ləutəs/ 荷花饰

louver /'lu:və/ (louvre) 百页窗，通气缝

low /ləu/ 低

lower /'ləuə/ 下部的，使低，放低，较低的，昏暗的；～ bound 下界；～ case 低格(字)，小写的，小写字母；～ case letter 小写字母

lozenge /'lɔzindʒ/ 菱形，菱形物

L. P (light pen) 光笔；(lighting panel) 照明配电盘

LPP (length of perpendiculars) 垂直距离

LSI (large scale integration) 大规模集成技术

L. side v. (left side view) 左侧视图

LT (location transition fit) 定位过渡配合

lth (lath) 水平尺

LTL (line-to-line) 两线间的，线到线的

LTR (letter) 字母，符号；信

LUB (lubricate) 润滑

lubricate /'l(j)u:brikeit/ 润滑

lubricator /'l(j)u:brikeitə/ 润滑器，油杯

lucidus /'lusidəs/ 光泽的

lug /lʌg/ 耳，突缘；接线片

lumber /'lʌmbə/ 木材，木料，制材

luminaire /l(j)u:mi'nɛə/ 光源

lunar /'lu:nə/ 月的，似月的，半月形的

lunate /'lu:nit/ 新月形的，半月形的

lune /lu:n/ 弓形，半月形(球面的)二角形；～ of a sphere 球面二角形

lunette /lu:'net/ (lunette window) 弦月窗

luniform /'lu:nifɔ:m/ 月形的

luster /'lʌstə/ (lustre /'lʌstə(r)/) 光泽，烛台；metallic ～ 金属光泽

lusterless 无光泽的

lustre /'lʌstə(r)/ (镀层的)光泽，亮光，闪光

lustreless /'lʌstəlɪs/ 无光泽的

lustrous /'lʌstrəs/ 有光泽的，光泽的

luxe /'luks/ (法语) deluxe /de'luks/ =warping /'wɔ:piŋ/ 精制的，精装的

L×W×H (length×width×height) 长×宽×高

M

M (module) 模数, 系数; (meter) 米, 公尺
MAG (magnet) 磁石, 磁铁
macadam /mə'kædəm/ 碎石路
mach. (machine) 机器
machine /mə'ʃi:n/ 机器, 机械; 机械加工, 机制, 机构; ~ all over 全部切削; ~ code 机器代码, 指令表; ~ drawing 机械画, 机械制图; ~ finish allowance 切削裕度; ~ language 机器(计算机)语言; ~ part 机械零件; ~ tool 机床, 金工机械; autographic ~ 自动绘图机; broaching ~ 拉床; ditto ~ 复印机, 复写器; drafting ~ 绘图机; drawing ~ 绘图机; engineering drafting ~ 工程绘图机; tracing ~ 描图机; whiteprint ~ 白印机, 晒图机
machinery /mə'ʃi:nəri/ (machines) 机器(总称), 机械, 机械装置, 机器制造; ~ joining* 机械联接; computing ~ 计算机
macroaxis /ˌmækrə'æksis/ 长轴
macroscopic(al) /mækrə'skɔpik(əl)/ 宏观的, 肉眼(或稍放大)可见的
macrostructure /mækrə'strʌktʃə/ 肉眼可见的, 宏观结构
mae (mean absolute error) 平均绝对误差
mag (magnitude) 大小, 数量, 数值; (magazine) 杂志
magazine /mægə'zi:n/ 杂志, 期刊; 箱, 盒, 料斗; (摄影)软片盒
magnetophone /mæg'ni:təufəun/ 磁带录音机
magnification /mægnifi'keiʃn/ 放大, 扩大
magnifier /'mægnifaiə/ 放大器, 扩大器, 放大镜
magnify /'mægnifai/ 放大, 扩大, 夸大; ~ing power 放大率; ~ M100 diameters 把 M 的直径放大100倍; ~ M200 times 把 M 放大(到)200倍
magni-scale 放大比例尺
magnitude /'mægnitju:d/ 长度, 大小, 尺寸, 数量, 等级; 值; 重要
main /mein/ 主要的, 主要部分, 主线, 干线; 总管, 管路, 电源; in the ~ 大体上, 基本上
mainframe /'meinfreim/ 主机(计)
maintain /mein'tein/ 维护, 保养; 保持
major /'meidʒə/ 长, 较大的, 较大范围的, 较多的; ~ axis 长轴, 主轴; ~ diameter 大径, 长径; ~ thread diameter 螺纹长径
make /meik/ 做, 制造, 构成, 组成, 引起, 制定; to ~ out 做成, 表示, 说明; 证明, 书写, 起草
making /'meikiŋ/ 制齐, 加工, 做; ~ a measurement 度量(尺寸)
malalign(e) ment /'mælə'lainmənt/ 不同轴度(性), 不平行度(性); 轴线不对称, 不成一直线, 相对位偏, 偏心率
male /meil/ 阳(螺纹); ~ screw 阳螺旋
malleable /'mæliəbl/ 可锻的, 韧性的; ~ cast iron 可锻铸铁
manifold /'mænifəuld/ 多样的, 复印, 复数
manifolder /'mænifəuldə/ 复印机, 复写机
manipulate /mə'nipjuleit/ 熟练地使用, 操作, 处理; ~ing the T-square 使用丁字尺
manual /'mænjuəl/ 手工的, 手控的; 手册,

说明书,教本,细则

manufacture /mænjuˈfæktʃə/ 制造,制造业,制成品

manufacturing /mænjuˈfæktʃəriŋ/ 制造的; ～ tolerance 制造公差; computer aided ～ 计算机辅助制造; computer intergrated ～ 计算机集成制造; machine ～ 机器制造

manuscript /ˈmænjuskript/ （工件的）加工图,手稿,底稿

many /ˈmeni/ ～ angled 多角的

many-sided 多边的,多角的

map /mæp/ 图,地图,天体图,天球图;绘制……的地图,测绘,用地图表示,拟定,安排,标记,映像,图像,变换; ～ board 图板; ～ border 图廓; ～ generalization 映像生成(计); ～ making 制图; ～ measure 量图仪,曲线仪; ～ of a mine 矿图,矿山图,矿区图; ～ projection 地图投影; ～ scale 地图比例尺; adjacent ～ 邻接地图; aerial ～ 航测图; aerial topographic ～ 航测地形图; aeronautic ～ 航空图; air ～ 航空图; areal ～ 地区图; base ～ （地质）工作草图; bit ～ 位映像,位图; bit display 位图显示器,高分辨率显示器; bit ～ representation 位图表示; block allocation ～ 存储块分配图; cadastral ～ 地籍图; coal mine geological ～ 煤矿地质图; coalseam floor contour ～ 煤层底板等高线图; coal topographic-geological ～ 煤田地形地质图; contour ～ 等高线地图; contour line ～ 等高线图; electronic ～ 电子地图; engineering ～ 工程地图; field ～ 矿区图; gas emission forecast ～ 瓦斯预测图; geographical ～ 地理图,地形图; geological ～ 地质图; hydrogeological ～ 水文地质图; hydrographic ～ 水位图; identical ～ 等角投影地图; integrate geological ～ of opencast 露天矿综合地质图; isobath ～ of coal seam 煤层等深线图; isomagnetic ～ 等磁线图; isothickness ～ of coal seam 煤层等厚线图; karnaugh ～ 卡诺图(一种简化开关函数的方格图); land ～ 区域图; landscape ～ 风景图,地景图; location ～ 井上下对照图; military ～ 军用地图; mine areahydro- geological ～ 矿区水文地质图; mine area topographic ～ 矿区地形图; mine area topographic and geological ～ 矿区地形地质图; mine geological ～ 矿山地质图; mine ～ 采掘工程平面图,矿山工程图; mineral ～ 矿物分布图; nautical ～ 航海图; navigation ～ 航海图; non geometrical conical ～ 非几何形的圆锥形地图; non-geometrical polyconic ～ 非几何形的多锥形地图; non-geometrical polar ～ 非几何形的极点地图; ordnance ～ 军用地图; outline ～ 略图; photo ～ 航空摄影地图; photogrammetric ～ 摄影地图; planimetric ～ 平面图; plat ～ 区域图; railway situation ～ 铁路周围情况图; reconnaissance ～ 踏勘地图,草测图; regional geological ～ 区域地质图; relief ～ 地貌图,地势图,立体地图,地形图; reserves estimation ～ 储量计算图; route ～ 线路图; route ordonnance ～ 路径图; scale ～ 比例图; screen memory ～ 显示屏幕存储映像; sketch ～ 草图,略图; spot ～ 点示图; station ～ 火车站图; statistical ～ 统计地图; stripping work ～ 采剥工程综合平面图; structural ～ 结构图; surface contour ～ 等高线地形图; survey ～ 测量图; topographic ～ 地形图; topographic ～ of mine field 矿田区域地形图; topographic ～ of (underground) mine field 井田区域地形图; valuation ～ 估价地图; ventilation ～ 通风系统图; ventilation pressure ～ 通风压力分布图; water level ～ 水位图

MAPL (manufacturing assembly parts list)

制造装配零件表
map-making 地图制图
mapper /'mæpə/ 测绘仪,测绘装置;制图人,绘图人
mapping /'mæpiŋ/ 绘图,测绘;映射,变换;对应,交换;映像,结疤,毛刺;aerial ~ 航空测量,航空绘图;aerial ~ work 航摄制图;computer-assisted ~ 电脑绘图;conformal ~ 保角变换,共形映射;field ~ 野外测图;survey ~ 测量绘图;topographic ~ 地形绘图
marble /'ma:bl/ 大理石
margin /'ma:dʒin/ 边缘,页边,空白;余地,余量,限界,差距,安全系数
marine /mə'ri:n/ 海运的,海的;船舶的
mark /ma:k/ 标记,戳,符号,标法,打印,记号;~ing machine 打印机;~ of conformity 合格标志;bearing ~ 方位标,安装方位标,指北针;bench ~ 水准点,标高标志;beyond the ~ 超出界限,过分;center ~ 定中心点;centring ~ 对中符号;check ~ 校对记号;corner ~ 角标志;datum ~ 基准点;detail drawing ~ 详图标志;detail index ~ 详图索引标志,详图索引号;finish ~ 加工符号,光洁度符号;fold ~ 折迭记号;foot ~ 英尺符号(');inch ~ 英寸符号(");land ~ 陆地标志;machining ~ 切削符号;pencil ~ 铅笔记号;punch ~ 冲点记号;reference ~ 参考点,基准点
marker /'ma:kə/ 标记,记号;~ clipping 标记剪裁
marking /'ma:kiŋ/ 标志
martempering /'ma:,tempəriŋ/ 分级淬火
masonry /'meisənri/ 石工;砖石;~ part of object 实体部分;~ plan 石工图;~ structure 石工结构
master /'ma:stə/ 主人,主要的,总的,院长,操纵者,师傅,硕士,熟练,精通,能手;~ check 校正,校对;~ diagram 总线图;~ drawing 样图;~ gauge 标准规,校准规;artwork ~ 照相原图,照相底图,原图掩膜;sketch ~ 草图底稿
mat /mæt/ 无光泽的,表面粗糙的,不光滑的,字模,垫子,席子
mate /meit/ 啮合,使紧密配合;~ well 啮合紧密
material /mə'tiəriəl/ 材料,原料,实体;building ~s 建筑材料;construction ~ 建筑材料,构造材料;drawing ~ 制图材料;flexible ~ 挠性材料;gasious ~s 气体材料;least ~ condition* 最小实体状态;liquid ~s 液体材料;maximum ~ condition* 最大实体状态;shape ~ 型材;solid ~s 固体材料
MATL (material) 材料
matrix /'meitriks/ 矩阵,模片;~ concatenation 矩阵连锁;accumulate transformation ~ 积累变换矩阵;evaluate transformation ~ 求变换矩阵;perspective projocction ~ 透视投影矩阵;transformation ~ 变换矩阵
matt (同 mat);~ surface 无光平面,粗面
max. (maximum) 最大,最大的
maxim /'mæksim/ 原理,原则,准则,谚语
maximal /'mæksiməl/ 最大的,极大的
maximum /'mæksiməm/ 极点,极限,最大值,最高值,最大的;~ admitted diameter of work 工件最大许可直径;~ material condition 最大材料条件
M/C (machine) 机械
M/CD (machined) 加工
M/CY (machinery) 机构
MD (mean deviation) 平均偏差
mean /mi:n/ 中间的,平均的;平均值,意是;~ line 中线,二等分线
meander /mi:ændə/ 曲折,弯曲;~ line 曲折线,蜿蜒

meandrine /miˈændri(:) n/ （或 meandroid）弯弯曲曲的,有螺旋形面的

means /mi:nz/ 方法,手段；工具,设备,装置

mean-square 均方；～ deviation 均方（偏）差；～ error 均方(误)差；～ root 均方根

meas. (measure) 测量,量度

measure /ˈmeʒə/ 测量,量器,计量；尺度,尺寸,比例尺；有……长(宽,高) ～ of precision 精密程序,精确度；angular ～ 角度法,度量角度；circular ～ 弧度法；shrinking ～ 收缩尺,放尺；tape ～ 卷尺,皮尺 ‖ *Example sentence:* A cube of steel measures 2cm on one side. 例句: 一块立方体钢,每边2厘米长。

measured /ˈmeʒəd/ 实测的,精确的；～ value 实测数值；to ～ 照尺寸,合调子

measurement /ˈmeʒəmənt/ 测量,测法；大小,尺寸,测定；surface roughness ～ 表面粗糙度测量；three-wire ～ 三线测(螺纹)法

measurer /ˈmeʒərə/ 测量器具；测量员；map ～ 量图仪

measuring /ˈmeʒəriŋ/ 测量；～ line 量线；～ plane 量度面；～ plane method 量度面法；～ point 量点；～ point method 量度点法(透视图)；～ tools 量具

mech. (mechanical) 机械的；(mechanics) 力学,机械学

mechanic /miˈkænik/ 机械的,机(械)工(人),机械员

mechanical /miˈkænikl/ 机械的,机械制的,机动的,力学的；～ drawing 机械制图,仪器画,机械画

mechanism /ˈmekənizəm/ 机械,机构,装置,机构学,机理；ratchet ～ 棘轮机构；servotype ～ 伺服机构

mechanization /mekənaiˈzeiʃən/ 机械化

medial /ˈmi:dəl/ 居中的,中间的,平均的

median /ˈmi:dən/ 中央的,中间的；中线；中位数；～ line 中线；～ of a triangle 三角形的中线

medium /ˈmi:diəm/ 中间,适中,中数,平均；导体,介质,手段；工具,方法；中等的,适中的,中型的；～ fits 中级配合；～ force fit 中压配合

med. s. (medium steel) 中碳钢,中硬钢

megascopic /megəˈskɔpik/ 肉眼可见的；放大了的,扩大了的

mem. (member) 构件,部件,组成部分；会员,成员

matingmember 配合件

memory /ˈmeməri/ 存储,记忆；read only ～ 只读存储器

meniscus /miˈniskəs/ 新月,新月形物；弯月形零件,半月板

menu /ˈmenju:/ 图样单；菜单

merge /mə:dʒ/ 合并

meridian /məˈridiən/ 子午圈,子午线；经线,经度；顶点,极点；～ plane 子午面

merotomize /meˈrəutəmaiz/ 分成几部分,分成几块

mesh /meʃ/ 网视,网孔；(齿轮的)啮合,使啮合,网格；～ coordinate 网格坐标；～ing gear 啮合齿轮；～ lines 网格线；be in (齿轮)互相啮合

message /ˈmesidʒ/ 信息

metacommand /metəkəˈmɑ:nd/ 中间命令

metafile /ˈmetəfail/ 元文件

metal /ˈmetl/ 金属,合金；Babbit ～ 巴比合金（含80%锡, 7%铜, 13%锑）

metaler /ˈmetelə/ (metaller) 板金工

meter /ˈmi:tə/ 表,仪,计；米(长度),公尺；～ screw 公制螺纹；～ taper 公制锥度

meth (method) 方法

method /ˈmeθəd/ 方法，办法；教学法（美国）；秩序，条理；~ of approxi mation 近似法；~ of auxiliary straight line 辅助直线法；~ of auxiliary straight plane 辅助平面法；~ of auxiliary spheric surface 辅助球面法；~ of construction 绘图方法；~ of determining the intersection point of line and surface 求线面交点的方法；~ of lapped projections 重叠投影法；~ of manufacturing 制造方法；~ of orthogonal projection 正投（射）影法；~ of parallel chords 平行弦法；~ of parallel displacement 平行移位法；~ of proportional square 比例方格法；~ of rabatterment 倒转法；~ of replacing planes of projection 更换投影面法；~ of revolution 旋转法；~ of revolution about aparallel axis 绕平行轴旋转法；~ of right cutting plane 正截面法；~ of rotation 旋转法；~ of substituting planes of projection 换面法；~ of three planes 三平面法；alternate ~ 变换法；alternate light and heavy line ~ 螺纹正规符号画法（美式）；alternative ~ 另一种方法；American ~ 美国画法（第三角法）；analysing line and surface ~ 线面分析法；approximate ~ 近似方法；approximate circle arc ~ 齿形近似画法；approximate polyconic ~ 近似锥面法；approximate polycylindric ~ 近似柱面法；auxiliary circle ~ 辅助圆法；auxiliary curve ~ 辅助圆法；auxiliary cutting elements plane ~ 辅助截素线平面法；auxiliary cutting plane ~ 辅助截平面法；auxiliary cutting plane parallel to projection plane ~ 辅助截平行面法；auxiliary cutting sphere surface ~ 辅助截球面法；auxiliary element ~ 辅助素线法；auxiliary plane ~ 辅助平面法，换面法；auxiliary projection plane ~ 辅助投影面法；auxiliary sphere ~ 辅助球面法；auxiliary view ~ 辅助视图法；‖ *Example sentence: By the* auxiliary view method *determine the intersection of the planes. Show visibility.* 例句：用辅助视图法确定两平面的交线，并判别可见性。basic angle ~ 基准角度法（圆锥度公差）；boxing ~ （画轴测图的）方箱法；breaking down ~ 形体（解剖）分析法；central projection ~ * 中心投影法；centre-to-centre ~ 中心连接法；change of position ~ 变换位置法；circle ~ for conjugate diameter 画椭圆的共轭径法；coincidence ~ 重合法；combination ~ 灭点法（透视图）；complementary-line ~ 余角法；complementary-angle ~ 余角法；concentric cirle ~ （画椭圆的）同心圆法；conventional ~ 习用方法，惯例；coordinate ~ 坐标法；cross section ~ 剖面法；curve plotting ~ 曲线画法；cutting ~ 切割法；cutting plane ~ 截平面法；cutting-sphere ~ 截球面法（求交线）；diagonal ~ 对角线法；direction view ~ 向视（图）法；direct-projection ~ 投影线法（透视图）；drafting ~ 制图法；edge view ~ （矢量）重影（边）视图法；edge wise view ~ 边视图法；eight-point ~ （已知共轭直径画椭圆的）八点法；element plane ~ 素线平面法；elipse-pin-and-string ~ 椭圆钉线法；enclosing-square ~ （画轴测图的）方框法；envelope ~ 包络线法（画抛物线）；European ~ 第一角法；first angle ~ * 第一角画法；first angle projection ~ 第一角投影法；four-center approximate ~ （画椭圆的）四心近似法；general ~ 一般方法；glass box ~ 透明盒法；gore ~ 柳叶法，球面经线展开法；Grant ~ 格兰脱近似画法（齿轮）；graphic ~ 图解法，图示法；graphical extrapolation ~ 图解外推法；grid ~ of polar coordinate system 极坐标网格法；grid ~ of rectangular coordinate system 直角坐标网格法；impression ~ 印模法，拓印法；lead wire ~ 铅丝法；line ~ 直线法；manufacturing ~ 制造方法；measuring point ~ 量点法；Monge's ~ 蒙日法；nested multiplication ~ 嵌套乘法；numerical-graphic ~ 数值图解法；oblique projection ~ *斜投影法；offset ~ 支距法（用坐标定物体上点的位置之法），坐标法；one-plan ~ 单面法；one-view ~ 单视图法；ordinate ~ 坐标法绘图；orthogonol projection ~ *正投影法；paralled plane ~ 平行面法；parallelepiped ~ （矢量）平行六面体法；parallelline ~ 平行线法；parallelogram ~ （矢量）平行四边形法；parallel plane ~ 平行面法；parallel projection ~ *平行投影法；perspective-plan ~ 透视平面图法；Piercing-point ~ 贯穿点法；pin and string ~ 钉线画（椭圆）法；plane ~ 平面法；point ~ 量点法；point by point comparative ~ 逐点比较法；point-view ~ 点视图法；polar triangle ~ 极三角形法；polyconic ~ 锥面法；polycylindric ~ 柱面法；profile projection ~ 侧面投影法；projection ~ *投影法；projection ~ A 第三角（投影）法；projection ~ E 第一角（投影）法；proportional ~ 比例法；radial-line ~ 辐射线法；replacing projection plane ~ 换面法；reversible ~ 逆求法；revolution ~ 旋转法；revolved-plan ~ 回转平面图法；right triangle ~ 直角三角形法；rolling ~ 侧滚法；rubbing ~ 拓印法；scaling ~ 比例法；selected-line ~ 选线法（交线求法）；setover ~ 跨距法，偏置法；simple and convenient developing ~ 简便展开法；sine-and-cosine ~ 正弦及余弦法；slitting-up ~ 全切开法；sphere ~ 球面法；sphere-gore ~ 球面三角狭条法；superposition ~ 叠加法；supplementary projection plane ~ 换面法；tangent ~ 正切法；tangent-sphere ~ 球切法；the ~ of resemble projection 相似投影法；third angle ~ *第三角画法；third angle projection ~ 第三角投影法；three-intersecting line ~ 三交线法；three-intersecting line ~ 三交线法；trammel ~ 椭圆规法；triangulation ~ 三角形法；true-length diagram ~ 真实长度图解法；two-view ~ 两视图法（用两视图求交点或交线之法）；Unwin ~ 恩文近似画法（齿轮）；vector ~ 矢量法；vector polygon ~ 矢量多边形法；visual-ray ~ 可见光线法，视线法（透视）；zone ~ 带状法，球面纬线展开法

metre /ˈmiːtə(r)/ (meter) /英/米（长度）；仪表

metric /ˈmetrɪk/ 公制的，米制的；度量的，测量的；~ measure 公制计量；~ scale 公制（比例）尺，米尺；~ size 公制尺寸；~ system 米制，公制；~ thread 公制螺纹

MFR (manufacture) 制造

MI (malleable iron) 可锻铸铁

MIC (micrometer) 分厘卡

micrify /ˈmaɪkrɪfaɪ/ 缩小尺寸，缩微，使变小

micro /ˈmaɪkrəu/ 微米，微小；测微计，千分表

microcomputer /ˈmaɪkrə(u)kəmˈpjuːtə/ 微型电脑，微型计算机

microfilm /ˈmaɪkrəfɪlm/ 缩微胶卷，用缩微法拍摄；computer output ~ing 计算机缩微输出

microgram /ˈmaɪkrə(u)græm/ 微克

micrograph /ˈmaɪkrəgrɑːf/ (micro.) 微型照片，微型图，微写器

micrographics /ˌmaɪkrəuˈgræfɪks/ 缩微图形学

micrography /maiˈkrɔgrəfi/ 显微绘图，显微照相(术)

microimage /ˌmaikrəuˈimidʒ/ 缩微图像

micro-inch 微英寸

micrometer /maiˈkrɔmitə/ 测微器，千分尺，分厘卡；slide ～ 滑动千分尺，游标千分尺

micron /ˈmaikrɔn/ 微米（10^{-6}米）

microprocessor /ˈmaikrəuˈprɔsesə(r)/ 微处理机

microscopic(al) /maikrəˈskɔpik(əl)/ 显微的，微观的，用显微镜可见的

microsize /ˈmaikrəsaiz/ 微小尺寸

microsoft /ˈməikrə,sɔːft/ 微软件

mid /mid/ (middle) 中间，中部，中央

middle /ˈmidl/ 中间，中部；中间的，中部的，中等的

midline /ˈmidlain/ 中线

midpoint /ˈmid,pɔint/ 中点

mid-section 中间截面，中间剖视

mil /mil/ 密耳，千分之一英寸 (=0.0254 mm=25.4μm)

millimetre /ˈmilimiːtə/ (millimeter) 毫米，公厘

millimicron /ˈmilimaikrɔn/ 毫微米

milling /ˈmiliŋ/ 铣

mimeograph /ˈmimiəgraːf/ 复写机，滚筒油印机

mimicked /ˈmimikt/ 模仿，仿制

min. (minimum) 最小，最小的；(minute) 分

mini /ˈmini/ 缩图，缩影，缩型；油印机，油印品，油印

miniature /ˈminətʃə/ 缩影图，缩样，雏型，袖珍画，微小绘画法，彩色画；尺寸小的，小型的；～ car 微型汽车；～ component 小型元件；in ～ 用缩图画，用缩图表示，缩图的，小型的

minicomputer /ˈminikəm,pjuːtə/ 微型电脑

minification /minifiˈkeiʃən/ (尺寸)缩小，缩小率

minify /ˈminifai/ 缩小尺寸，弄小，使缩小

minimum /ˈminiməm/ 最小的，最小量

minor /ˈmainə/ 较小的，较少的，短的；～ axis 短轴；～ diameter 短径

minuscule /ˈminəskjuːl/ 小写字母；很小的，很不重要的

minus /ˈmainəs/ 负的，减去的，减少

minute /ˈminit/ 分(钟)，精细的，微小的

mirror /ˈmirə/ 镜子，反映

mirror-image 镜像

mirroring /ˈmɪrərɪŋ/ 镜像变换(计)

misalignment /misəˈlainmənt/ 直线不重合度，位移；偏差，角度误差；不同轴度，轴线(中心线)不重合度，不同心度，不平行度，非直线性

miscellaneous /misiˈleiniəs/ 混杂的，各种各样混在一起的；多方面的，有各种特点的；～ bolt 有各种特点的螺栓

missing /ˈmisiŋ/ 遗漏，缺失，空白；缺少的；～ lines 缺少线条；～ views 缺少视图

miter /ˈmaitə/ (mitre) 斜接，斜接面；～ wheel 斜齿轮，圆锥齿轮；～ joint 斜面接合，斜削接头

mk. (mark) 标记，刻度，特征

ML (machine language) 机器语言；(material list) 材料单；(mid-line) 中线

MLP (machine language program) 机器语言程序

mm (milimeter) 毫米

M. M. C (maximum material condition) 最大材料条件，最大实体状态

mmp (metric module pitch) 米制模数(齿轮)节距

mock /mɔk/ 制造模型，模仿的，模拟的

mock-up 仿真

mod. (modulus) 模数，模量，系数，率

mode /məud/ 方式；alphaphotographic ~ 字图点式传输，字图照相法显示；analog plot ~ 模拟绘图方式

model /ˈmɔdl/ 模型，标本，式样，缩图；geometric ~ling 几何造型，几何建模；graphic ~ 立体图，三维图；open system interconnection reference ~ 开放系统互联参照模型；solid ~ 实体模型；solid ~(l)ing 实体构形；surface ~ 表面模型；surface ~(l)ing 表面构形；wire frame ~ 线框模型

MO-DEM (modulator-demodulator) 调制解调器

modernization /mɔdənaiˈzeiʃn/ 现代化

modification /mɔdifiˈkeiʃn/ 变更，变换，改变，更改，改造，修改；限制；sign ~ 符号变换

modifier /ˈmɔdifaiə/ 更改者，修改者；修饰语

modify /ˈmɔdifai/ 使变更，使变换，修改；改变

modulation /mɔdjuˈleiʃn/ 调制，转变，变换

modulator /ˈmɔdjuleitə/ -demodulator 调制解调器

module /ˈmɔdju:l/ 模数，系数；模件，组件，模块，组件，程序片，存储体；optical ~ 光学组件

modulo /ˈmɔdjuləu/ 模，模数；系数，模量；殊余数，按模计算

modulus /ˈmɔdjuləs/ 模数，系数，模件，比率；axial ~ 轴向模数；end ~ 端面模数；section ~ 截面模数；scale ~ 标值（图尺）系数

Mohshardness 莫氏硬度

Mohs scale (of hardness)莫氏硬度（计）~ number 莫氏硬度数值

moiety /ˈmɔiəti/ 一半，二分之一，一部分

mold /məuld/ (mould) 模型，样板，曲线板，性质，造模；~ line 模线（即圆角之理论尖角线）；casting ~ 铸型

molding /ˈməuldiŋ/ 造型；嵌线（建筑物上的装饰线）；compression ~ 压塑

monogamy /məˈnɔgəmi/ 一一对应，一对一

monitor /ˈmɔnitə/ 监视器

montage /mɔnˈtɑ:ʒ/ 剪辑

moony /ˈmu:ni/ 月亮（状）的，新月形的；圆的；月光似的

mortar /ˈmɔ:tə/ 砂浆，胶泥

mosaic /məuˈzeik/ 镶嵌法（建筑图中用淡色线条描绘室内之地板花纹，家具等），镶嵌图；aerial (photographic) ~ 空中（航空）照片嵌拼地图；airphoto ~ 航摄相片镶嵌图

motion /ˈməuʃn/ 运动；circular ~ 圆周运动；rotary ~ 旋转运动；straight line ~ 直线运动

motor /ˈməutə/ 电动机，发动机

motorization /məutəraiˈzeiʃən/ 机械化

mould /məuld/ 曲线板，模型，样板，性质，造型；~ing practice 翻砂

mounting /ˈmauntiŋ/ 装配，固定，座架；~ surface 装配面

mouse /maus/ 鼠标器，鼠形器

movable /ˈmu:vəbl/ 活动的，可移动的

movement /ˈmu:vmənt/ 移动，运动

mst. (measurement) 测量

mtl. (mat., matl., material) 材料

multidimensional /ˌmʌltidiˈmenʃnl/ 多面的，多维的

multidirectional /ˌmʌltidiˈrekʃənl/ 多方面的

multilayer /ˌmʌltiˈleiə/ 多层的

multimedia /ˌmʌltiˈmi:djə/ 多媒体

multiparty /ˌmʌltiˈpɑ:ti/ 合用，同一

multiple /ˈmʌltipl/ 并联，复接；多倍的，复合的，多样的，多重的，倍数；~ spline 花键；~ thread 多头螺纹

multiple-series 复联,串并联
multiplication /mʌltipliˈkeiʃn/ 放大,增倍;相乘,乘法
multiply /ˈmʌltiplai/ (按比例)放大,乘,增加数量
multiview /ˈmʌltivju:/ 多面视图; ‖ *Example sentence:* A multiview *drawing is a systematic arrangement of orthographic views on a single plane (the drawing paper).* 例句:多面视图是各正投影视图在单一平面(图纸)上有规则地排列。
mural /ˈmjuərəl/ 墙,壁(上的),壁饰(画)
mutual /ˈmju:tjuəl/ 共同的,互相的;～-perpendicular axes 互垂直轴线;～-perpendicular edges 互垂直边;～-perpendicular planes 互垂直平面;～ positions 相互位置;～ly perpendicular lines 相互垂直的直线
mylar /ˈmaila/ 聚酯薄膜

N

N "美国式样螺纹"的代号;(north) 北; N11° 5′ E 正北11° 5′ 偏东
name /neim/ 名,名称;～ of part 零件名称;～ plate 铭牌; array element ～ 数组元素名
nappe /næp/ 片,叶
naught /nɔ:t/ 零,无
nautical /ˈnɔ:tikəl/ 航海的,海上的;～ scale 海图比例尺;～ chart 航海图
naval /ˈneivəl/ 海军的,军舰的;～ brass 船用黄铜
NB (American Standard Buttress Threads) 美国标准锯齿螺纹
NBS (National Bureau of Standards) 美国国家标准局; (New British Standard, Imperial Wire Gauge) 英国新标准线规(帝国线规)
NC (American National Coarse Thread Series) 美国粗(牙)螺纹系列;(Numerical Control) 数字控制
ND (not dark enough) 不够黑
near /niə/ 靠近,接近;近,差不多,近似的,仿制的
nearness /ˈniənis/ 接近,近似,密切
neatline /ni:tlain/ 图表边线,准线
NEC (national electrical code) 美国电气规程
neck /nek/ 颈,刻颈,退刀槽,收缩; shaft ～ 轴颈
needle /ˈni:dl/ 针,指针;～ point 针尖(圆规);～ point leg 固定针脚,针脚
NEF (American National Extra Fine Thread Series) 美国特细螺纹系列
negative /ˈnegətiv/ 底片,负数,负的;阴性的;否定的,反面的;～ sign 负号;～ thread 阴螺纹; Vandyke brown ～ process 范戴克晒图法
neighbo(u)r /ˈneibə/ 邻近的,邻接的,与……接邻
network /ˈnetwə:k/ 网络; mesh ～ 网格的网络
NETWT (net weight) 净重
new-string 新字符串
NF (National Fine Thread Series) 美国细(牙)螺纹系列
NH (not heavy enough) 不够粗

nib /nib/ 笔头, 笔尖; 字模; 模孔; 修尖

nickel /'nikl/ 镍, 镀镍

nil /nil/ 零, 无

nipple /'nipl/ 螺纹接管, 接头, 喷嘴; close ~ 紧密插口接头

No. (number) 第……号, 号码, 数

node /nəud/ 节点, 结点结; 重点交点; 轨迹相交点; ~ name 节点名称; connect ~ 连接点; linkage ~ 联节点

nom. (nominal) 公称的, 名义的

nomenclature /nə'menklətʃə(r)/ 名称, 术语; 名词汇编; 命名法; blade ~ 叶片几何参数

nominal /'nɔminl/ 名义, 标称, 名称; 标称的, 名义的, 公称的; ~ diameter 名义直径; ~ error 名义误差; ~ line width 标准行宽; ~ size 名义尺寸

nomogram /'nɔməgræm/ 诺漠图; 列线图; 图解

nomograph /'nɔməgra:f/ 列线图解, 计算图表, 图解, 诺漠图

nomographs /'nɔməgra:fs/ 图算, 算图

nomography /nəu'mɔgrəfi/ 诺漠图, 列线图, 图算法, 计算图表学

non- /nɔn/ (字首) 表示"非""不""无"

nonagon /'nɔnəgən/ 九边形

noncircular /ˌnɔn'sə:kjulə/ 非圆的, 非圆形的; ~ curve 非圆曲线

non-coplanar 不共面(的), 异面(的)

nondevelopable /ˌnɔndɪ'veləpəbl/ 不可展; ~ rulod surface 不可展直纹曲面

non-graphical 非图形的

non-intersecting 不相交的, 交叉的

nonparallel /ˌnɔn'pærələl/ 不平行的

non-planar 空间的, 非平面的, 曲面的

NON-STD (non-standard) 非标准

non-trace 非迹线; ~ plane 非迹线平面

non-transparent 不透明的

norm /nɔ:m/ 标准, 规格, 定额, 模方, 规范; fundamental ~s 基本规范; technical ~s 技术标准, 技术规范

normal /'nɔ:məl/ 标准的, 正规的; 垂直的, 正交的, 法线的, 法向的; 法线, 规定; ~ axis 垂直轴线; ~ circular pitch 规定周节 (弧线长); ~ circular thickness 规定弧线齿厚; ~ cross section 正剖面, 垂直剖(断)面; ~ direction 法线方向; ~ line 法线; ~ module 法向(面)模数; ~ pitch 法向节距; ~ to a curve 曲线的法线; common ~ 公法线; surface ~ 曲面法线

normalization /ˌnɔ:məlai'zeiʃn/ 标准化, 规格化, 归一化; (热处理)解除内应力, 正(常煅)火, 常化

north /nɔ:θ/ 北, 北方; ~ point 指北针

nose /nəuz/ 尖头, 凸耳; 前端

not /nɔt/ 不; ~ less than 不下于, 不小于; ~ more than 不超过, 至多; ~ to seale 不按比例, 不用比例尺

notation /nəu'teiʃn/ 符号; 标志法, 表示法; 标记, 注释

notch /nɔtʃ/ 切口, 凹口, 槽口; 缺口, 凹坑; 将……切口; 刻 V 形槽; 给……开槽; ~ curve 切口曲线, 阶形曲线

notched, notchy /nɔtʃt, 'nɔtʃi/ 带有切口的, 刻有凹槽的

notching /'nɔtʃiŋ/ 切口, 开槽; 阶梯式, 下凹的; ~ curve 下凹(阶梯)曲线

note /nəut/ 记录, 笔记; 草稿, 符号; lecture ~s 讲课(演讲)草稿

nowel /ˌnəuel/ (铸造)下型箱

NP (name plate) 铭牌; (normal pitch) 公称节距, 标准间(节)距

NPS (American Standard Straight Pipe Thread) 美国标准直管螺纹

NPSC (American Standard Straight Pipe Thread in Coupling) 美国标准接头直管

螺纹

NPT (American Standard Taper Pipe Thread) 美国标准锥管螺纹

NS (American National special thread series) 美国标准特殊螺纹级

n. t. (number of teeth) 齿数

NTS (not to scale) 不（未）按比例（绘制），不按比例

nt. wt. (net weight) 净重

num. (number) 数，数字，号码

number /ˈnʌmbə/ 号码，编号；数，数目，数量，总计；～ of part required 所需件数；～ of starts 螺纹头数；～ of teeth 齿数；～ of threads 螺纹扣数；～ of turns（螺纹）圈数；assembly drawing ～ 装配图图号；code ～ 序号，代号；dash ～ 零件编号，（各部分）细号；descriptive ～ 注释，说明；detail drawing ～ 零件图号；drawing ～ 图号；issue ～ 顺序编号；item drawing ～ 零件图号；item ～ 件号；line ～ （计算机语句）行（的编）号；odd ～ 奇数；parts ～ 件号；prefered ～s 优先数；random ～ 随机数；round ～s 取整数；sequential ～ 有序数；serial ～ 序号；statement ～ 语句标号

numbering /ˈnʌmbəriŋ/ 编号；～ of bars 钢筋编号

numeral /ˈnjuːmərəl/ 数字，号码；数的，表数的；～ order 编号次序；Arabic ～s 阿拉伯数字

numerator /ˈnjuːməreitə/ 计算者，（分数的）分子

numerical /njuːˈmerikəl/ 数字的，数值的，用数表示的；～ -graphic method 数值图解法

nut /nʌt/ * 螺母，螺帽；上螺母；acorn ～ * 盖形螺母；back ～ 逆螺母，防松螺母；blank ～ 粗制螺母；bolt ～ 螺栓的螺母；bright ～ 精制螺母；butterfly ～ 翼形螺帽（元宝螺帽）；cap ～ * 盖形薄螺母；castellated (castle) ～ 槽形螺母；channel ～ 槽形螺母；check (lock) ～ 防松螺母；check 锁紧螺帽；die ～ 扳牙，钢板螺母；finished ～ 精制螺母；flat ～ * 扁环螺母；foundation ～ * 地脚螺母；gland ～ 压紧螺母；grip ～ 夹紧螺母，固定螺母；hexagon ～ * 六角螺母；hexagon ～ with collar* 六角凸缘螺母；hexagon ～ with flange* 六角法兰面螺母；hexagon slotted ～ * 六角开槽螺母；hexagon thin ～ * 六角薄螺母；hexagon thin castle ～ * 六角冠状薄螺母；hexagon weld ～ * 焊接六角螺母；jam ～ 压（锁）紧螺母，防松螺母；lifting eye ～ * 吊环螺母；light castle ～ 轻型槽形螺母；light-thick ～ 轻厚螺母；octagon ～ * 八角螺母；pentagon ～ * 五角螺母；12 point flange ～ * 十二角法兰面螺母；round ～ 圆螺母；slotted round ～ * 开槽圆螺母；square ～ * 方螺母；square ～ withcollar* 方凸缘螺母；square weld ～ * 焊接方螺母；thin ～ 薄螺母；two-face view of ～ 螺母见二面视图；thumb ～ 元宝螺母，蝶（翼）形螺母；triangle ～ with collar* 三角凸缘螺母；three-face view of ～ 螺母见三面视图；washer faced hexagon ～ * 六角垫圈面螺母；wing ～ * 翼形螺母

NW (north west) 西北

n. wt. (net weight) 净重

O

OA (O-A, overall) 总（尺寸），外轮廓的（尺寸）

oad (overall dimension) 总尺寸，外形尺寸，全部尺寸

object /'ɔbdʒekt/ 物体，实物；对象，目标，课程；～ lesson 实物教学，具体例子；～ line 外形线，可见轮廓线；cylindrical ～ 圆柱体

objective /ɔb'dʒektiv/ 目标，目的；客观的，真实的，（显微镜等的）物镜

object-line 轮廓线，外形线，地形线，等高线

oblate /'ɔbleit/ 扁的，扁圆的，扁球状的；～ ellipsoid 横轴椭圆球；～ spheroid 扁球体，椭球体

oblique /ə'bli:k/ 斜的，倾斜的，斜面的，斜角的，非直角的；成45°角地，斜投影；～ cone 斜圆锥；～ cutting 斜剖；～ cylinder 斜圆柱体；～ elliptical cylinder 斜椭圆柱面；～ helicoid 斜螺旋面；～ hexagonal pyramid 斜六棱锥体；～ illumination system 斜照式（地形图上的山坡线具有日光及坡度形象）；～ line 斜线，一般位置线；～ parallelepiped (on) 斜平行六面体；～ pentagonal prism 斜五棱柱体；～ perspective 斜透视；～ projection 斜投影；～ roctangular 斜长方体；～ surface 斜面，一般位置面；～ view 斜视图，辅视图；‖ *Example sentence: Oblique lines* that appear parallel in two or more principal views are parallel in space. 例句：倾斜直线在两个或两个以上的主要视图中平行，他们在空间也平行。

obliquity /ə'blikwəti/ 斜度，倾斜

obliterate /ə'blitəreit/ 擦去，涂去，除去，消去，清除

oblong /'ɔblɔŋ/ 长方形，阔椭圆形；长方形的，长椭圆形的

OBS (observation) 观察；(observer) 观察者

obscure /əb'skjuə/ 遮蔽；阴暗的，含糊的

observable /əb'zə:vəbl/ 可看到的，显著的；值得注意的

observation /ɔbzə'veiʃn/ 观察，观测；注视，监视；言论，意见

observe /əb'zə:v/ 观察，观测；注视，遵守；评述

observer /əb'zə:və/ 观察者，观察员；评论员；观察器

obtain /əb'tein/ 得到，找出

obtuse /əb'tju:s/ 纯（角）的，圆头的，不尖的；～ angle 钝角

obvious /'ɔbviəs/ 明显的，显著的；～ly 明显地，显然

OC (on center) 对准中心

OCB (oil circuit breaker) 油断路器（电）

occupy /'ɔkjupai/ 处于；占有

occur /ə'kə:/ 发生，出现；存在，想起，发现

occurrence /ə'kʌrəns/ 发生，出现；存在，所在地

oct (octagon) 八角形，八边形；(october) 十月

octahedron /'ɔktə'hedrən/ (octahedra) （正）八面体；regular ～ 正八面体

octant /'ɔktənt/ 挂限，圆周的八分之一，八分仪

octet(te) /ɔt'tet/ 八隅（体），八角体，八位

二进制数字
octagon /'ɔktəgən/ 八角形, 八边形
ocular /'ɔkjulə/ 用眼的, 凭视觉的 ~ estimate 目测法; ~ estimation 目测法
OD (outside diameter) 外(直)径; (outside dimension) 外形尺寸
odd /ɔd/ 奇数的, 单数的, 不成对的, 零散的, 多余的
odd-numbered (spokes) 单数的(辐条)
odontograph /əu'dɔntəgra:f, -græf/ 画齿规, 齿面描记器; ~ method 画齿法; three point ~ 三点画齿规
odontoid /ɔ'dɔntɔid/ 齿形的, 牙状的
OF (outside face) 外表面
off-bear 移开, 拿走, 移去
off-center (ed) 偏心的, 不对称的
off-gauge 不标准的
office /'ɔfis/ 室, 普通; drawing ~ 绘图室
off-line 脱机
off-round (faux-round) 立椭圆
offset /'ɔ:fset/ 偏斜, 偏置, 偏移, 平移, 位移偏距; 失调; 调整偏差不均匀性; 胶印; (坐标)支距(法)量, 型值(船舶); ~ angle 偏角, 偏斜; ~ construction 支距法作图; ~ cutting 阶梯截平面, 阶梯剖; ~ method 支距法(用坐标定点之位置法); ~ printing (橡皮)印刷, 胶印; ~ sections 偏置剖面, 移位剖面; table of ~s 型值表
offsetting /'ɔ:f,setiŋ/ 偏心距, 倾斜, 偏置法
off-size 偏差, 偏移; 偏角; 尺寸不合格, 不合尺寸, 非规定尺寸
OG (ogee) S 形曲线; (ogival) 蛋形的, 尖顶的
ogee /'əudʒi:/ S 形曲线, S 形曲线的; ~ arch 双弯拱; ~ curve 双弯曲线, S 形曲线
OGL (out going line) 引出线, 引线
oil /ɔil/ 油, 石油; ~ colour 油画颜料; ~ painting 油画; ~ stone 油石

oiler /'ɔilə/ 油杯, 加油器
oil-hardening 油淬火
oil-quenching 油淬火
OK (okay, okeh, okey) 对, 是, 好, 行; 全对, 正确, 同意, 批准
old-string 老字符串
O-member (Observers member) 观察员成员(国)
omit /o'mit/ 省略, 遗漏, 忽略, 去掉
one-half 一半; ~ size 半尺寸
one-point 一点; ~ perspective sketch 一点透视草图
one-size 同一尺寸的, 同一大小的
one-third 三分之一; ~ size1/3尺寸, 1比3的尺寸
one-to-one 一(对)一的; ~ correspondence 一一对应
one-way 单向的, 单程的, 单面的
on-gauge 合格的, 标准的
on-line (电脑)联机, 联用, 联成
onto (同 on to)向……方面, 到……上
onwards /'ɔnwədz/ 向前
opaque /əu'peik/ 不透明的, 不透光的; ~ body 不透明体
open /'əupən/ 开, 展开, 伸直; ~ system 开放系统
open-end(ed) 开口的, 无底的, 无终止的; ~ wrench(或 ~ spanner)开口扳手
opening /'əupniŋ/ (接头的)口, 孔, 洞; 开, 打开, 开度; 开放, 道路; 开始; ~ of cock 旋塞开度; air intake ~ 进气口; centre ~ 中心孔; exhaust ~ 排气口; root ~ 焊缝底距
operate /'ɔpəreit/ 计算, 运算, 运转, 开动, 操作, 工作
operating /'ɔpəreitiŋ/ 工作的, 操作的; ~ position 工作位置; ~ security 操作安全; ~ space 操作空间; ~ system 操作系统

operation /ɔpəˈreiʃn/ 计算，运算；运转，操作，运用；作业，效果，说明，工序；～ charts 运用图，运转图；Boolean ～ 布尔运算；file ～ 文件操作；machining ～ 加工顺序

opposite /ˈɔpəzit/ 相对的，对立的，相反的；相应的，对应的

optic /ˈɔptik/ （光学仪器中的）镜片

optical /ˈɔptikl/ 光学的，光导的；眼睛的；～ module 光学组件

optimal /ˈɔptiməl/ 最佳的，最优的，最适当的

optimization /ɔptiməiˈzeiʃn/ 优化

option /ˈɔpʃn/ 选择，挑选；选择方案；at one's ～ 随意；make one's ～ 进行选择

optional /ˈɔpʃənl/ 任意的，随意的

OR (outside radius) 外半径

orb /ɔ:b/ 球，球体，天体；轨道；球形的，圆的

order /ˈɔ:də/ 次序，序列，序；阶，级；指令；订货单；～ of drawing 绘图次序；～ of inking 上墨次序；～ of stroke 笔划顺序；operational ～ 运算顺序

ordinal /ˈɔ:dinl/ 依次的，次序的；～ number 序数

ordinate /ˈɔ:dinət/ 纵坐标；axis of ～ 纵坐标轴；vertical ～ 纵坐标

ORFC (orifice) 孔，小洞

organ /ˈɔ:gən/ 机构，机关，构造；元件，部件；报刊

organigram /ɔ:ˈgænɪgræm/ 构造示意图

organization /ɔ:gənaiˈzeiʃn/ 编制，构造；组织，机构

oricycle /ˈɔ:rəˌsaikl/ 极限圆

orient /ˈɔ:riənt/ (orientate) 定向，定方位；方位

orientation /ɔ:rienˈteiʃn/ 定向，定方位；方位，方向性

orifice /ˈɔrifis/ （小）孔，小洞，眼，管口，阻尼（节流）孔；blast ～ 鼓风口；exhaust ～ 排气口；square ～ 方口；triangular ～ 三角口

origin /ˈɔridʒin/ 原点，起源，由来；～ of coodinates 坐标原点；～ of vector 矢量原点

ornament /ˈɔ:nəmənt/ 花饰；loop ～ 绳环花饰

ortho- （词头）正，原；直（线）；垂直，正形

orthogon /ˈɔ:θəgɔn/ 矩形，长方形

orthogonal /ɔ:ˈθəgənl/ (rightangled) 直角（投影），相互垂直的，正交的，直角的，矩形的；～ cone 正交的锥面；～ coodinate 正交坐标；～ projection 正（交）投影；～ section 正（交）截面

orthogonality /ɔ:θəgəˈnæliti/ 相互垂直，正交（性），垂直度；～ relation 正交关系

orthograph /ˈɔθəgrɑ:f/ 正视图，正投影图，正射图，投影图

orthographic(al) /ɔ:θəˈgræfik(əl)/ 正交的，直角的，正射的，直线投射的；～ projection 正投影，正交射影；～ sketch 正投影草图；‖ *Example sentence: Orthographic projection is a method of representing an object by a line drawing on a projection plane that is perpendicular to parallel projectors.* 例句：正投影是在与投影线垂直的投影面上画出线条来表示物体的方法。

orthography /ɔ:ˈθəgrəfi/ 正交射影，正射投影；正射法，正投影法；正投影图，正视图；正字法，表音法

orthohexagonal /ˌɔ:θəˈheksəgənl/ 正六方体的

orthonormal /ɔ:θəuˈnɔ:məl/ 正规化的，标准化的

OS (over size) 尺寸过大，过大，超差；加大（尺寸）；(operating system) 操作系统

oscillograph /ɔˈsiləgrɑ:f/ (oscilloscope) 示波器，快速过程（脉冲）记录仪

osculatory /ˈɔskjuˌlətəri/ 切线，接触线，接触的，共同的

OTO (out-to-out) 外廓尺寸，全长，全宽，外到外

o/u (over and under) 上下

outboard /ˈautbɔːd/ 外侧

outcrop /ˈautkrɔp/ 露头

outer /ˈautə/ 外部的，外层的，外面的，表面的；外线，边线

outfall /ˈautfɔːl/ 出口

outgoing /ˈautgəuiŋ/ 引出；～ line 引出线

outlet /ˈautlet/ 引出线，引出端，出口管；出口，排出口

outline /ˈautlain/ 外形，轮廓；外形线，轮廓线；略图，草图；概要概括，提纲，大纲，略述；剖面；画……的轮廓，作……的草图，提出……的要点；‖*Example sentence:* The outline *of a view will always be visibility.* 例句：一个视图的外形轮廓线总是可见的。～ assembly drawing 外形装配图；～ drawing 轮廓图；～ map 略图；～ of proccess 生产过程简图；～ section 轮廓剖面；～ sketch 略图；an ～ of (draw ～s of) 画……的轮廓，作……的草图；double ～ 双轮廓线；fictitious ～ 虚拟轮廓线；front ～ 正视图，前视图，垂直投影；in ～ 只画轮廓；tooth ～ 齿廓，齿外形

out-of-round 不圆的

out-of-shape 形状不规则的

out-of-size 尺寸不合规定的

output /ˈautput/ 输出；产品，产量；～ device 输出设备；～ primitive 输出原语(计)

outreach /autˈriːtʃ/ 伸出去，展开

outset /ˈautset/ 开头，开始

outside /autˌsaid/ 外部(的)，外面(的)，外界(的)；外表，外观，表面；在外；除去；达到极点的；～ diameter 外径；～ drawing (view) 外形图

outspread /autˈspred/ (使)伸开，(使)展开，伸开的，展开的

outstretch /autˈstretʃ/ 拉长，伸长，扩张

out-to-out 总(外廓)尺寸，总长(宽)度，全长(宽)

outward /ˈautwəd/ 外部的，外表的，外形的，表面的；明显的，可见的；外部，外表，外形；表面上，外表上；向外，在外；～ly 外表上，表面上

oval /ˈəuvəl/ 椭圆，椭圆形的，卵形的

ovality /əuˈvæliti/ 椭圆度(性)，卵形度

ovaloid /ˈəuvəlɔid/ 卵圆面

overall /ˈəuvərɔːl/ 全部的，全面的，总的；轮廓的，综合的；～ design 总体设计；～ dimensions (size) 总尺寸，外形尺寸，最大尺寸；全部尺寸，全长尺寸；～ height 总高度；～ length 总长，全长

overarm /ˈəuvəraːm/ 横梁

overdimensioned /ˈəuvəˌdaiˈmenʃənd/ 超尺寸的，全尺寸的

overgauge /ˈəuvəgeidʒ/ 超过规定尺寸的，等外的

overhasty /ˈəuvəˈheisti/ 太草率的，太急速的

overhaul /ˈəuvəhɔːl/ 仔细检查，大修，修理

overhead /ˈəuvəhed/ 架空的，在头以上的；(电脑的)辅助操作；～ projector 教学(架空)投影仪

overlap /əuvəˈlæp/ 搭接，重叠；巧合

overlay /ˈəuvəlei/ 涂，镀，覆盖；外罩

overlook /əuvəˈluk/ 俯视，过目；漏看，忽略，忽视

overmeasure /ˈəuvəˈmeʒə/ 留量

overpitch /ˈəuvəpitʃ/ 夸大，夸张

overrun /əuvəˈrʌn/ 超过，占据，出头；越程槽

oversee /ˈəuvəˈsiː/ 省略，错过；管理，监督

overshoot /ˈəuvəʃuːt/ 曲线的凸起；越过；射过头，过冲

oversize /ˈəuvəsaiz/ 过大的，加大尺寸的，

比普通尺寸大的，超过尺寸的，带有余量的尺寸；非标准尺寸

overt /'əuvə:t/ 外表的；明显的；展开的，公开的

overview /'əuvəvju:/ 观察，概述；溢出

oviform /'əuvifɔ:m/ (ovoid) 卵形的，卵形体

oxidation /ˌɔksi'deiʃən/ 氧化

oxygon(e) /'ɒksɪgən/ 锐角三角形

oxygon(i)al /'ɒksɪgənl/ 锐角的

ozalid /'ɔzəlid/ 氨熏；~ paper 氨熏晒图纸；~ print 氨熏晒图

P

P (pitch) 节距，螺距，齿距，间距，行距；步

PA (pressure angle) 压力角

pack /pæk/ 部分，组件；（一）束，（一）包；填塞

package /'pækidʒ/ 组件，插件，标准部件，包装，装箱；包；software ~ 软件包

pack-harden 全部淬火，整体淬火，碳化及皮硬

packing /'pækiŋ/ 填料，包装，密封；~ gland 填料压盖；~ piece 垫片；~ ring 垫圈

pad /pæd/ 衬套，垫圈，极低之凸台，垫板；sandpaper ~ 砂纸板

paint /peint/ 画，描绘，描写，叙述；油漆，涂料，颜料；red lead ~ 红铅油漆

painting /'peintiŋ/ 涂色，涂漆；色标，镀；油画；mural ~ 壁画；oil ~ 油画；wall ~ 壁画

paintwork /'peint,wɜːk/ 油画，油漆工

pair /pɛə/ 副，对偶；配对，线对；一对，一付；~ of compasses 圆规；homologous ~ 对应线对

pallet /'pælit/ 棘爪；托板；夹板

pan /pæn/ 盘，头盖；under ~ 底盘

pane /pein/ 窗格玻璃，(棋盘图案式的)长方格，长方块，(螺母等的)边，面；a 6-~d nut 六角螺母

panel /'pænl/ 配电盘，区段；distribution board 配电盘；lighting ~ 照明配电盘 (LP)；power ~ 动力配电盘

panning /'pæniŋ/ 漫游(计)

panoptic /pæ'nɔptik/ (用图)表示物体全貌的

pantograph /'pæntəgrɑːf/ 缩放仪，伸缩器，比例画图器，放大器；~ ratio 缩放比；electronic ~ 电子伸缩绘图器；skew ~ 斜角缩放仪

pantography /pæn'tɔgrəfi/ 缩放图法

pantometer /pæn'tɔmitə/ 经纬(万能)测角仪

paper /'peipə/ 纸；论文，文件；记录；纸的；~ fastener 图纸扣；~ location 纸上定线，图上定线；~ size 纸张大小；~ tape 纸带；abrasive ~ 砂纸；Alexandria ~ 亚历山大绘图纸(以埃及亚历山大港命名的一种中等绘图纸)；asbestos ~ 石棉纸；asphalt ~ 油纸；atlas ~ 印图纸，绘图纸；autopositive ~ 正图纸(晒图)；blue print ~ 兰晒纸，晒图纸；blue-lined ~ 兰格子的纸；bristol ~ 图案纸，上等板纸；calendered ~ 铜板纸；carbon ~ 复写纸，炭纸(俗名)；carborundum ~ （金刚）砂纸；carte ~ 地图纸；cartridge ~ 图画纸，厚纸；colored ~ 色纸；commercial graph ~ 方格纸；coordinate ~ 坐标纸，方格纸；copying ~ 复写纸，复印纸；correspondence ~ 高级书写纸；cover ~ 书面纸；craft ~ 牛皮纸；cross-section ~ 横剖面纸，方格纸，坐标纸；cyclic ~ 旋转抛物线面；demy ~ 小图纸；depth of ~ 抛物线深度；design ~ 图纸；detail ~ 底图纸，画详图用的描图纸；drafting ~ 绘图纸；drawing ~ 绘图纸；dull side of ~ 图纸无光泽面；egg-shell ~ 光厚绘图纸；emery ~ 金刚砂纸；enamel ~ 铜版纸，蜡图纸，印图纸；ferroprussiate ~ 蓝图纸，晒图纸；fine drawing ~ 精致绘图纸；fish ~ 青壳纸（商名），钢纸；flint (glass) ~ 粗砂纸；glazed printing ~ 道林纸；graph ~ 方格纸，毫米纸；grid ~ 方格纸，网格纸；gummed ~ 胶纸；helicgraphic ~ 晒图纸；India ~ 薄而坚韧的印刷纸，凸版（字典）纸；isometric ~ 正等轴测格纸；isometric sketching ~ 等角格子纸；kraft ~ 牛皮纸；line ~ 横格纸；manifold ~ 复印纸，打字纸；Manila ~ 马尼拉硬纸板；map ~ 地图纸；mining coordinate ~ 坐标方格网；multiple cycle ~ 复循环格纸；oil ~ 油纸；ozalid ~ 氨熏（正象）晒图纸；photo-sensitive ~ 感光纸；plain ~ 无格纸；plotting ~ 方格绘图纸，比例纸；print ~ 有格纸；printing ~ 印图纸，晒图纸；profile ~ 纵剖面（格）纸，断面纸；rag ~ 废纸；ruled ~ 方格纸，坐标纸；sand ~ （金钢）砂纸；scale ~ 比例方格纸；scetion ~ (cross section ~) 方格纸；semilog ~ 半对数坐标纸；sized ~ 涂胶纸；sketching ~ 草图纸；soft-sized ~ 吸水纸；square(d) ~ 方格纸；square-lined ~ 方格纸；tracing ~ 描图纸；transfer ~ 复制图纸；Vandyke ~ 范代克晒图纸；vellum ~ 牛皮纸；virgin ~ 白纸；Whatman ~ 瓦特曼纸，绘图纸；white manifold ~ 复印纸；‖*Example sentence:* This paper is well written except for a little miscalculation. 例句：这篇论文写得很好，只是有少量的误算。

paper-tape 纸带；~ coil 纸带卷；virgin coil 无孔纸带卷

papyrograph /pəˈpaiərəgra:f/ 复写器，复写板

papyrus /pəˈpaiərəs/ 纸莎草纸（古埃及人绘图的一种纸。纸莎草是一种禾状水生植物，盛产于尼罗河一带，其茎为木质，呈钝三角形，有的粗如手腕，高可达三四米，古埃及人用以造纸，叫作纸莎草纸，是古埃及主要的书写和绘画材料）

PAR (parallel) 平行的(线)；(para., paragraph) 段，节，短文，短评

parabola /pəˈræbələ/ 抛物线(型)；~ by intersection method 交线法绘抛物线；~ segment 抛物线弓形面；~ spandrel 抛物线半余面；~ template 抛物线模板；depth of ~ 抛物线深度；half-cubic ~ 半立方抛物线

parabolic(al) /pærəˈbɔlik(əl)/ 抛物线的；~ envelope 抛物线包络；~ spiral 抛物线蜗线

paraboloid /pəˈræbəlɔid/ 抛物面，抛线体，次抛物线；~ of revolution 回转抛物面；hyperbolic ~ 双曲抛物线体

paraboloidal /pəræbəˈlɔidl/ 抛物面的

para-curve 抛物线

parallel /ˈpærəlel/ 平行，平行线；并联，并行，等距，带；纬圆，纬度；滑板；~ line 平行线；‖*Example sentence: Two horizontal, two frontal, or two profile lines that appear to be parallel* in twos *may or may not be actually parallel in space.* 例句：两条水平线、两条正平线或两条侧平线，在两个主要视图

中都平行,但两线在空间可能平行也可能不平行。~ rule 绘图用平行移动尺;~ to each other 相互平行;~ with 和……平行;;H-~ 平行于 H 的面(水平面);P-~ 平行于 P 的面(侧平面);V-~ 平行于 V 的面(正平面)‖ *Example sentence*: AB is a line *parallel with* CD. 例句: AB是一根和CD平行的线段。

paralleled /ˈpærəleld/ 并行的,并联的

parallelepiped /ˌpærəleləˈpaiped/ (parallelepipedon /pærəˌleliˈpaipidən/) 平行六面体

paralleling /ˈpærəleliŋ/ 并联

parallelism /ˈpærəlelizm/ 平行度,类似,对应;相同,比较;平行性‖ *Example sentence: Parallelism of lines is a property that is preserved* in orthographic projections. 例句: 平行直线在正投影中具有保持平行的性质。

parallelogram /pærəˈleləgræm/ 平行四边形;~ of forces 力的平行四边形

parallelometer /ˌpærələˈlɔmitə/ 平行仪

parallel-serial 并—串联,复联

parameter /pəˈræmitə/ 参数;altitude ~ 高度参数

parapet /ˈpærəpit/ 女儿墙;~ wall 女儿墙

parenthesis /pəˈrenθisis/ 括弧,插句

par exemple /pa:r egˈzã:mpl/ (法语)举例,例如

parkerising /ˈpa:kəraiziŋ/ (coslettising) 磷化处理

part /pa:t/ 部分,组成部分,零件;分开,分割;~ list 零件目录;~ of...一部分;~s explosion 零件分解图;a ~ of...一小部分;adjacent ~ 邻接件;Army-Navy standard ~s (美国)陆海军用标准零件;assembled ~s 装妥零件;commercial ~s 商用零件(市售零件);component ~ 组合零件,部件;exchangeable ~ 可换零件;family of ~s 零件族;interchangeable ~ 可互换零件;machine ~ 机件;machined ~ 已切削部分;mating ~ 配合件;moving ~ 运动机件;non-standard ~ 非标准零件;quickly-worn-out ~ 易损件;solid ~ 实心件;spare ~ 备用件;standard ~ 标准件;survey and drawing of ~s 部件测绘;threaded ~ 螺纹紧固件;unmachined ~ 不(未)机械加工部分

part. (partial) 部分的,局部的;(participating) 参与

parti /pa:ˈti/ (scheme) 图解

partial /ˈpɑ:ʃəl/ 部分的,局部的,不完全的,单独的;偏的;~ assembly drawing 零件装配图,装配分图;~ automatic 半自动的;~ views 部分视图,局部视图

participating /pa:ˈtisipeitiŋ/ 参与

particular /pəˈtikjulə/ 特殊的,特别的

particulars /pəˈtikjuləs/ 详细资料,详细数据

partition /pa:ˈtiʃn/ 分隔,分类,分区

part-number 件号

party /ˈpa:ti/ 党,派;参与者;一组,一批;~ line 政党的路线,合用线,共同线,分界线

pass /pa:s/ 孔型,轧槽;经过,穿过,越过,使通过,作出;‖ *Example sentence: Pass a line AB parallel to CD and intersecting the skew lines EF and GH.* 例句: 作直线AB平行于CD并与交叉直线EF、GH相交。

pass. (passage) 通道,走廊

passameter /pæˈsa:mitə/ 外径指示规

passivation /ˌpæsiˈveiʃən/ 钝化

password /ˈpa:swə:d/ 口令

paste /peist/ 贴,粘起来;糊,浆糊;abrasive ~ 消字膏

pat. (pattern) 模型,样本,模范

path /pa:θ/ 轨迹线,路径,轨道,跑道;cam

PATT (pattern) 图案, 图型, 模型, 样本

pattern /ˈpætən/ 图, 图形, 图表, 方向图;（特性）曲线; 模式, 模型, 样式, 样品; 图像, 图案, 花样; 晶格, 结构; 制度, 规范; 仿造, 构图, 加花样; ～ draft 拨模斜度; ～ generation 模式发生（计）; ～ layout 图案设计; ～ primitive 图形元; ～ recognition 图形识别; ～ taper 拨模斜度; black and white ～ 黑白图形; camera-scanning ～ 照相机扫描图形; conductive ～ 导电图形; dash ～ 虚线图形; dot ～ 光点图形, 点模式; geometric ～ 几何图案; texture ～ 网纹图

pause /pɔːz/ 暂停

pawl /pɔːl/ 棘爪, 卡子

PC (percent) 百分比; (personal computer) 个人计算机; (piece) 零件; (pitch circle) 节圆, 基圆

P/C (parts catalog) 零（部、配、元）件目录

PCD (pitch circle diameter) 节圆直径

PCS (pcs, pieces) 零件

P. D. (pitch diameter)（齿轮）节圆直径,（螺纹）中径

PDES (product data exchange standard) 产品数据交换标准

pdo (production design outline) 产品设计草图

P. D. P. (profile datum plane) 侧基准面

PE (picture element) 画面元素; (pixel picture element) 象素

peak /piːk/ 最高点, 尖点; 山顶, 山峰, 波峰; 尖端, 减少, 缩小

peculiar /piˈkjuːljə/ 特殊的, 特有的, 独特的

peculiarity /piˌkjuːliˈærəti/ 特性, 特质, 奇特

pedal /ˈpedl/ 垂足的, 足部的,（脚）踏板, 脚的

pediment /ˈpedimənt/ 人字形的, 人字墙

peg /peg/ 楔

pelage /ˈpeilidʒ/ 砂纸, 砂皮, 毛皮

pellucid /pelˈjuːsid/ 透明的, 透彻的

pen /pen/ 笔, 写头; 充填带; 风眼; 墨水笔, 钢笔; ～ attachment 笔附件, 鸭嘴笔接脚; ～ choice 选笔; ～ for lettering 书写用钢笔; ～ holder 笔座; ～ legs 鸭嘴笔脚; ～ number 笔号; ～ setter 笔架; ～ speed 笔速; ～ straight line 直线笔; ～ stylus 电子感应笔, 笔头; ball-point ～ 圆珠笔; black ～ 黑色笔; blade of ～ 鸭嘴笔叶片; blade ruling ～ 鸭嘴笔; border ～（画轮廓用的）绘图笔,（专画边框粗线用的）边框笔, 图框笔; bow ～ 上墨弹簧圆规; contour ～ 等高线笔, 曲线笔, 轮廓笔; curve ～ 曲线笔; detail ～ 特号笔; dotted line ～ 虚线笔; dotting ～ 虚线笔; double line ～ 双线笔, 复线笔; drawing ～ 绘图笔, 划线笔, 鸭嘴笔; drop ～（绘）点笔, 点圆规, 骤落笔, 坠笔圆规; fountain ～ 自来水笔; fountain ruling ～ 自来水直线笔（绘图笔）; jackknife ruling ～ 扳开式鸭嘴笔; light ～ 光笔, 光写入头; light ～ attention 光笔中断; light ～ tracing 光笔跟踪; lining ～ 直线笔; needle ～ 绘图笔, 针笔; needle-in-tube ～ 针笔; plot ～ 绘图笔; railroad ～ 双线笔; railway ～ 双线笔; reed ～ 芦苇笔; rivet ～ 铆钉圆规（小圆规）; ruling ～ 直线笔, 鸭嘴笔 (pen straight line); sketch ～ 草图笔; sonic ～ 声笔（声音输入笔）; special lettering ～ 特制书法用笔; stencil ～ 铁笔（刻蜡纸用）; stylographic ～ 管形笔; Swede ～ 瑞典笔; swivel ～ 曲线笔, 转笔; tank ～ 储水笔, 槽头笔

pencil /ˈpensl/ 铅笔, 画笔; 用铅笔画,

用铅笔写；束；～ cloth 铅画布，(擦)铅笔画的绒布；～ compasses 有铅笔头的圆规；～ eraser 擦铅笔画的橡皮；～ing 铅笔绘图；～ line 铅笔线；～ of lines 直线束；～ of planes 面束；～ of circles 圆束；～ of conics 圆锥曲线束，二次曲线束；～ pointer 磨铅笔之物；～ point of compass 圆规用铅笔尖；～ removable 自动铅笔；～ sharpener 铅笔刀；～ tracing 加深用铅笔; bow ～ 铅笔弹簧圆规, 带铅笔的小圆规; colored ～ 颜色铅笔; draftsman's ～ 制图铅笔; drawing ～ 绘图铅笔; hard ～ 硬铅笔; line ～ 线束; mechanical ～ 活动铅笔; order of ～ing 铅笔绘图顺序; plain point of ～ 锥形铅笔尖; plot ～ 绘图铅笔; removable ～ 活动铅笔; soft ～ 软铅笔; using the ～ 铅笔的用法; wood-cased drawing ～ 普通铅笔；

pencil-case 铅笔盒

pencil-work 铅笔图

penetrate /ˈpenitreit/ 贯穿, 穿透, 渗透, 透过, 透视

penetration /peniˈtreiʃn/ 贯穿, 穿透, 渗透; 穿透深度

penetrator /ˈpenitreitə/ 贯穿者, 穿透物, 侵入者; (硬度试验)压头

penholder /ˈpenhəuldə/ (钢)笔杆

penstock /ˈpenstɔk/ 钢笔杆; 管道, 闸门

pentagon /ˈpentəgən/ 五角形, 五边形

pentagonal /penˈtægənl/ 五角的, 五角形的

pentagram /ˈpentəˌgræm/ (pentacle) 五角星形

penumbra /piˈnʌmbrə/ 画面明暗交界处; 黑影周围的半阴影; 半影(太阳黑子周围的较淡部分)

perambulator /ˈpræmbjuleitə/ 测距仪, 间距规; 手推车

percent /pəˈsent/ 百分之……, 按百计, 每百; 100 ～ bar 百分比条图

perceptible /pəˈseptibl/ 显而易见的, 觉察得出的

perception /pəˈsepʃn/ 感觉, 感受; 体会, 理解(力); stereo ～ 立体感(觉)

PERF (perforate 穿孔)

perfect /ˈpəːfikt/ 正确的, 精确的; 完美的, 精通的

perforate /ˈpəːfəreit/ 穿孔, 钻孔, 打孔

perforation /ˌpəːfəˈreiʃn/ 穿孔, 孔眼, 洞

perform /pəˈfɔːm/ 完成, 执行; 进行; 表演

performance /pəˈfɔːməns/ 特性, 性能; 表现, 表演; 实行, 完成; 成品; 动机, 操作, 特性曲线; ～ chart 工作特性图

perimeter /pəˈrimitə/ 周长, 圆周, 周边; 视野计

periphery /pəˈrifəri/ 圆柱表面, 圆周, 周线, 周围, 周界

permissible /pəˈmisəbl/ 容许的, 许可的

permission /pəˈmiʃn/ 允可, 许可

permit /pəˈmit/ 容许, 许可

permutation /pəːmjuˈteiʃn/ 变更, 交换, 取代, 排列

permute /pəˈmjuːt/ 变更, 交换; 置换

PERP (perpendicular) 垂直

perpendicular /ˌpəːpənˈdikjulə/ 垂直, 正交; 垂直的; 垂线, 垂面; ‖ *Example sentence: If a line is perpendicular to a plane, it is perpendicular to every line in the plane.* 例句: 如果直线垂直于平面, 它必垂直于平面内的每一条直线。～ alignment 垂直度; ～ bisector 中垂线; ～ lines 垂线; ～ planes 垂面; ～ to a given plane 与一已知平面垂直; after ～ 尾垂线(船); be out of (the) ～ 倾斜; forward ～ 艏垂线(船)

perpendicularity /ˌpəːpəndikjuˈlæriti/ (线)垂直度, 垂线, 直立, 垂直性

perspective /pəˈspektiv/ 透视, 远景; 前

程；透视的，透视图；~ collineation 透视变换；~ correspondence 透视对应；~ drawing 透视图；~ fore-shortening 透视收缩画；~ grid 透视格；~ plan 透视平面图；~ plan method 透视平面图法；~ mapping 透视变换；~ on curved surface 曲面透视；~ projection 透视投影；~ scale 透视比例尺；~ view 透视图，远景图；aerial ~ 空中透视；angular (two-point) ~ 角透视（两点透视）；architectural ~ 建筑透视；birds eye ~ 鸟瞰（透视）图；center of ~ 透视图中心；doubly ~ 二重透视；isometric ~ 等角透视；linear ~ 直线透视（图）；military ~ 军用透视；normal ~ 正透视；oblique (three-point) ~ 斜透视（三点透视）；one point ~ 一点透视；panoramic ~ 全景透视；parallel (one-point) ~ 平行透视（单点透视）；reversed ~ 倒象的透视图；secondary ~ 次透视；three point ~ 三点透视；two point (angular) ~ 两点透视（角透视）

perspectivity /pəːspekˈtiviti/ 透视性
perspicuity /pəːspiˈkjuːiti/ 明晰，清楚
pertusion /pəˈtjuːʃən/ 穿孔，孔，眼
PF (plain face) 光面；(profile) 侧面，侧面图，轮廓

phantom /ˈfæntəm/ 假想，虚似，幻象，影子，无形的；部分剖视图（的）；~ line 假想线（双点划线）；~ section 假（幻）想剖面（不取掉外形，而将剖面线画成虚线的剖面）；~ view 部分剖视图（经过透明壁的内视图）

phase /feiz/ 形象，形态；相位，阶段；approval ~ 检验阶段；development ~ 开发阶段；distribution ~ 发布阶段；revision ~ 修改阶段；storage ~ 存储阶段

ph brz (phos bro) (phosphor bronze) 含磷青铜

photo. (photograph) 照片，照相；照相的；光，光电，光敏；wire ~ 有线传真，有线传真照片，有线传真装置

photocartograph /ˌfəutəuˈkaːtəgraːf/ 摄影测图仪
photocartography /ˌfəutəuˌkaːˈtɔgrəfi/ 摄影制图
photo-charting 摄影制图
photoduplicate /fəutəuˈdjuːplikeit/ 照相复制，照相复制本
photoduplication /ˈfəutəuˌdjuːpliˈkeiʃən/ 照相复制法
photogram /ˈfəutəgræm/ 传真电报；黑影相片，物影照片
photogrammetry /fəutəˈgræmitri/ 摄影测量学，摄影制图法；aerial ~ 航空摄影测量制图（学）
photograph /ˈfəutəgraːf/ 照片，摄影，照相
photographs /ˈfəutəgraːfs/ 照相法
photolithography /ˈfəutəliˈθɔgrəfi/ 照相石印
photomapping /ˈfəutəumæpiŋ/ 摄影制图；aerial ~ 航空摄影制图
photoprint /ˈfəutəprint/ 影印（画），照相复制品
photosensitive /ˈfəutəˈsensitiv/ 光敏的，（能）感光的；~ paper 感光纸
photosensitize /fəutəˈsensitaiz/ 使具有感光性，使光敏
photostat /ˈfəutoustæt/ 直接影印机；直接影印制品；用直接影印机复制；影印法；~ print 照相印图
photostereograph /ˌfəutəˈstiəriəgraːf/ 立体测图仪
physical /ˈfizikəl/ 物理（学）的，实际的，物质的，有形的；~ construction 机械结构；~ design 结构设计

PI (point of intersection) 交点，交叉点
pick /pik/ 截齿，镐；读出，接收……信号

pictogram /ˈpiktəɡræm/ 象形图

pictograph /ˈpiktəɡrɑːf/ 象形文字；统计图表

pictorial /pikˈtɔːriəl/ 插画；绘（图）画的，有插图的，图示的，有插画的；画板；图景；形象，写生画；立体图，想象；～ display 图像显示；～ drawing 形象画；～ intersections 交线的立体图；～ projection 立体图；～ representation 写生画法；～ sketching 写生草图

pictorialization /pikˌtɔːriəlaiˈzeiʃən/ 用图画表示

pictorialize /pikˈtɔːriəlaiz/ 用图画表示

pictorially /pikˈtɔːriəli/ 用插图，图画

picture /ˈpiktʃə/ 画，图画，图像，图片，形象，概念电影；～ data bases 图像数据库；～ distortion 图形畸变；～ element 图（象元）素，象素；～ file 图形文件；～ gallery 画廊；～ plane 画面；～ processing 图像处理；～ receiver 电视接收机；～ recording 录像；～ regeneration 画面重新生成；～ signal 图像信号；～ size 图像尺寸；～ window 陈画窗；automatic ～ transmission 自动图像传输；automatic ～ transmission system 自动图像传输系统；black and white ～ 黑白图像；composite ～ signal 复合图形信号；deferring ～ changes 延迟图形变化；hard ～ "硬"图像；motion ～ 电影；moving ～ 影片；radio ～ 电视；soft ～ "软"图像；three dimensional ～ 立体图像

PID (piping and instruments diagram) 管路及仪表布置图

pie /pai/ 馅饼，馅饼状物；～ chart 圆瓣图（统计图），（或称100 percent circle）百分比图

piece /piːs/ 件，部分；connecting ～ 连接件；mating ～ 配合件；thickness ～ 厚薄规；transition ～ 变形接头

pierced /piəsd/ 穿入，贯穿

pig /piɡ/ 金属块，生铁；～ copper 铸铜；～ iron 铸铁，生铁；casting ～ 铸铁

pilage /ˈpilidʒ/ 砂布，砂纸

pin /pin/ * 销，针，栓，钉；conical ～ * 圆锥销；cotter ～ * 开口销；cuphead ～ 圆头销；cylindrical ～ * 圆柱销；dowel ～ 定位销，暗钉；drawing ～ 图钉；fixing ～ 固定销；general cylindrical ～ * 普通圆柱销；generaltaper ～ * 普通圆锥销；grooved ～ * 槽销；locating ～ 定位销；map ～ 图钉；round ～ 圆柱销；safety ～ * 安全销；split ～ * 开口销；stop ～ 止动销，定位销；straight ～ * 圆柱销；taper ～ * 圆锥销，斜销

pin-hole 小孔，塞孔，销孔

pinion /ˈpinjən/ 小齿轮

pink /piŋk/ 刺，扎，戳，穿小孔；彩红的，淡红的

pin(-) point /ˈpinpɔint/ 针尖，一点点；定点的，极精确的，准确地定位

pipe /paip/ 管，缩管；（钢表面的）皮；～ coil 盘管；～ fitting 管接头；～ flange 管子法兰盘；～ line 管系，管路；～ organ chart 管风琴图（象管风琴形的铅直式统计图表）；copper ～ 铜管；run ～ 管路，管线；down ～ 雨水管，水落管；four-way ～ 四通管；rain-water ～ 雨水管；Y-～ Y形三通管

pipework /ˈpaipwɜːk/ 管道系统，管道工程

piping /ˈpaipiŋ/ 管路，管道系统；缩孔；～ drawing 管系图

π plane π 平面

pit /pit/ 坑，槽，洞孔，凹窝，砂（缩）孔

pitch /pitʃ/ * 螺距，* 齿距，* 节距，齿节，节圆；顶尖，高峰，竖起，倾斜，投掷，屋面斜度，坡度，俯仰角；～ angle 节锥半角；～ circle 节圆；～ cone 节圆锥；～ cylinder 节圆柱；～ diameter 节圆直径，螺纹中径；～ line* 中径线，节线；～ of centers 顶尖高度，

轴心高度；～ of holes 孔间距；～ of roof 屋面斜度，屋顶开势；～ of screw 螺距；～ of stairs 楼梯斜势；～ of teeth 齿距；～ of thread 螺距；～ point 节点；～ surface 面；～ wheel 互相啮合的齿轮；throat diameter 节圆喉径；8-～ thread series 8螺距级（美国三个特级螺纹之一，每英寸8个螺纹）；12-～ thread series 12螺距级（美国三个特级螺纹之二，每英寸12个螺纹）；16-～ thread series 16螺距级（美国三个特级螺纹之三，每英寸16个螺纹）；axial ～ 轴向节距；circular ～ 圆周齿节，周节；circumferential ～ 圆周齿节，周节；coarse ～ 大（粗）螺距；deviation in ～ * 螺距偏差；diameter ～ 直径节距，径节；diametral ～ (diametric ～)（齿轮的）径节；linear ～ 直线节距；normal ～ 法向节距；screw ～ 螺距；tooth ～ 齿节，齿距；trans-verse circular ～ 端面弧线节距

pitchwheel /pɪtʃwiːl/ 相互啮合的齿轮
pivot /ˈpivət/ 旋转中心，支点；枢轴；中心点，参考点
pixel /ˈpiksəl/ 象素，象元，点
PL (pipe line) 管路；(plate) 板
place /pleis/ 地方，地点，地位，位置；放置，配置；～ of projection 投影位置；～ing the paper 安放（贴）图纸；to take the ～ of…… 代替……
placement /ˈpleismənt/ 位置，部位，方位；布置，放置
plain /plein/ 简单的，平凡的，普通的，明白的；平坦的；张开
plainness /ˈpleinnis/ 不平度
plan /plæn/ * 平面图，轮廓图，水平投影，示意图，图，图样，设计图；图表，计划，规划，设计，方案，概要；～ area 直线面，设计面积；～ of site 场地图，总布置图；～ of wiring 线路图，布线图；～ sketch 草图，设计图，简图，平面图；～ view 顶视图，俯视图，平面图；～ with contour line 等高线图；anchor arrangement ～ 锚泊设备布置图；anchor mooring and towing arrangement ～ 系泊布置图；approved ～ 承招图，核准图样；architectural ～ 建筑平面图；arrangement ～ 配置图；arrangement of life saving appliances 救生设备布置图；arrangement ～ of small hetchover and manhole cover 小仓盖及人孔盖布置图；awning ～ 天幕图；basement (floor) ～ 地下室平面图；block ～ 总（区域）平面图，区划图，配置图；body ～ 横断面，（船）正面图，横剖型线图；bottom plug arrangement ～ 船底塞布置图；building ～ 建筑平面图；bulkhead ～ 舱壁结构图；bulwark ～ 舷墙图；capacity ～ 船容图；central ～ 中心平面图；composite ～ 综合平面图；construction ～ 施工布置图，施工平面图，基本结构图；cruciform ～ 十字形平面图；deck ～ 甲板平面图；deck covering ～ 甲板敷料图；detail ～ 详细图；disposition ～ 配置平面图；draining arrangement ～ 排水工程布置图；draught-tube ～ 尾水管层平面图；electrical ～ 电路布置图；excavation engineering ～ 采掘工程平面图；excavation finish ～ 采掘竣工图；excavation ～ 采掘计划图；excavation progress ～ and form 采掘进度计划图；excavation system ～ 开拓系统图；excavation system ～ of opencut 露天开拓系统图；falsework ～ 脚手架图；first floor ～ 第二层平面图（英国），第一层平面图（美国）；floor ～ 楼平面布置图，地面布置图，地板平面图；flow ～ 流程图；foundation ～ 基础平面图，基座图；frame body ～ 肋骨型线图；framing ～ 结构平面图，布置图；船体结构图；general ～ * 总布置图，总平面图，总计划，计划概要；general arrangement

～ 总平面布置图; general location ～ 区域总平面图; general surface ～ 地面总平面图; ground ～ 底面图,下层平面图; ground floor ～ 底层平面图(英国),首层平面图; half breadth ～ 半宽水线图; hold ～ 船底图; horizontal ～ 平面图,水平投影; horizontal excavation ～ 水平开拓掘进计划图; hull block division ～ 分段划分图; hydroelectric machine's floor ～ 水力发电机层平面图; hydro-junction's ～ 水利枢纽布置图; inverted ～ 仰视图; irrigation project ～ 灌区规划图,灌区平面布置图; key ～ 索引图; lines ～ *型线图; location ～ 位置平面图; longitudinal ～ 前视图,纵剖面图,垂直投影; lot ～ 地段图,地区图; main horizontal roadway ～ 水平主要巷道平面图; master ～ 总平面图,总布置图,总图,总体规划; masonry ～ 石工图,泥水工程图; midsection complex geological ～ 中段复合地质平面图; midship section ～ 中横剖面图; mine arrangement ～ 工业场地平面图; minesite ～ 井上下对照图; mine yard ～ 工业场地平面图; moulding bed ～ 胎架图; opencut division ～ 露天矿分层平面图; opencut pre-production ～ 露天矿基建终了平面图; ore piece excavation ～ 矿块采掘计划图; perspective ～ 透视图; piling ～ 桩位布置图; piling fundation ～ 桩位图; piping ～ 管路图; plot ～ 总(区域)平面图; profile ～ 侧面图; project ～ of mine area 矿区规划图; raised ～ 正面图,投影图; regional ～ 区域图; reinforced concrete single story factory framing ～ 钢筋混凝土单层厂房结构布置图; reinforced concrete structure framing ～ for high-rise and multistory building 钢筋混凝土高、多层结构布置图; road ～ 道路平面图; roof ～ 屋顶平面图; rough ～ 设计草图,初步计划; safety pillars ～ 保安矿柱图; sectional ～ 部分平面图; second floor ～ 第二层平面图(美国); shaft bottom ～ 井底车场平面图; sheer ～ 侧视图,侧面图(船壳的),纵剖型线图; shell expansion ～ 外板展开图; site ～ 总平面图; sketch ～ 草图,示意图; star-shaped ～ 星形布置图(建筑); stem post ～ 首柱图; stern post ～ 尾柱图; structure ～ 结构图; structure ～ of section 分段结构图; structure ～ of block 总分段结构图; structure ～ of stem section 首段结构图; structure ～ of stern section 尾段结构图; theoretical lines ～ 理论线图; trunk bulk-head ～ of super structure and deck house 上层建筑及甲板室围壁图; turbines floor ～ 水轮机层平面图; utilities trench ～ 地沟平面图; valley project ～ 流域规划图; working ～ 工作程序图,施工图; year-end integrate ～ of opencut 露天矿年末综合平面图; zinc plate arrangement ～ 锌板布置图

planar /ˈpleɪnə/ 平面的,一个平面上的
plane /pleɪn/ 平面,面;平的,水平;飞机;投影; ‖ *Example sentence: A plane is a surface such that a straight line connecting any two points in that surface lies wholly within the surface.* 例句: 平面是指这样一个面,该面上任意两点连成的直线都重合在该面内。～ curve 平面曲线; ～ figure 平面图; ～ of contact system 接触网平面图; ～ of contact system design 接触网平面设计图; ～ of coordinates 坐标面; ～ of flexure 挠曲面; ～ of homology 透射平面; ～ of projection 投射面,投影面,射影面; ～ of reference 参考面,基准(平)面; ～ of rotation 转动面; ～ of rupture 折断面; ～ of symmetry 对称面; ～ of tangency 切平面; ～ of the paper 图纸平面; ～ of vision 视平面; ～ on general position 一般位置平面; ～ perpendicular

to the vertical ～ 正垂面；～ position indicator (P. P. I) 平面示位图；～ section 剖面；～ surface 平面；～ tangenies 切平面；affine ～ 仿射平面；auxiliary ～ 辅助平面，辅视平面；auxiliary projection ～ 辅助投影面；auxiliary sectional ～ 副断面，辅助平面；axial ～ of symmetry 对称轴面；axonometric ～ 轴测投影面；basic ～ 基础平面；basic projection ～ 基本投影面；bit ～ 位平面；cardinal ～ 基面；central ～ 中心面；centre ～ 中线面（船体）；collinear ～s 共线面；coordinate ～ 坐标面；critical ～ 临界面；cut ～ 剖面；cutting ～ 切割平面，剖切（平）面，截（平）面；datum ～ 基准面；double inclined ～ 双斜面；directing ～ 准平面；direction of ～ 平面方向；frontal ～ 正面，正立面；正平面（平行于正立投影面的面）；frontal projection ～ 正立投影面；frontal receding ～ 主视图为一直线之平面；general ～ 任意平面；ground ～ 地平面，水平投影；horizon ～ 水平面；horizontal ～ 水平面；horizontal projection ～ 水平投影面；ideal ～ 理想平面；inclined ～ 斜面；infinite ～ 无限平面；isometric ～ 等角面；limiting ～ 极限平面；longitudinal vertical ～ 纵直平面；meridian ～ 子午面；midstation ～ 中站面；non-trace ～ 非迹线平面；normal ～ 法向面；oblique ～ 斜平面，一般位置平面；offset (cutting) ～ 转折切割面；parallel ～s 平行的平面；perpendicular ～ 垂直面；perpendicular ～ of the projection ～ 投影面垂直面；perspective ～ 透视面；picture ～ 画面；points and lines in the ～ 平面内的点和直线；‖ *Example sentence: If a line is known to be in a plane, then any point on that line is in the plane.* 例句：如果直线在平面内，直线上的点也都在平面内。principal coordinate ～s 主坐标平面；profile ～ 侧面，侧平面（平行于侧立投影面的面）；profile ～ of projection 侧投影面；profile projection ～ 侧投影面；projecting ～ 投射面；projection ～ 投影面；reference ～ 参考面，基准面；secant ～ 切断平面；secondary ～ of projection 辅助投影面；section ～ 剖切面，截平面；special positional ～ 特殊位置平面；starting ～ 起画面（画轴测投影时，最初画的物体上的面）；symmetrical ～ 对称平面；tangent ～ 切平面；tapered ～ 斜平面；transverse vertical ～ 横向坚直面；vertical ～ 竖直面；view ～ 投影平面；viewing ～ 观察平面；

planeness /ˈpleinis/ 平面度

planform /ˈplænfɔːm/ 平面的，平面图，平面形状

planimegraph /ˌpleiniˈmegrəf/ 比例规，缩图器

planimetry /plæˈnimitri/ 平面几何，测面（积）学，测面法

planish /ˈplæniʃ/ 抛光，磨光；辗平

planometer /pləˈnɔmitə/ 平面规，测平器

plaster /ˈplɑːstə/ 灰，泥；～ floor 抹灰地板；granitic ～ 水刷石

plastic /ˈplæstik/ 塑料，塑胶

plat /plæt/ 区域图；～ map 区域图；～ of a survey 测区图；～s of mineral claims 矿区图；city ～ 市区图；industrial ～s 工厂配置图

plate /pleit/ 板，盘；中厚板，薄板；钢板；电镀；bent- ～ 弯板；bottom ～ 底板；buckle ～ 凹凸板（压有凹槽增加抗弯强度）；curved ～ 曲线板；flat ～ 平板；gusset ～ 角撑板，角牵板；heavy ～ 厚板；surface ～ 平板，面板

plating /ˈpleitiŋ/ 镀涂，镀涂层；brass ～ 镀黄铜；chromium ～ 镀铬；copper ～ 镀铜；hard chromium ～ 镀硬铬；nickel ～ 镀镍；

tin ~ 镀锡; zine ~ 镀锌

play /plei/ 余裕, 宽余; 活动; 间隙; 玩; 作用

pliability /plaiə'biliti/ 可弯性

plot /plɔt/ 曲线, 图表, 图, 图样, 地图; 绘图板, 测绘板; 设计, 绘图, 画曲线; 作图, 作图表示 (方程式), 于图上定位置, 测定 (点、线) 的位置; 标绘, 标绘图, 区分, 划分, 策划, 计划; ~ ...against...画出……对……的曲线; ~ pen 绘图笔; building ~ 房屋地区图; check ~ 校验图(计); colour jet ink ~ 彩色喷墨绘图机; graphical ~ 图表, 图像; ground ~ 地基图; layout ~ 设计图的绘制; polar ~ 极坐标图; test ~ 初步方案图

plot-observer 测绘员

plot(o)mat /plɔt(əʊ)mæt/ 自动绘图机

plotter /'plɔtə/ 绘图仪, 标绘器, 绘迹器, 绘图机, 笔绘仪; 标绘员, 标板; 计划者; ~ close 关绘图机; ~ open 开绘图机; ~ step size 绘图机步距; aircraft ~ 飞机绘图器, 航行绘图尺; data ~ 数据绘图仪; digital ~ 数字绘图机; dot matrix ~ 点阵绘图仪; drum ~ 滚筒式绘图机; electrical ~ 静电绘图仪; electrostatic ~ 静电绘图机; field ~ 绘图板; flat bed ~ 平台式绘图机, 平板绘图仪; graph ~ 制(绘)图仪; incremental ~ 增量绘图机; ink jet ~ 墨喷绘图机; pen ~ 笔式绘图机; personal ~ 个人绘图仪, 小型绘图仪; photo ~ 光绘图机; raster ~ 光栅绘图机

plotting /'plɔtiŋ/ 点绘(绘点连线), 点绘曲线, 绘曲线, 绘图, 测绘; 标图, 标定; 标航路; ~ angle 测绘角度; ~ a curve 画一条曲线; ~ board 曲线板, 绘图板, 图形输出板; ~ curve 测绘曲线; ~ head 绘图头; ~ method 画法, 测绘方法; ~ ofice 绘图室; ~ paper 方格绘图纸, 比例纸; ~ scale (绘图)比例尺; ~ tablet 绘图板, 图形输入板; automation ~ 自动绘图; magnetic tape ~ system 磁带绘图系统; map ~ 地图绘制

plug /plʌg/ 插头, 塞; ~-in 插入

plumb /plʌm/ 铅锤, 线铊, 垂线, 垂直的; ~ line 铅垂线

plumbline /'plʌmlain/ 铅垂线, 用铅垂线检查……的垂直度

plummer-block /'plʌməblɔk/ 止推轴承

plummet /'plʌmit/ 测锤, 垂线

plus /plʌs/ 加号, "+" 号, 正号, 正数, 零上, 正的, 加

plywood /'plaiwud/ 胶合板

P-member (participating member) 参加者成员(国)

P/N (part number) 零件号, 部件号

PO (plotting office) 绘图室; (project office) 设计办公室

pofo (point of origin) 原点

point /'pɔint/ 点; 刀刃; 端, 末端, 尖端; ~ of contact 切点; ~ of contraflexure 反弯点; ~ of convergence 集点; ~ of crossing 交叉点; ~ of curve 曲线起点; ~ of division 分点; ~ of infinity 非固有点, 无穷远点; ~ of inflexion 拐点, 弯曲点; ~ of intersection of tangents 切线交点; ~ of piercing 穿过点; ~ of sight 视点; ~ of support 支点; ~ of tangency 切点; ~ of weld 焊接点; absolute zero ~ 绝对零点 (开始计数的基点); base ~ 基点, 原点; brilliant ~ 亮点; cardinal ~ 基点; center ~ 中心点; chisel ~ 楔形尖 (铅笔); collective ~ of traces 迹线集合点; common ~ 公共点; cone ~ 锥端; conjugate vanishing ~s 共轭灭点; connecting ~ 连接点; consecutive ~s 连续点; control ~ 控制点, 基准点; critical ~ 转向点; culminating ~ 最高点, 顶点, 转折点; curve ~ 曲线起点; cutting ~ 刀口;

datum ~ 基点; dividing ~ 分界点; driving ~ 动点; end ~ 终点; external ~ 尖端(螺纹攻的); fixed ~ 定点,固定点; flat ~ 平端; focal ~ 焦点; initial ~ 始点; intermediate ~ 中间点; internal ~ 定心孔(螺纹攻的); intersection ~ 交点; jig ~ 基点; locating ~ 定位点; measuring ~ 量点; mid ~ 中点; moving ~ 动点; needle ~ 针脚,针尖; nodal ~ 节点,交点,会聚点; north ~ 指北针; observation ~ 视点,观 测点; opposite ~ 对点; piercing ~ (贯)穿点; pitch ~ 节点; projection coincided ~ 重影点; reference ~ 参考点,基准点; spacer ~ 针头; special position ~ 特殊点; starting ~ 起点; station ~ 站点,驻点(观察点); tangent ~ 切点; tool ~ 刀锋; tracing ~ 描绘点; trammel ~ 大圆规针尖; transition ~ 转变点,过渡点; vanishing ~ 灭点; view ~ 视点; visible ~ 可见点; working ~ 工作点,施力点; zero ~ 零点,原点

point-blank 在一条直线上,直截了当(的)
pointer /'pɔintə/ 指针,指示器; character ~ 字符指针,当前指针; hand lead ~ 笔附磨尖器,卷笔刀
pointing /'pɔintiŋ/ 勾缝
point-view 直线的端视图
polar /'pəulə/ 极,极线,极面; 南极的,北极的; 极性的; ~ chart 极坐标图; ~ coordinate 极坐标; ~ curve 配极曲线; ~ point 极点
polarity /pəu'lærəti/ 配极对应; absolute ~ 绝对配极
pole /'pəul/ 杆,柱; 极点,顶点
polish /'pɔliʃ/ 润饰; 光泽,擦亮,抛光,打光; 光泽表面
polycylinder /ˌpɔli'silində/ 多圆柱体,多柱面
polygon /'pɔligən/ 多边形,多角形,封闭折线; inscribed ~ 内接多边形; regular ~ 正多边形
polyhedra /ˌpɔli'hedrəl/ 多面体
polyhedral /ˌpɔli'hedrəl/ (polyhedrical) 多面体的
polyhedron /ˌpɔli'hedrən/ 多面体
polyline /'pɔlilain/ 折线
polymarker /'pɔlimɑːkər/ 多点标记
polymeter /'pɔlimitə/ 多测计
pommel /'pʌml/ 圆头,球端
Poncelet 庞司勒(公元1788—1867)法国著名的几何学家,射影几何的创始人,有"射影几何学的开山祖师"之称
pony /'pəuni/ ~-size 小型的,小尺寸的
population /ˌpɔpju'leiʃn/ 总体,全体; 人口,人数
popup /'pɔpʌp/ 弹出
porous /'pɔːrəs/ 多孔的
porrect /pə'rekt/ 伸出,延长; 伸出的,平伸的
port /pɔːt/ 管口,港口; 终端站; 孔,入口
portion /'pɔːʃn/ 部分,段,份; 分配,分布; circular ~ 圆形部分; full divided ~ 全刻度部分; visible ~ 可见部分
portray /pɔː'trei/ 画,描绘,描写; 扮演
pos. (positive) 正的
posit. (position) 位置,情况; (positive) 正的
position /pə'ziʃn/ 位置,地位,状态,情况; 安置,定位; ~ and size dimensions 位置与尺寸大小; ~al distance 位置距离; ~ altolerance 位置公差; ~ing 配置; arrange ~ 安放位置; functioning ~ 习用位置; general ~ 一般位置; machining ~ 加工位置; mid~ 中间位置; relative ~ 相对位置; secondary ~ of side view 侧视图之第二位置; service ~ 工作位置; special ~ 特殊位置; true ~ 位置度; working ~ 工作位置
positive /'pɔzitiv/ 正的,阳性的,确定的,

实的；正象的，正常，正面，阳极，正片；～ print（白底的）蓝图；～ sign 正号

postiche /pɔsˈtiːʃ/ （法语）多余的（添加物），伪造的，假的

postprocessor /ˈpəust,prəsesə/ 后处理程序

potentiate /pəˈtenʃieit/ 加强，使更有效力

power /ˈpauə/ 动力，电源

P. P (power panel) 动力配电盘

P-parallel 平行于 P 面的面（侧平面）

PPI (plane position indicator) 平面示位图

practice /ˈpræktis/ 画法，做法，作业，练习，实习，实践，实验，实际操作，习惯；～ problem 习题；accepted ～ 常例，习惯做法；poor ～ 拙劣的做法；regular ～ 常规做法；‖ *Example sentence:* The practice of projection drawing of points. *例句:* 点的投影图画法。

prc (part requirement card) 零件规格卡片

precede /priˈsiːd/ 优先，优于；领先，先于；占先；高于，在……之前；位于……之前；M is ～d by N. M 以前是 N, N 在 M 之前

precedence /priˈsiːdəns/ （同 precedency）领先，优先，占先，优先权，优先性；～ of line 线的占先（几类线型重合时画那一个）

precise /priˈsais/ 精的，精密的；明确的，正确的，准确的；～ measurements 精确的尺寸（量度）

precisely /priˈsaisli/ 确切地，正确地，明确地

precision /priˈsiʒən/ 精度，精密度；～ measure 精确度量；manufacturing ～ 制造精度

preface /ˈprefis/ 前言，序，引言，开始

prefer /priˈfəː/ 宁可，宁愿，更喜欢；推荐，介绍；提升，提拔

preferable /ˈprefərəbl/ 更好的，更可取的；it is ～ to use……最好使用

preferably /ˈprefərəbli/ 引出，呈出；更可取地

preferred /priˈfəːd/ 择优的，优先的；～ plan 最佳规划，最佳方案；M is ～d by N. 最好是用 N。

prefix /ˈpriːfiks/ 字首

prep /prep/ 预备功课，家庭作业；预习，预备的

preparation /prepəˈreiʃn/ 准备；～ for drawing 制图之准备

prepare /priˈpɛə/ 准备；配备；作业，制订

preprocessor /priːˈprəusesə/ 预处理器，前处理程序

prescribe /prisˈkraib/ 规定，指示，叙述

prescript /ˈpriːskript/ 规定的，指示的，命令的

present /ˈprezənt/ 呈现，表现；现在的，目前的；介绍，提出；～ section 本节

presentation /prezənˈteiʃn/ 显示，展示，呈现，表现，表示；图像，描述；介绍，提出；赠送，礼物；～ drawing 表意图；dismounted ～ 拆卸画法；isometric graphical ～ 等角图法；oblique graphical ～ 斜视法；uncovered ～ 掀开表示法

preserved /priˈzəːvd/ 保持，保存，保护，收藏

press /pres/ 出版社；印刷，印刷机；印刷物；冲压；～ fit 压配合，压入配合

pressure /ˈpreʃə/ 压力；～ angle 压力角；working ～ 工作压力

previous /ˈpriːvjəs/ 在前的，上述的，初步的

prick /prik/ 扎，穿，刺，用小点或小记号标出，刺孔；～ punch mark 定中心点

pricker /ˈprikə/ 刺孔针，冲子

primarily /ˈpraimərili/ 第一，主要是，首先是；原来，本来

primary /ˈpraiməri/ 初次的，原始的

prime /praim/ 最初的，原始的；（字码右上角的）撇号（′），带撇号的字母如 A′、a′；

加注,加小撇;最高,第一部分

primitive /ˈprimitiv/ 原素,原语基元,图元;最初的,原始的

prin. (principal) 主要的,重要的;(principle) 原理,原则

principle /ˈprinsəpl/ 原理,原则,法则;～ of operation 工作原理;～ of work 工作原理;basic ～ 基本原理;contour ～ 外形原则;correlation ～ 相关原则;envelope ～ 包容原则;independent ～ 独立原则;manafacturing position ～ 加工位置原则;maximum material ～ 最大实体原则;working ～ 工作原理;working position ～ 工作位置原则

print /ˈprint/ 印刷,晒图,晒印,复制,出版,印行;打印;出版物,印刷品;照片,图片;assembly ～ 装配图,组合图;black-line ～ 黑线图;blue ～ 蓝图,蓝晒;blue ～ apparatus 晒蓝图器;blue ～ing 晒蓝图;blue ～ing machine 晒图机;blue ～-paper 晒图纸;detail ～ 零件蓝图;figure ～ing 正楷书写;brown ～ 棕色图;offset ～ing(橡皮)印刷,胶印;ozalid ～ 氨熏晒图;photographic ～ 影印;photostat ～ 照相印图;positive ～ (白底的)兰图;white ～ 白色图,印正象

printer /ˈprintə/ 印相机,晒图机,印刷机,复印机,打印机;band ～ 带式打印机;colour ～ 彩色印片机;dot ～ 点式打印机;dot matril ～ 点阵打印机;graphic matrix ～ 图形点阵打印机;laser ～ 激光打印机;line ～ 行(式打)印机;standard line ～ 标准行式打印机;xerographic ～ 静电复印机

prism /ˈprizəm/ 棱柱,棱形;oblique ～ 斜角柱;oblique pentagonal ～ 斜五角柱;oblique rectangular ～ 斜矩形柱;pentagon ～ 五角柱;regular ～ 正棱柱;triangular ～ 三棱柱,三角柱;truncated right rectangular ～ 截头直四棱柱

prismatic(al) /prizˈmætik(əl)/ 棱柱(形)的

prismoid /ˈprizmɔid/ 平截头棱锥体,棱柱体

pro (拉丁语)为了,按照;正面;～ rata 按比例,成正比例的

prob. (problem) 问题,习题

problem /ˈprɔbləm/ 问题,难题,疑问;习题,几何作图题;self-testing ～ 自我测试题

proc. (proceedings) 会刊,学报;(process) 工序过程

procedure /prəˈsiːdʒəˈ/ 步骤,程序;工序,方法,过程

proceed /prəˈsiːd/ 进行;继续进行;开始,着手,作出

process /ˈprəuses/ 过程,程序,工序,工艺规程;方法,处理,加工;工艺过程,操作方法;凸起;～ control 程序控制;continued bisection ～ 连续二等分法;data ～ing 数据处理;diazo ～ 影印法;graphical ～ 绘图程序;interactive image ～ing 交互图像处理;manufacturing ～ 制造程序;predefined ～ 既定处理

processor /ˈprəusesə/ 处理机;micro ～ 微处理机

production /prəˈdʌkʃn/ 生产,制造;～ dimension 制造尺寸

profile /ˈprəufail/ 侧,侧面;侧面图,断面图,外形,轮廓,型面;截面,断面;纵断面,纵截面;分布图;画侧画图,画纵剖面图,画轮廓;齿形轮廓,外廓线(船舶),叶型;靠模,仿形;～ angle 齿形角;～ datum plane 侧基准面;～ inelevation 注有标高的纵断面图;～ inplane 水平剖面图;～ line 侧线(侧平面内之线),侧平线;～ meter (表面光洁度)轮廓仪;～ of any line 线轮廓度;～ of any plane 面轮廓度;～ of any surface 面轮廓度;～ of geological prospecting 地质剖面图;～ of hydrogeological prospecting 水文地质剖面图;～ paper 纵剖面图格纸;～ plane 侧平面;～ plate 仿形样

板；~ projection 侧投影；~ trace 侧面迹线；active ~ 有效齿廓；basic ~ * 基本牙型；constructed ~ 示意剖面图；cross ~ 横剖面；cross-sectional ~ 横剖面（图），横断面（图）design ~ * 设计牙型；flight ~ 飞行剖面图，航迹；geological ~ of exploratory line 勘探线地质剖面图；horizontal ~ 水平剖面，水平断面；inboard ~ 内部剖面图；involute ~ 渐开线齿形；longitudinal ~ 纵剖面（图）；master ~ 标准剖面图；thread ~ 螺纹牙型；tooth ~ 齿廓，齿形；transversal ~ 横剖面，横断面

profiling /ˈprəufailiŋ/ 轮廓法，绘轮廓图

profilograph /prəuˈfiləɡrɑːf/ (profilometer) 表面光度计，轮廓曲线仪，纵断面测绘器（测定平整度用），测面计（量光洁度），测平仪

program(me) /ˈprəuɡræm/ 程序，程式，程序表，图表；计划，方案，提纲，大纲，说明书，设计程序；~ed control 程序控制；~ structure 程序结构；~ unit 程序单元；application ~ 应用程序；assembly ~ 装配程序，安装程序，汇编程序；computer ~ 计算机程序；main ~ 主程序；master ~ 主程序；object ~ 目的程序；source ~ 源程序

program-controlled 程序控制

program(m)ing /ˈprəuɡræmiŋ/ 程序设计，编制程序；~ language 程序设计语言；~ procedure 程序设计步骤；automatic ~ 自动编程；machine code ~ 手编程序；manual ~ 手工编程；the ~ of drafting 绘图程序的编制

progressive /prəˈɡresiv/ 分级；发展的

project /ˈprɔdʒekt/ 投射，投影，射影，放映，把……投影在……上，作……的投影图，画出，设计，计划，设想，凸出，草图，方案；~ engineer 设计工程师 ‖ *Example sentence: sentence: Project a vertical line from point M, upward to N. 例句: 从 M 点向上画一条垂线到 N。*

projecting /prəˈdʒektiŋ/ 投影，射影；投影的；射影的；设计；~ line 投影线，投影线；~ plane* 投影面，投射面；horizontal ~ plane 水平投影面；profile ~ plane 侧面投射面

projection /prəˈdʒekʃn/ * 投影，投射，发射，射影，射出，放映；投影法；投影图，抛射物，凸出，凸出部；~ drawing 投影制图，投影图；~ lantern 幻灯，映画器；~ line 投影线；~ of boundary 边界线的投影；~ of contour element 轮廓线的投影；~ of curved surface 曲面的投影；~ of line 线的投影；~ of point 点的投影；~ on a plan 平面上的投影；~ plane 投影面；~ receiver 投影式电视接收机；~ screen 银幕，投影屏；~ transformation 投影变换；~ weld 突焊；3rd (third) angle ~ 第三角（象限）投影；American ~ 美国制投影（第三角投影）；arbitrary ~ 任意投影；authalic ~ 等积投影；auxiliary ~ 辅助投影；axonometric ~ *轴测投影，三向图，不等角投影；bounded ~ 有界投影；cabinet ~ 斜二等轴测投影，半斜投影，减半投影；cavalier ~ 等斜轴测投影，斜投影，卡华里尔投影；center of ~ 投影中心；central ~ 中心投影，中心射影，几何投影法；characteristic of ~ 投影特性；clinographic ~ 斜射影法；coincidence of ~s 重影性；computer general pictorial ~s 计算机立体投影；conformal ~ 保形射影，正形射影，保角射影，保角投影；保形射影法；conical ~ 圆锥形投影；corresponding ~ 对应投影；cylindrical ~ 圆柱形投影；dimetric ~ 正二等轴测投影（正二测）；dual central ~ 双心投影；English ~ 第一角投影，第一角法；English system of ~ 第一角投影，第一角

法; equidistant ～ 等距离投影; European ～ 第一角投影; Europe system of ～ 第一角法; first-quadrant ～ 第一角象限投影; frontal ～ 正面投影; frontal oblique axonometric ～ 正面斜轴测投影; frontal plane of ～ 正面投影面; geometric ～ 几何投影法; globular ～ 球状投影; gnomonic(al) ～ 心射切面投影; homolographic ～ 相应投影; horizontal ～ 水平投影, 标高投影; horizontal oblique axonometric ～ 水平斜轴测投影; horizon-tal plane of ～ 水平投影面; inclined ～ 斜投影; indexed ～ * 标高投影; in ～ with X 与 X 在同一投影; isometric (al) ～ 等角投影, 等距射影, 正等轴测投影; law of ～ 投影规律; map ～ 地图投影; method of central ～ 中心射影法, 中心投影法; method of orthographic ～ 正射法, 正投影法; multiple-plane ～ 多面投影; oblique ～ * 斜投影, 斜射影; oblique axonometric ～ 斜轴测投影; oblique dimetric ～ 斜二等轴测投影（斜二测）; one-plane ～ 单面投影; ore body vertical ～ 矿体纵影图; orthogonal ～ * 正投影, 正射投影, 正交射影, 直角投影; orthographic ～ 正投影, 正射投影, 正交投影; parallel ～ 平行投影; perspective ～ * 透视投影, 透视射影; pictorial ～ 立体图; polar ～ 极投影; polyconic ～ 多圆锥投影; polyhedral ～ 多面投影; primary auxiliary ～ 一次变换投影面, 初次辅助投影; pro-file ～ 侧面投影; profile plane of ～ 侧面投影面; rectangular axonometric ～ 正轴测投影; reflective ～ * 镜像投影; resemble ～ 相似投影图; single-plane ～ 单面投影; spherical ～ 球投影; stereographic ～ 体视投影, 球极平面投影, 赤平极射投影; successive auxiliary ～ 连续辅助投影, 二次变换投影面; theorem ～ 理论投影; third-angle ～ 第三角投影; third-quadrant ～ 第三角象限投影; three dimensional ～ 三维投影; trimetric ～ 三度投影; true ～ 真实投影; upright ～ 垂直投影, 垂直剖面图, 侧视图; vector ～ 矢量投影图; vertical ～ 垂直投影; worm's-eye ～ 蛙式投影; zenith ～ 方位投影

projectionist /prəˈdʒekʃənist/ 投影图绘制者, 地图绘制者, 电影放映员, 电视播放员

projections /prəˈdʒekʃənz/ 投影线, 投射线, 射影线

projection-type 投影式的

projective /prəˈdʒektiv/ 射影, 射影的, 投影的, 投射的, 发射的; ～ correspondence 射影对应; ～ equivalence 射影等价; ～ geometry 射影几何（学）, 投影几何（学）; ～ group 射影群; ～ plane 射影平面; ～ transformation 射影变换

projectivity /prədʒekˈtiviti/ 射影, 直射, 射影变换, 射影对应（性）; anti～ 反射影变换; direct ～ 正射影变换; elliptic ～ 椭性射影变换

projector /prəˈdʒektə/ 投影机, 投影仪, 放映机, 发射器, 映画器, 幻灯机（制图的）投射线, 投影线; ～ line 投射线; ～ contour ～ 轮廓投影仪; over-head ～ 架空（教学）投影仪

projecture /prəˈdʒektʃə/ 投影, 凸出

prolate /ˈprəuleit/ 伸长的, 扁长的, 长球状的, 扩大的; ～ ellipsoid 长椭球, 长球状的, 长椭圆面; ～ spheroid 扁长球体

prolong /prəˈlɔŋ/ 延长, 拉长, 引伸

prolongation /prəuləŋˈgeiʃn/ 拓展, 拉长, 延长（部分）

prompt /prɔmpt/ 提示符; 提示

proof /pru:f/ 证明, 证据, 试验, 试验对的, 合乎标准; be above ～ 合乎标准; below ～ 不合标准; under ～ 不合标准的, 不合格的

proof-read 校对, 校读; ～er 校对员

prop (propeller, propellor) 螺旋浆, 推进器; 推进者

property /ˈprɔpəti/ 性质, 性能, 特性; 特征, 参数, 所有权; concentration ～ 积聚性

proport. (proportional 或 proportionnel) 成比例

proportion /prəˈpɔːʃn/ 比, 比例, 比率; 一部分; 相等, 使适宜; 匀称, 相称; 配合, 大小, 尺寸, 容积, 面积; ～ M to N 按 N 来定 M; definite ～ 定比; drawing ～ 制图用尺寸比例; easy drawing ～ 习用比例; the ～ of four to one 四与一之比; in ～ as (in ～ to) 按……的比例, 越……越; inverse ～ 反比; odd ～ 罕用比例

proportional /prəˈpɔːʃənl/ 比例的, 相称的, 有比例的, 成比例的; ～ divider 比例分规; ～ method 比例法

proportionality /prəpɔːʃəˈnæliti/ 定比性

proportioning /prəˈpɔːʃəniŋ/ 比例法

proposal /prəˈpəuzəl/ 建议, 提议; ～ drawing 建议图

proposition /prɔpəˈziʃn/ 定理, 命题, 讨论题; 建议, 主张

prorata /prəuˈreitə/ (拉丁语)按比例, 成比例的

PROT (protractor) 量角器, 分度规

protract /prəˈtrækt/ 绘制(平面图等), 延长, 拖长(时间等)

protractor /prəˈtræktə/ 量角器, 分度规, 半圆角度视; 延长器; angle ～ 量角器, 量角规, 分角规, 斜角规; draftsmen's ～ 绘图量角器; semicircular ～ 半圆量角器; sine ～ 正弦量角器, 正弦尺; three-arm ～ 三臂分度器; universal bevel ～ 万能量角器, 组合角尺; v-edge ～ V 型量角器; vernier ～ 量微半圆角度规, 游标分度器

provide /prəˈvaid/ 提供, 规定; 装备; 定出

proving /ˈpruːviŋ/ 校对, 勘探

provision /prəˈviʒən/ 准备, 予备; 设备, 供给; 规定, 条件; safety ～ 安全设备; technical ～s 技术条件

proximal /ˈprɔksiməl/ 近似的, 接近的, 最近的

prtsc (print screen) 打印屏幕内容

pt. (part) 部分; (point) 点; (point of tangency) 切点

pts (parts) 部分, 零件, 份; (points) 点

pub. (public) 公共的

pulley /ˈpuli/ 皮带轮, 滑轮; band ～ 带轮; belt ～ 皮带轮; double curved arm of ～ 皮带轮的双弯幅; step ～ 塔轮

pulse /pʌls/ 脉冲, 脉动; 半圆周; program ～ 程序脉冲

pump /pʌmp/ 泵, 唧筒; ～ body 泵体; centrifugal ～ 离心泵; gear (rotary) ～ 齿轮泵; piston ～ 活塞泵; plunger ～ 柱塞泵; tri-rotor ～ 三转子泵; vane ～ 叶轮泵

punch /pʌnts/ 打眼, 穿孔; 打出的孔眼, 切口; 冲压机; center ～ 中心冲

puncher /ˈpʌntʃə/ 穿孔机

puncture /ˈpʌŋktʃə/ 刺孔, 扎孔, 小孔, 刺痕; ～ plane 有孔平面

purlin(e) /ˈpəːlin/ 檩条

P & W (Pratt & Whitney)(键名)

P & W key (Pratt and Whitney key) 圆头平键

pyramid /ˈpirəmid/ 棱锥, 金字塔; 锥形, 棱锥体, 四面体; frustum of ～ 棱锥台; oblique ～ 斜棱柱; regular ～ (right ～) 正棱锥体; right triangular ～ 正三棱锥; spherical ～ 球面棱锥体; square ～ 正四棱锥; triangle ～ 三角锥; truncated ～ 截头棱锥体, 棱锥台

pyramidal /piˈræmidl/ 棱锥体形的

Q

q. (quenching) 淬火

QC (quality control) 质量管理

QED (quod erat demonstrandum)（拉丁语）已如所示,证(明)完(毕)

QPL (qualified parts list)（合格）零件一览表；零件目录；(qualified products list)（合格）产品一览表,商品目录

Qr. (quire) 一刀纸（24张）；(quarter) 四分之一

Qt. (quarter) 四分之一

Q'ter (quarter) 四分之一,一刻钟,季度

qu. (quarter) 四分之一,一刻钟,季度；(question) 问题

qua. (qualitative) 定性的；(quality) 特性,性质,质量

quad /kwɒd/ 四边形,方形；象限,四分仪；四倍的,由四部分组成的；扇形体,扇体齿轮

quadrangle /kwɔ'dræɡl/ 四角形(的),四边形(的),方形(的)

quadrangular /kwɔ'dræɡjulə/ 四棱柱；四边形(的),方形(的)

quadrant /'kwɒdrənt/ * 分角,象限,四分之一圆；象限仪,四分仪,扇形齿轮；换向器；~ designation 象限名称；gear ~ 月牙轮

quadrate /'kwɒdrit/ 正方形,长方形；四等分；正方形的,长方形的

quadratic /kwə'drætik/ 二次的,方形的；~ curve 二次曲线；~ surface 二次曲面

quadratrix /kwə'drætriks/ 割圆曲线

quadrature /'kwɒdrətʃə/ 正交,求面积,平方面积,求积分

quadric /'kwɒdrik/ 二次的,二次曲面(的)；~ cone 二次锥面；~ crank mechanism 曲柄连杆机构,四连杆机构；~ cylinder 二次柱面；~ surface 二次曲面；unruled ~ 非直纹二次曲面

quadrilateral /kwɒdri'lætərəl/ 四边形,四边的,四边形的；四方面的；cyclic ~ 圆内接四边形；gauche ~ 挢 (lie) 四边形,扭四边形

quadscreen /kwɒdskri:n/ 四方屏幕

qualification /kwɒlifi'keiʃən/ 技能,熟练程度,资格

quality /'kwɒliti/ 性质,本质,质量；surface (finish) ~ 表面性质,表面质量

quantity /'kwɒntiti/ 量,数量；大量；vector ~ 矢量,向量

quarter /'kwɔ:tə/ 四分之一,四等分；~ scale1/4比例,（1/4″=1′）；~ size1/4 尺寸

quartering /'kwɔ:təri/ 四等分,四开,成直角的

quatrefoil /'kætrəfɔil/ (quaterfoil /'kætrəfɔil/) 四叶形；四花瓣的花朵,四叶饰；~ knot 四叶式结

quench /kwent/ 淬火,急冷,使淬硬；熄灭,抑制；~ aging 淬火（后自然）时效；~ hot 高温淬火；harden ~ 淬硬；isothermal ~ 等温淬火；local ~ 部分淬火；oil ~ 油淬火；water ~ 水淬火

quincunx /'kwinkʌks/ 梅花式,五点形

quinquangular /kwi'kwæɡjulə/ 五角(形)的,五边形的

quit /kwit/ 退出,离去,清理

qunty. (quantity) 量,数量

R

R (radius) 半径;(rolling) 滚制;(rough finish) 粗糙的,未加工的;近似的,粗略的

RA (Ra. Rockwell hardness A-scale) 洛氏硬度A

rabbet /ˈræbit/ 槽,口,凹部;插孔,塞孔

rack /ræk/ 架,齿条;chart ～ 图架;gear ～ 齿条

RAD (radian) 弧度;(radius) 半径

rad. (radial) 半径的;径向;(radical) 根,茎;(radius) 半径

radial /ˈreidiəl/ 径向,半径的,径向的,放射的,辐射(状)的星形;～ ball bearing 径向球轴承;～ line 径向线,辐射线

radian /ˈreidiən/ 弧度

radiate /ˈreidieit/ 辐射,辐射状的

radii /ˈreidiai/ (radius 的复数)半径

radius /ˈreidiəs/ 半径,辐条,放射状;倒圆,切成圆角;～ angle 圆心角;～ at bend 曲率半径;～ of corner 圆角半径;～ of curveture 曲率半径;～ of curve 弧线半径;～ of gyration 回转半径;～ of rounded crest* 牙顶圆弧半径;～ of rounded root* 牙底圆弧半径;～ of turn 旋转半径;bend ～ 弯曲半径;connecting ～ 连接半径;heel ～ 外半径(肘管);interrupted dimension line of ～ 截短半径尺寸线;pitch ～ 节圆半径;revolving ～ 旋转半径;root ～ 根部半径;rotation ～ 旋转半径

rafter /ˈrɑːftə/ 椽条

rag /ræg/ 毛刺,飞边;除去毛刺;刻纹,压花,滚花;擦布,擦拭材料

ragged /ˈrægid/ 不平的,参差不齐的,锯齿形的

rail /reil/ 轨,栏杆,钢轨;lead ～ 导轨;steel ～ 钢轨

railroad /ˈreilrəud/ (railway) 铁路;～ pen 铁路笔(有两个鸦嘴笔头,用以画双线);～ situation map 铁路四周情况图

rake /reik/ 倾斜,倾角,斜度,坡度;front ～ 前角,前倾角;side ～ 侧倾角,侧斜角

ral. (right and left) 左右

ramp /ræmp/ 斜面,滑行台

random /ˈrændəm/ 任意的,随便的,不规则的;不一律的;随机的;～ process 随机过程;～ variable 随机变数

range /reindʒ/ 排列,并列,与……平行;距离,范围,区域;分类;to ～ from...to...从……开始,到……为止;分布;在……到……的范围内

rank /ræk/ 排,横列,次序;把……列成横列,把……等分

rapid /ˈræpid/ -graph 快速绘图器

raplot /ˈreiplɔt/ 等点绘图法

raster /ˈræstə/ 光栅;～ scan 光栅扫描;～ unit 光栅单位

ratch /rætʃ/ (ratchet /ˈrætʃit/) 棘轮,齿弧;～ and pawl 机轮与爪,棘轮机构;～ gear 棘轮装置

rate /reit/ 比值,比例;等级,程度,标准,值得,应予以;～ of change chart 变率图;～ special attention 值得特别注意;accuracy ～ 精确度

rating /ˈreitiŋ/ 等级;参数,特性,规格;～ curve 关系曲线;～ plate 铭牌;surface

smoothness ～ 表面光洁度等级
ratio /ˈreiʃiəu/ 比例，比，比率，比值；系数；传动比；～ chart 比率图；～ of foreshortening for any axis 轴向缩短率；～ of gear（齿轮转）速比；～ of slope 坡度比率；aspect ～ 纵横比；cross ～ 交叉比；definite ～ 定比；reduction ～ 缩小比例；same size ～ 1:1比例；speed ～ 速比
raw /rɔː/ 生的，未加工的，半加工的，不完善的；～ material 原材料
ray /rei/ 光线，射线；～ tracing 光线追迹，声线描迹；cone of ～ 射线锥；diverging ～s 发散光线；parallel light ～s 平行光线
RB (Rb, Rockwell hardness B-scale) 洛氏硬度 B；(roller bearing) 滚子轴承
RC (reinforced concrete) 钢筋混凝土
Rc (Rc, Rockwell hardness c-scale) 洛氏硬度 C
R. D. (root diameter) 根圆直径；(round) 圆形
RD (rd, round) 外圆角
RdHd (round head) 圆头，半圆形
re (reference) 参考，基准，坐标
reach /riːtʃ/ 延伸，伸出；伸到，到达；伸展，距离
react /riˈækt/ (re-act) 重作
read /riːd/ 读，阅读，读入；～ing view 读图
reader /ˈriːdə/ 读者，校对者，审稿者；读数器，指示器，读入机；card ～ 卡片读入机；paper tape ～ 纸带读入机
reading /ˈriːdiŋ/ 读数；读出；accurate ～ 准确读数；red ～ 标尺读数；vernier ～ 游标读数
real /riəl/ 真实的；实型，实数
real-line 实线
realm /relm/ 范围，区域，领域，王国
ream /riːm/ 铰孔，扩孔，铰
rear /riə/ 后部，背面，后方的，背后的；竖起，建立；～ auxiliary 后辅视图；～ elevation 背立面图；～ face 背面；～ projection 背面投影；～ view 后视图
rearmost /ˈriəməust/ 最后（面）的
rebound /riˈbaund/ 弹回，跳回；～ hardness 肖氏硬度
recede /riˈsiːd/ 退却，向后倾斜；缩减，降低，缩短
receding /riˈsiːdiŋ/ 退隐，退隐线，退隐轴；面之边视图；～ line 退隐线；～ surface 退隐面（垂直面），（向后）倾斜面
recess /riˈses/ 凹口，切口，凹座，凹槽，环槽，退刀槽；暂停
reciprocity /ˌresiˈprɑsəti/ 相互关系，相互作用，相关性，互换性
recombination /ˈriːkɔmbiˈneiʃn/ 复合；再化合
recommend /rekəˈmend/ 推荐，介绍；建议
reconnaissance /riˈkɔnisəns/ 踏勘，探测
record /riˈkɔːd/ v. 记录；/ˈrekɔːd/ n. 记录；drafting room ～ 制图室记录
rect. (rectangle) 矩形，长方形；(rectangular) 直角的，矩形的
rectangle /ˈrektæŋgl/ 长方形，方形，矩形，直角；enclosing ～ 外切矩形
rectangular /rekˈtæŋgjulə/ 长方形的，矩形的，成直角的，方格式的；～ axis 直交轴；～ coordinate 直角坐标；～ paralelepiped 长方体；直角平行六面体；～ prism 矩形棱柱；～ pyramid 四棱锥；～ section 矩形剖面
rectifiable /ˈrektifaiəbl/ 可矫正的，可求长的，可用直线测度的；～ curve 可求长的曲线，有长曲线
rectifier /ˈrektifaiə/ 整流器
rectify /ˈrektifai/ 修改，纠正，矫正；求（曲线）的长度；～ing an arc 修改圆弧；to ～ an arc 展直弧形
rectilineal /rektiˈliniəl/ (rectilinear /rekti

liniə/) 直线的, 直线组成的, 构成直线的, 直线性的, 用直线围着的; 直的; 直角坐标, 直导线; ～ chart 直角坐标图; ～ generators 直纹母线; ～ scale 直尺

rectilinearity /ˌrektiˌliniˈæriti/ 直线性

rectitude /ˈrektitjuːd/ 正直, 笔直, 正确

recto /ˈrektəu/ 纸张的正面, 书籍的右页(单数页); 与 verso 相对

redented /riˈdentid/ 锯齿形的, 齿的

redesign /ˈriːdiˈzain/ 重新设计

redraw /riːˈdrɔː/ 重画

reduce /riˈdjuːs/ 减少, 变形, 使变为; 把……归纳; 简化

reducer /riːdjuːsə/ 减缩器, 减缩接头; rectangle-to-circle ～ 天圆地方接头

reduction /riˈdʌkʃn/ 简化, 压缩, 压下, 减缩, 减少, 缩小; ～ of area 面积缩小

redundance /riˈdʌndəns/ (redundancy) 多余的东西, 过多, 重复, 剩余度

redundant /riˈdʌndənt/ 多余的, 重复的

reduplicate /riˈdjuːplikeit/ 重复, 加倍

reed /riːd/ 芦苇; ～ pen 芦苇笔

reef /riːf/ 折叠

REF (reference) 参考, 基准

refacing /ˈriːˈfeisiŋ/ 光面

refer /riˈfəː/ 参考; 查询; 系指, 称为

reference /ˈrefrəns/ 参考, 参照, 查阅, 基准, 标准; 标准的, 基准的; 坐标, 定位, 核对位置; ～ books 参考书; ～ line 基准线; ～ plane 基准面, 基准平面; ～ surface 基准面

refine /riˈfain/ 精制, 精炼; 改进, 改善

REFL (reference line) 基准线, 参考线, 零位线

reflect /riˈflekt/ 反射, 反照, 反映, 表现; ～ed view 反射图

reform /riˈfɔːm/ 重作, 改编; 革新; 变换, 改造

Reg. (regulations) 规则

regard /riˈgɑːd/ 视为, 看作, 考虑

regardless /riˈgɑːdlis/ 不视为, 不考虑, 未考虑; ～ of feature size 未考虑特征尺寸

regional /ˈriːdʒənl/ 地方(性)的, 局部的; ～ coding 局部编码

regular /ˈregjulə/ 规则, 规则的, 规定的, 正规的

reinforce /riːinˈfɔːs/ 加强, 增强, 补充

reinforcement /riːinˈfɔːsmənt/ 加强, 加固; 钢筋; ～ concrete 钢筋混凝土; steel ～ 钢筋

reink /ˈriːiŋk/ 重新涂墨水于, 重加墨水于

related /riˈleitid/ 有关的, 相关的, 有联系的; ～ angle 相关角

relation /riˈleiʃn/ 关系式, 关系曲线, 比例关系; ～ curve 相关曲线

relationship /riˈleiʃnʃip/ 关系; ～ of space 空间位置关系; assembly ～ 装配关系

relative /ˈrelətiv/ 相对的, 成比例的, 关联的, 相关的; ～ position 相对位置; ～ to 相对于, 与……相对应

release /riˈliːs/ 释放

relevant /ˈrelevənt/ 有关的, 相关的; 贴体的, 成比例的, 相应的

relief /riˈliːf/ (形状)起伏, 地形; 浮雕; 解除; 消除; 降压; (刀具的)后角, 背面; 立体的, 安全的; ～ map 地形图, 立体地图, 地势图; ～ television 立体电视; end ～ 主后角, 端切口

relief-map 地势图

REM (remark) 注释

remains /riˈmeinz/ 其余

remake /riːˈmeik/ 重制, 重做; 重制物; 摘要, 要点

remark /riˈmɑːk/ 备考, 附注, 评语, 意见, 注视, 观察, 表示

reminder /riˈmaində/ 暗示, 提示; 提醒者;

as a ~ 提示一下

remove /ri'mu:v/ 移置, 搬开, 移动; 拆去, 清理, 切掉; ~ burrs 清理毛刺; ~d section 移出剖面, 详细剖面

rename /ri:'neim/ 给以新名

rendering /'rendəriŋ/ 翻译; 初涂, 抹灰, 打底, 渲染图, 透视图, 示意图, 复制图

rendezvous /'rɔndi,vu:/ 会合点, 会合, 集合

repaint /ri:'peint/ 重画, 重新涂(漆); 重画的部分, 重涂漆的

repair /ri'pɛə/ 修理, 修补, 改正, 修正; ~ charts 修理图

repeat /ri'pi:t/ 重复; 反复; 复制, 代替; 出现; 重做

repetition /repi'tiʃn/ 重复, 反复, 重做; 复制品, 副本; 反复再现

replace /ri:'pleis/ 替换, 取代, 更换, 代替, 复原, 移位; ~ M by (or with) N 用 N 来代替 M

replacement /ri'pleismənt/ 替换, 更换; 取代, 代替; 换位, 复位; ~ of M by (或 with) N 用 N 来代替 M

replicate /'replikeit/ 复制品; 复制, 复现(计)

replot /ri:'plɔt/ 重画, 重制(图表)

represent /repri'zent/ 描绘, 描述, 表示, 体现, 阐述, 回忆; 作; ...to oneself 想象出; ‖ *Example sentence*: Represent a plane *that contains CD and is parallel to AB*. 例句: 过直线CD作一平面平行于直线AB。

representation /reprizen'teiʃn/ 表示法, 表现, 体现, 表示; 显示, 图像; 画法; ~ of develop 展开画法; ~ of fair surfaces 流线型表面的画法; ~ with geometric elements 几何元素表示法; ~ with trace 迹线表示法; assembled ~ 集中表示法; block ~ 方块图表示法; boundary ~ 边界表示法; conventional ~ 惯用表示法; conventional breaks ~ 折断画法; detached ~ 分开表示法; development ~ 展开画法; diagrammatic ~ 图示法; disassembling ~ 拆卸画法; dismantlement ~ 拆卸画法; graphic (al) ~ 图解(表示)法, 图示法, 用图表示; imaginary ~ 假想画法; lapped ~ 重叠画法; multiline ~ 多线表示法; omissive ~ * 省略画法; orthographic ~ 正视表示法; overstate ~ 夸大画法; perspective ~ 透视表示法; pictorial ~ 直观图, 立体图; pseudo-pictorial ~ 非透视立体表示法; schematic ~ * 示意画法; semi-as-sembled ~ 半集中表示法(电); simplified ~ * 简化画法; single-line ~ 单线表示法; special ~ 特殊表达法; specified ~ * 规定画法; symbolic ~ 符号表示法; thread ~ 螺纹标准图表

representative /repri'zentətiv/ 典型的, 有代表性的, 样品, 代表

reprint /'ri:'print/ 再版, 翻印, 转载

reproduce /ri:prə'dju:s/ 复制, 仿制; 再生, 再现; 还原; ~ to scale 按比例复制

reproduction /ri:prə'dʌkʃn/ 复制, 仿制; 再生, 再现; 还原; black-and-white ~ 黑白复印; scale ~ 按比例复制

reqd. (required) 规定的, 要求的, 需要的

require /ri'kwaiə/ 要求, 命令, 需要; ~d line 所求线

requirement /ri'kwaiəmənt/ 要求, 必要条件, 需要, 需要量; manufacturing ~ 制造条件; technical ~ 技术要求

rescaling /ri:'skeiliŋ/ 尺度改变, 改比例

reseat /'ri:'si:t/ 研磨, 研配, 修整

resemblance /ri'zembləns/ 类似, 相似

reservoir /'rezəvwa:/ 储存器, 容器

reset /ri:'set/ 复位

residual /ri'zidjuəl/ 剩余的

residue /'rezidju:/ 剩余, 余数, 其余, 其他

resistance /ri'zistəns/ 阻力，电阻；阻尼；～ welding 电阻焊；buckling ～ 抗弯强度；shear ～ 抗剪强度；compressive ～ 抗压强度

resolution /rezə'lu:ʃn/ 分解法，解；决定；分辨率；～ of force 力的分解；～ of vector 矢量分析

resort /ri'zɔ:t/ 求助，凭借，采取

respect /ris'pekt/ 关系；方面，考虑，尊重；遵守；with ～ to 关于……，对于……

respective /ris'pektiv/ 各自的，各个的；～ly 各自地，分别地

response /ris'pɔns/ 答复，签应，回答；特性曲线；对应

rest /rest/ 其余

restart /ri'sta:t/ 再启动

resultant /ri'zʌltənt/ 合量，合力；组合的，合成的；～ of forces 合力

retain /ri'tein/ 保留

reticule /'retikju:l/ 十字线，标线，分度线

return /ri'tə:n/ 返回

rev. (reverse) 反向的，相反的，反的；(revolution) 旋转；～ bend 反向弯曲

reveal /ri'vi:l/ 展现，露出，揭示

reversal /ri'və:səl/ 反向，反的；～ of curvature 反曲率

reverse /ri'və:s/ 颠倒，反转；相反的，反向的，反面；～ curve 反向曲线

review /ri'vju:/ 检查，观察；审查，审阅，杂志，评论，评审

revision /ri'viʒən/ 校阅，校核，校订，修订；～ of dimension 复校尺寸

revolution /revə'lu:ʃn/ 旋转，周期；革命，改革；～ about a parallel axis 绕平行轴旋转；～ of a line 直线的旋转；～ of a plane 平面的旋转；～ of a point 点的旋转；～ of an object 立体的旋转；complete ～ 全转；primary (single) ～ about a normal axis 绕垂直轴一次旋转；secondary (succesive) ～ about anormal axis 绕垂直轴二次旋转；surface of ～ 回转面

revolve /ri'vɔlv/ 旋转，循环，回转；～d-plan method 回转平面法；～d sections 旋转剖面；‖*Example sentence: As a line* revolves *about an axis, all-points of the line revolve through the same angle.* 例句: 当直线绕轴旋转时，直线上所有的点都旋转了同样角度。

R. F. S (regardless of feature size) 不管特征（定型）尺寸如何

R-gauge R 规，圆弧规

RH (right hand) 右旋，右手；向右的，右边的；(right head) 右首；(round head) 圆头

RHC (Rockwell hardness C scale) 罗氏硬度 C

RHN (Rockwell hardness number) 罗氏硬度（数）

rhomb /rɔm/ 菱形，斜方形

rhombohedron /rɔmbə'hedrən/ 菱形（六面）体

rhomboid /'rambɔid/ 长菱形，斜长方形，平行四边形

rhombus /'rɔmbəs/ 菱形，斜方形

rhs (right hand side) 右方，右边

rhythm /'riðəm/ 变化的和谐，有规律的重复发生，节奏

rib /rip/ 肋；～ stiffener 加强肋；reinforcing ～ 加强肋；wheel ～ 轮肋

Richter （圆规名）弓形圆规

ridge /ridʒ/ 山脊，屋脊，波峰，背

right /rait/ 右，直，正，对；正确的，恰当的；右向；～ angle 直角；～ cone 直维，正圆维；～ of way 通路；～ prism 正棱柱体；～ section 正剖面，正截面；the ～ side of paper 纸的正面；right-angled 成直角的；right-hand 右旋；right-lined 直线的

rim /rim/ 缘，边；colar ～ 轮凸缘，轮圈

ring /riŋ/ * 挡圈, 环, 轮, 圈; 环面; ～ for shoulder* 轴肩挡圈; "E" ～ * 开口挡圈; snap ～ * 弹性挡圈

ringent /'rindʒənt/ 开口的, 张口的

rip /rip/ 切开, 撕开

rise /raiz/ 升高, 高地, 斜坡, 升度; 高, 楼梯的级高; 矢高

rivet /'rivit/ 铆钉, 铆接; ～ head 铆钉头; ～ hole 铆钉孔; ～ pen 铆钉圆规 (小圆规); ～ pitch 铆钉间距; ～ spacing 铆钉间距; brazer head ～ 扁头铆钉; button head ～ * 半圆头铆钉, 纽头铆钉; cone head ～ * 平锥头铆钉; countersunk head ～ 沉头铆钉; field ～ 现场铆接; flat head ～ * 平头铆钉; globe head ～ 圆头铆钉, 球头铆钉; headless ～ * 无头铆钉; mushroom ～ 平圆头铆钉; snap head ～ 半圆头铆钉, 半球头铆钉; sunk ～ 埋头铆钉

riveting /'rivitiŋ/ * 铆钉联接, * 铆接

RL (reference line) 基准线

RMS (root mean square) 均方根值, 均方的

RMVD (removed) 拆去的, 移去的

RMVG (removing) 拆去, 移去, 移置

ro. (recto) 右页

roadway /'rəudwei/ 道路, 路面; 行车道, 铁道线

rock /rɔk/ 石头, 石块, 摇动

rocket /'rɔkit/ 火箭

Rockwell 洛氏 (硬度); ～ apparatus 洛氏硬度计; ～ B 洛氏 B 级硬度; ～ C 洛氏 C 级硬度

rock wool /'rak, wʊl/ 石棉, 玻璃纤维

rod /rɔd/ 杆, 条; connecting ～ 连杆; welding ～ 焊条

roller /'rəulə/ ＊ 滚子

ROM (read-only memory) 只读存储器

Roman /'rəumən/ 西文罗马字

roof /ru:f/ 屋顶, 屋面, 顶部; ～ framing 房屋构造; ～ truss 屋架; arched ～ 拱形屋顶; common ～ing 普通屋面; flat (gable) ～ 平 (人字) 屋顶; hip ～ 四坡屋顶

room /ru:m/ 房, 室; chart ～ 图表室; drafting ～ 制图室; drawing ～ 绘图室; meeting ～ 会议室; model ～ 模型室, 模型车间; recording ～ 资料室; tool ～ 工具室, 工具车间

root /ru:t/ * 牙底, 根 (螺纹根), 齿根, 根部; ～ cone (齿) 根锥; ～ diameter (齿轮) 根圆直径; ～ of thread 螺纹牙底; minimum ～ radius 最小螺纹根弧半径

root-mean-square (value) 均方根值

rot. (rotation) 旋转, 转动

rotary /'rəutəri/ 回转, 转动的; anti-clockwise ～ (counter-clockwise ～) 逆时针方向旋转的; clockwise ～ 顺时针方向旋转的; left-handed ～ 左向旋转的; right-handed ～ 右向旋转的

rotate /rəu'teit/ 旋转, 回转, 转动; 使旋转

rotation /rəu'teiʃn/ 旋转, 转动; ～ axis 转动轴 (线), 旋转轴 (线); ～ of axes 轴的旋转; ～ of corrdinate plane 坐标面的旋转; angle of ～ 旋转角; angular ～ 角位移

rough /rʌf/ 粗糙的, 初步的, 毛坯; 要略, 草图, 画……的轮廓; ～ draft 草图, 略图, 示意图; ～ing 画草图; ～ plan 草图, 初步设计; ～ turn 粗加工, 粗车

rough(-)cast 粗涂, 打底 (子), 草拟, 粗制的, 毛坯

roughness /'rʌfnis/ 粗糙度, 不平度; ～ parameters 粗糙度参数; ～ of surface 表面粗糙度; surface ～ 表面粗糙度, 表面不平度; ～ -width cut-off 粗糙度截取长度

round /raund/ 外圆角; 圆形的, 球形的, 完全的; 周围, 一周; ～ed corner 圆角; ～ head 圆头; fillet and ～ 内圆角与外圆角

rounding /'raundiŋ/ 倒圆, 倒角, 外圆角修

整, 圆的

roundlet /'raundlit/ 小圆, 小的圆形物

roundness /'raundnis/ 圆度, 球度, 圆形, 圆, 球形; average ～ 平均圆度; out of ～ 椭圆度, 不圆度

roundnose /'raʊnd'nəʊz/ 圆头

route /ru:t/ 路线, 线路; ～ chart 路线图; ～ location on paper 纸上走线

routine /ru:'ti:n/ 常规, 惯例, (计算机的) 程序

row /rəu/ 排, 行; in a ～ 成一长行, 成一排

RP (refernece plane) 参考基准面

RPM (rpm, revolutions per minute) 每分钟转数

RPS (rps, revolutions per second) 每秒钟转数

Rr (rear) 后方; (railroad) 铁路

R. sidev. (right side view) 右侧视图

rte. (route) 线路

Rto. (ratio) 比, 比例, 比率

rubber /'rʌbə/ 橡皮, 橡胶, 粗锉, 磨石; India- ～ 擦墨橡皮

ruckle /'rʌkl/ 弄皱, 折叠

rude /ru:d/ 粗糙的, 未加工的

rugged /'rʌgid/ 不平的, 有皱纹的; 粗暴的

rule /ru:l/ 规则, 章程; 画线, 尺, 规, 比例尺; 用尺画(线), 画线板, 破折号, 标准, 规定, 惯例; 定则, 律, 法则; ～ depth gage 深度规; ～ of thumb 奏成法, 概算法; ～ out 划去, 取消; a sheet of paper ruled with rectangular coordinates 一张画有直角坐标的纸; adjustable ～ 自由曲线规; caliper ～ 卡尺, 测径器; carpenter's ～ 木工尺; common ～ 普通尺; comparing ～ 比例尺; copy ～ 仿形尺, 放大尺; depth ～ 深度尺; dotted ～ 点线规; drawing ～ 绘图尺; em ～ 长破折号; empirical ～ 经验规则; en ～ 短破折号; evaluation ～ 判别法则; flat ～ 平尺; flexible ～ 卷尺, 盒尺; folding ～ 折尺; foot ～ 英制尺; gnomonic ～ 心射投影尺; guiding ～ 样板, 量规; keyseat ～ 键槽尺; measuring ～ 量尺; parallel ～ 平行尺(绘图仪器); production ～ 形成规律; projection ～ 投影规律; shrink ～ 收缩尺; shrinkage ～ 缩尺; slide ～ 计算尺, 滑尺; stainless steel ～ 不锈钢尺; steel ～ 钢尺; steel folding ～ 钢皮折尺; three-edged ～ 三棱尺; three-square ～ 三棱尺; transformation ～ 变换规则, 变换法则; triangular ～ 三棱尺; wave ～ 曲线尺; wood ～ 木尺; zigzag ～ 曲尺

ruled /ru:ld/ 规则的; ～ line 尺子线(用尺子画的线); ～ surface 规则表面, 直纹曲面

ruler /'ru:lə/ 直尺, 直规, 尺, 画线板; curve ～ 曲线板, 曲线尺; parallel ～ 平行尺, 一字尺

ruling /'ru:liŋ/ 画线, 格线; 刻度, 量度; 支配; ～ of a ruled surface 直纹面的母线; ～ pen 直线笔, 鸦嘴笔; semilogarithmic ～ 单对数格线

run /rʌn/ 划, 描(线); 运行, 进行; 刊印; 平距(水平距); 画, 跑, 开动; 突出, 伸向; 贯穿; 流出, 流量; ; ～ book 运行资料, 题目文件, 备忘录; ～ curve 运转曲线 ‖ *Example sentence:* Run a line across the sheet. *例句:* 在纸上画一条线。

run(-)out /'rʌnaut/ 转角, 跳动, 径向跳动, 摆动度; 跑出; 伸出; 完成; 偏斜; ～ line 出线(表示小圆角的), 过渡线; ～ of thread 退刀纹, 螺纹尾部; circular ～ 圆周偏转; cycle ～ 圆跳动; total ～ 全跳动

rupture /'rʌptʃə/ 破断, 破折, 破裂, 折断, 挠曲

ry. stn (railway station) 火车站

S

S. (south) 南; S 11° 5′ E 南偏东11° 5′; (surfaceness) 表面粗糙度
S. A. (sectional area) 横断面积
sable /ˈseibl/ 深褐色, 黑色
SAC (Standardization Administration of China) (中国) 国家标准化管理委员会
SAE (Society of Automotive Engineers) 美国汽车工程师学会; (number) (美国汽车工程师学会) 编号
sagitta /səˈdʒitə/ 矢 (数学)
same /seim/ 同样的, 同一的
sample /ˈsa:mpl/ 样品, 模型, 实例, 试样, 样本
sampler /ˈsæmplə/ 样板, 规 (具), 模型
sand /sænd/ 砂
sandcloth /sændkˈlɒθ/ 砂布
sanding /ˈsændiŋ/ 砂纸打光, 喷砂清理; blast ～ 喷砂处理
sandpaper /ˈsændpeipə/ 砂纸, 用砂纸擦; ～ board 砂纸板
sandpapering /ˈsænd,peipəriŋ/ 砂纸打磨
sash /sæʃ/ 框格, 窗框, 门框
satisfy /ˈsætisfai/ 满足, 使满足; 达到, 符合
save /seiv/ 保存, 存
sawtooth /ˈsɔtu:θ/ 锯齿, 锯齿形的
sc (scale) 刻度, 标度, 比例尺; (screw) 螺钉; (steel casting) 铸钢件; (subcommittee) 分技术委员会
scalar /ˈskeilə/ 纯量的, 无向量的; 梯状的, 分等级的
scalariform /skəˈlærifɔ:m/ 梯子状的, 阶状的
scalary /skæˈləri/ 如梯的, 有阶段的
scale /skeil/ * 比例; 刻度, 标值; 图尺, 标尺; 尺度, 比例尺, 缩尺, 测尺; 用缩尺 (按比例) 制图; 等级; 阶梯; 氧化皮, 鳞皮; ～ down (把……) 按比例缩小; ～ equation 标值 (图尺) 方程; ～ factor 比例因子; ～ing 定比例 (计); ～ of chord 弦线量角度尺; ～ of reduction 缩小比例尺; ～ reproduction 按比例复制; ～ up (把……) 按比例扩大; ～'s blade 比例尺的画线边; architect's ～ 建筑师比例尺; architectural ～ 建筑上用的比例尺; arithmetic ～ 算术标度; arithmetical horizontal ～ 算术分度水平坐标; a shop drawing on the ～ of one hundredth 百分之一比例的加工图; be drawn to ～ 按比例绘制, 用比例尺测量; celluloid-edge ～ 赛璐珞边尺; chain ～ (engineer fully divided scale) 全部刻度的尺; civil engineer's ～ 土木工程师比例尺; contracted ～ 缩小比例; contraction ～ 缩尺; diagonal ～ 斜线尺; division by using ～ 比例尺分格法; double-bevel ～ 双斜面比例尺; drafting ～ 绘图比例尺, 制图尺; drawing ～ 绘图比例尺, 制图尺; drawing to ～ 照比例绘制; dual ～ 双比例尺 (同一视图上); duplex ～ 复式比例尺; eighth ～ = 1/8″=1′ 比例尺; engineer's ～ 工程用比例尺, 工程师比例尺; enlarged ～ 放大比例尺; exaggerated ～ 扩大比例尺; flat ～ 平尺; flat bevel ～ 平斜尺, 刀口平尺; folding ～ 折尺; foreshortened ～ 深度缩小比例尺; full ～ 实尺, full divided ～ 全刻

度尺; full size ～ 足尺; 足尺(比例), 全尺寸; functional ～ 函数标值(图尺); graduated ～ 分度尺; graphic ～ 图上比例尺, 图示比例尺; grey ～ 灰度等级; half(inch) ～ 1/2=1′ 之比例尺; imperial ～ 英制比例尺; isometric ～ 等角投影尺; linear ～ 直尺; logarithmic ～ 对数比例尺, 对数标值图尺, 计算尺; map ～ 地图比例尺; mechanical engineering's ～ 机械工程比例尺; metric ～ 米尺, 公制(比例)尺; nautical ～ 海图比例尺; notto ～ 未按比例绘制, 不按比例; open divided ～ 两端细分刻度尺; opposite bevel ～ 双斜尺; plotting ～ 制图尺, 绘图比例尺; proportional ～ 比例尺; quarter ～ 1/4比例尺; quarter ～ 1/4″=1′ 之比例尺, reduced ～ 缩尺; semilogarithmic ～ 半对数标值(图尺); single-bevel ～ 单斜面比例尺; smal ～ 缩小比例尺, 缩尺, 小比例尺的; steel ～ 钢(比例)尺; structure's ～ 结构用比例尺; three-square ～ 三棱尺; to ～ 按比例(尺); to the(a) ～ of M to (for) N. 按 M 比 N 的比例; to the same ～ 按同一比例(尺); triangular ～ 三棱尺; unnecessary ～ 不必按比例; vast ～ 大比例; vernier ～ 游标尺

scaled /skeild/ 成比例的, 有刻度的

scaled-down 按比例缩小, 分频

scalene /'skeili:n/ 不等边三角形(的); ～ obtuse 钝角三角形; ～ triangle 不等边三角形, 不规则三角形; ～ right ～ 直角三角形

scale-paper 坐标纸, 方格纸

scaling /'skeiliŋ/ 定标, 定比例, 按比例描绘, 比例变换, 缩尺; ～ down 按比例缩小; ～ method 比例法; ～ system 计算图, 设计图; ～ up 按比例扩大

scaling-off 定尺寸(根据图纸的比例尺)

scallop /'skɔləp/ 扇形

scan /skæn/ 扫描, 搜索, 检查; raster ～ 光栅扫描

scan-and-paint 扫描和绘画

scanner /'skænə/ 探伤器, 扫描器, 光电继电器; 扫描程序

scanner-recorder 图示记录器

scanning /'skæniŋ/ 扫描

scantling /'skæntliŋ/ 略图, 样品; 少量, 一点点

scarf /ska:f/ 修整, 清理

SCD (specification control drawing) 技术要求控制图纸

SCEM (schematic) 图解的, 简图

scenography /si'nɔgrəfi/ 透视图法

SCH (sclerosmpe hardness) 回跳(肖氏)硬度

schedule /'ʃedju:l, 'skedju:l/ 图表, 程序, 简图; 工艺制度; 编目, 目录, 一览表, 表; 清单; ～ number 序数, 号码; bar ～ 钢筋表, 钢筋成型图; bending ～ 钢筋表; design ～ 计算简图; door ～ 门户一览表; nozzle ～ 管口表; technical characteristic ～ 技术特性表; weight ～ 重量表

scheduling /'ʃedju:liŋ/ 制表, 编目录

schema /'ski:mə/ (schemata) 大纲; 概要; 图解, 略图; 一览

schematic /ski(:)'mætik/ 图解的, 纲要的, 示意的, 概略的; 简图, 略图; ～ drawing (diagram) 简图, 略图, 示意图, 原理图; functional ～ 作用原理图; simplified ～ 简图, 略图, 原理图

schematically /ski'mætikli/ 示意地, 大略地; 用示意图, 用图解法

scheme /ski:m/ 图表, 图解, 线路图, 设计图, 电路, 计划, 方案, 简图, 示意图, 原理图, 分类表, 系统, 形式, 方式; ～ of wiring 布线图, 电器安装图, 线路图; block ～ 方块图, 方框图; coding ～ 编码方案; flow ～ (工艺)流程图, 操作程序图; functional

～ 作用原理图, 方框图; hydraulic ～ 液压系统图; kinematic ～ 传动系统图, 运动系统图; original ～ 原始方案; wiring ～ 接线图, 布线图

sciagraph /ˈsaiəɡrɑːf/ 投影图, 房屋纵断面图, X 射线照片, 制投影图

sciagraphy /saiˈæɡrəfi/ (skiagraphy) 投影法, X 射线照相术; 房屋纵断面图

scission /ˈsiʒən/ 切断, 切开, 分离

sclerometer /skliəˈrɔmitə/ (同 scleroscope) (肖氏) 硬度计, (回跳) 硬度计, 测硬器

scope /skəup/ 指示器, 显示器, 范围, 余地; 眼界, 视野

scotch /skɔtʃ/ 制止, 阻止, 切口; ～ drafting tape 制图带, 胶纸带; ～ tape 胶带（黏贴用透明胶带）

scour /ˈskauə/ 擦净, 净化

scr. (screw) 螺钉, 螺丝, 螺杆

scrape /skreip/ 刮

scraper /ˈskreipə/ 刮刀, 刮除机

scratch /skrætʃ/ 擦, 刮, 草率书写, 涂写, 乱画, 刻线, 标线; ～ paper 草稿纸

scratchboard 刮画纸板（一种画图纸）

scratcher /ˈskrætʃə/ 擦刮器

screed /skriːd/ 模板, 样板

screen /skriːn/ 屏幕; ～ point area 屏幕指示区

screw /skruː/ * 螺钉, 具有螺旋之物; ～ bolt 螺栓, 螺杆; ～ down 用螺钉拧住, 用螺旋拧紧; ～ eye 螺孔; abutment ～ 止动螺钉; adjustable stop ～ 可调整止动螺钉; anchor ～ 基础螺钉, 地脚螺钉; attachment ～ 紧固螺钉, 装配螺钉; backing ～ 止动螺钉; backing-up ～ 止动螺钉; ball point set ～ 球端定位螺钉; bench ～ 台用虎钳; bihexagonal head ～ * 十二角头法兰面螺栓; bleed ～ 带孔螺钉（放气用）; butterfly- ～ 元宝螺钉; button head cap ～ 圆头螺钉; cap ～ 有帽螺钉, 内六角螺钉; check ～ 压紧螺钉, 止动螺钉, 定位螺钉; cheese head ～ 圆头螺钉; clamp ～ 夹紧螺钉; closing ～ 螺纹规, 螺纹塞; collar ～ * 凸缘螺钉; connection ～ 连接螺钉; counter sunk ～ 埋头螺钉; coupling ～ 连接螺钉, 夹紧螺钉; cover ～ * 面板螺钉; cross recessed panhead ～ * 十字槽盘头螺钉; cup point set ～ 尖端止动螺钉; drive ～ 传动螺杆; eared ～ 翼型螺钉; Edison ～ 圆（爱迪生）螺纹, 螺丝灯头的螺纹; endless ～ 蜗杆; expansion ～ 可调螺钉; external ～ （阳）螺钉; eye ～ 环首螺钉; fastening ～ 紧固螺钉; female ～ 内螺纹; fillister head ～ 凹槽螺钉; fixed ～ 固定螺钉; fixing ～ 固定螺钉, 定位螺钉; flat ～ 平头螺钉; flat counter-sunk head ～ 平顶埋头螺钉; flat-headed ～ 平头螺钉; flat leaf ～ * 扁平头螺钉; foot ～ 地脚螺钉; forcing ～ 紧固螺钉; grub ～ 埋头螺钉, 平头螺钉, 木螺丝; half-round ～ (button-headed ～) 半圆头螺钉; headless ～ * 无头螺钉; hexagonal head ～ 六角头螺钉; hexagon head tapping ～ * 六角头自攻螺钉; hexagon head wood ～ * 六角头木螺钉; hexagon socket head cap ～ * 内六角圆柱头螺钉; holding ～ 紧定螺钉, 紧固螺丝; hollow ～ 空心螺钉; inner ～ 阴螺旋; inside ～ 螺母; instrument-headed ～ 半圆头螺钉; joint ～ 连接螺钉; knurled thumb ～ * 滚花高头螺钉; lag ～ 木螺旋, 方头螺钉; lifting eye ～ * 吊环螺钉; lock ～ 锁紧螺钉; machine ～ 机螺钉; male ～ 阳螺旋（纹）; millededge thumb ～ 铣边翼形螺钉; minus ～ 一字槽头螺钉; nut ～ 螺母, 螺帽; pinching ～ 固定螺钉, 夹紧螺钉; plug ～ 螺塞; pointed ～ 尖端螺钉; 12 point flange ～ * 十二角头法兰面螺栓; pointless ～ 平端螺钉; positioning ～ 定位螺钉, 调整螺钉; raised head ～

凸头螺钉; self tapping ～ 自攻螺钉; set ～ * 紧定螺钉, 止动螺钉; setand looking ～ 止动及锁紧螺钉; shoulder ～ * 轴位螺钉; slotted cheese head ～ * 开槽圆柱头螺钉; slotted countersunk flat head ～ * 开槽沉头螺钉; slotted pan head ～ * 开槽盘头螺钉; slotted raised countersunk ovalhead ～ * 开槽半沉头螺钉; slotted round head ～ * 开槽圆头螺钉; slotted ～ 有槽螺钉; socket ～ 凹头螺钉; socket head cap ～ 内六角螺钉; square ～ 方纹螺旋; square head ～ * 方头螺钉; stop ～ 制动螺钉; tapping ～ * 自攻螺钉; thread cutting ～ * 自切螺钉; three-start ～ 三头螺纹; thumb ～ 翼形螺钉; tommy ～ * 旋棒螺钉; T-slot ～ T 形槽螺钉; wing ～ * 翼形螺钉; wood ～ * 木螺钉; worm ～ 蜗杆

screw-gear 螺旋齿轮
screw-pitch 螺距, 螺节
screw-thread (thread) 螺纹
screwy /ˈskruːi/ 螺旋形的, 扭曲的
scribe /skraib/ 作家, 划线器; 划线, 缮写
scriber /ˈskraibə/ 划线器, 划线针
scribing /ˈskraibiŋ/ 划线
script /skript/ 底稿, 手稿
scrolling /ˈskrəuliŋ/ (屏幕显示)推进, 滚动, 上卷
scrutinise (scrutinize) /ˈskruːtinaiz/ 考查, 审查, 核对, 细看
S. E. (standard error) 基本误差
seal /siːl/ 密封, 闸, 隔离层, 绝缘(装置); 图章, 印记; ～ing equipment (device) 密封装置; oil ～ 油封; rubber oil ～ 橡胶油封; rubber ring ～ 胶圈密封; staffing box ～ 填料函密封
sealing /ˈsiːliŋ/ 封, 封接, 填缝, 填充物; ～ device 密封装置
sealed /siːld/ 封闭的

seam /siːm/ 缝; 接合面; button punched dovetail ～ 燕尾缝; standing ～ 压痕接缝
search /səːtʃ/ 检索, 测试, 寻找
SEC (section) 截面, 剖面
secant /ˈsiːkənt/ 正割, 割线, 割平面; ～ plane 截面, 正割面
sec. ar (sectional area) 截面积
secondary /ˈsekəndəri/ 第二的, 副的, 辅助的, 代表, 代理的; 二次的
second-order 二级的, 二阶的
sect. (section) 区域, 部门; 部分, 节, 段; 剖面, 断面, 截面, 截口
section /ˈsekʃn/ * 断面(图), 剖面图, 剖面, 截面(图), 切断, 剖开, 切下的部分, 切片, 一段, 一部分, 处, 科, 室, 股, 组, (机器的)零件, (文章的)节, (条文的)款, 项, (管子的)一段; ～ area 截面区域, 断面区域; ～ drawing 截面图, 断面图; ～ elevation 立剖面(图), 立面剖视(图); ～ line 剖面线; ～ liner 剖面线器; ～ lining 剖面线法(画剖面线的方法); ～ of the rail 钢轨截面; ～ (on line) A-A A-A 剖; ～ paper 制图用的格纸; ～ plane 剖切面, 切割面; ～ steel 型钢; ～ view 剖视(图); airfoil ～ 机翼断面; aligned ～ 校(转)直剖, 旋转剖; angular ～ 斜剖视; architectural ～ 建筑剖视图; assembly ～ 组合剖面; auxiliary ～ 辅助剖面; axial ～ 轴向截面; bench ～ 横截面法; bent offset ～ 用旋转剖切法画的剖视; bent-tube ～ 弯管部分; body ～ 机身, 床身, 基本部分; borehole columnar ～ 钻孔柱状图; broken-out ～ 破碎剖视, 断裂剖视, 局部剖视(图); built-up ～ (compound ～) 组合截面(建筑), 组合部分; center ～ 中心截面; circular ～ 圆形剖面; coal-seam correlation ～ 煤层对比图; coincidental ～ 重合截面; columnar ～ 柱状剖面; composite ～ 组合剖(视), 组合剖面;

compound ～ 复合剖（视）；conic ～ 圆锥截面，圆锥曲线，二次曲线；conic ～s 圆锥曲线，二次曲线，割锥线；conventional ～ 习用剖视，剖视的习惯画法；cross ～ 横断面，横截面，横断面图，横截画图；cut ～ 截口；dangerous ～ 危险断面；detailed ～ 详细截面；diagrammatic ～ 图解截面；dotted ～ 虚字剖面线；echelon ～ 用阶梯剖切法画的剖视；excavated ～ 开挖断面；fracture ～ 断裂剖面；fragmentary ～ 断裂剖面，局部剖面；full ～ 全剖视；full ～ed view 全剖视图；full ～ view 全剖视图；general ～ （房屋）全剖视图；geological ～ 地质剖面图；geological columnar ～ 地质（钻孔）柱状图；geometric cross ～ 几何横断面；half ～ 半剖视（图）；half round ～ 半圆剖（视）；horizontal ～ 水平剖（视）；horizontal geological ～ 水平地质剖面图；insufficient ～ 危险断面，不足部分；intermediate ～ 中部，中段；interpolated ～ 内插截面（重合截面）；isometric ～ 等角剖视；lateral ～ 横向（侧）剖面（图）；lengthwise ～ 长度方向剖面，纵剖面；local ～ 局部剖视（图）；longitudinal ～ 纵剖面，纵剖面图，纵断面图；main cross ～ 主（横）剖面；meridian ～ 子午截口，沿子午线之截面；normal ～ 正剖面，垂直剖面，法向剖面；oblique ～ 斜剖（视）；offset ～ 移位剖，阶梯剖，转折剖；part ～ 局部剖视图；partial ～ 局部剖视图；phantom (ghost) ～ 假想（幻影、阴影）剖视；plane ～ 平面截口；principal ～ 主剖面，主断面；radial ～ 径向截面；rear ～ 后部；rectangular ～ 矩形剖面；removable ～ 可拆卸部分；removed ～ 移出断面图；revolved ～ 旋转断面，重合断面；right ～ 正截面，主截面，正截口；roadway ～ 道路断面；round ～ 圆形剖面；separate ～ 个别截面，逐次断面，详细截面（用数个截面表示某处造型轮廓之变化）；shearing ～ 剪切面积；shift ～ 移动截面，多个截面，详细截面；spherical conic ～ 球锥剖面；staggered ～ view 阶梯状剖视图；steep ～ 阶梯剖；successive ～ 依次相续的断面；tail ～ 尾部；thin ～ 细薄剖面，窄截面；transverse ～ 横断面，横剖面，横断面图；trapezoid cross ～ 梯形截面；vertical ～ 垂直剖面，竖断面，竖截面；vertical geological ～ 垂直地质剖面图；wall ～ 墙壁剖面；working ～ 工作剖面；Z- ～ Z 形剖面

sectional /ˈsekʃənl/ 截面的，断面的，剖视的，局部的，部分的，部门的，章节的；由部分组成的；～ assembly 剖视图，断面图，截面图；～ view* 剖视图

sectioning /ˈsekʃəniŋ/ 剖面法，剖切；～ convention 习用剖切法；outline ～ 沿边剖面线；sliced ～ 截面；taper ～ 斜剖面法；type of ～ 剖面法种类

section-paper 方格纸

sections /ˈsekʃənz/ （法语）（不表示其他轮廓的）截面图，断面，断面；～ in two parallel planes 两平行面的断面；type of ～ 断面种类

sector /ˈsektə/ 扇形（物），区段，部分；象限

security /siˈkjuəriti/ 安全性；installation ～ 安装的安全性

seeming /ˈsiːmiŋ/ 外观上的，表面上的；外观，表面

seg. (segment) 弓形

segment /ˈsegmənt/ 扇形体，扇形块；段，线段，节，部分；图，图段，图块；弓形圆的一部分，片段，程序段；～ analysis 线段分析；～ attributes 图段属性；～s connection 线段连接；～ transformation 图段变换 associate ～ 图段；calling ～ 调用程序段；close ～ 关闭图段；connecting ～ 连接线

段；create ~ 建立图段；delete ~ 删除图段；detectable ~ 可检测图段；extremity of a ~ 线段之端点；given ~ 已知线段；insert ~ 插入图段；major ~ 优弓形；master ~ 主程序段；medium ~ 中间线段；minor ~ 劣弓形；open ~ 打开图段；procedure ~ 过程子程序（过程段）

segmental /seg'mentl/ (segmentary) 弓形，扇形的，圆缺的，球缺的，分割的，零碎的，辅助的

select /si'lekt/ 选择，精选

selection /si'lekʃn/ 选择，精选；选址（计算机的）；~ of instrument 仪器选择

selective /si'lektiv/ 选择的，局部的，部分的，分别的，优先的；~ hardening 局部硬化，局部淬火

semi- /'semi-/ 半，部分的，部分地；~ axis 半轴；~ circle 半圆；~ finished 半加工的，半光制的

semichord /semɪ'kɔːd/ 半弦

semicircle /'semisəːkl/ 半圆

semicircular /ˌsemi'səːkjulə/ 半圆形的

semicircumference /ˌsemisə'kʌmfərəns/ 半圆周

semicylinder /ˌsemi'silində/ 半柱面

semidiameter /ˌsemidai'æmitə/ 半径

semiellipse /ˌsemi'lips/ 半椭圆

semiellipsoid /ˌsemi'lipsɔid/ 半椭圆体

semielliptic /ˌsemi'liptɪk/ 半椭圆（式）的

semifin /semɪ'fain/ (semifinished /'semi'finiʃt/) 半加工，半精加工

semifinished /ˌsemi'finiʃt/ 半精加工的

semilog /ˌsemi'lɔg/ (semilogarithmic) 半对数；~ chart 半对数图表；~ paper 单对数坐标纸

semilogarithmic /ˌsemilɔgə'riθmik/ 半对数，单对数；~ ruling 单对数格线

semilunar /ˌsemi'ljuːnə/ 半月形的，新月形的，月牙形的

semi-major 长半；~ radius 长半轴

semi-minor 短半；~ radius 短半轴

semi-refined 半精制的

semisphere /'semisfiə/ 半球

semi-symmetric 半对称的

separate /'sepəreit/ 分开，独立的；~ element 分开各部分，形体分析；~ section 独立的剖视图

separator /'sepəreitə/ 分隔符；field ~ 区域分隔符（字段分隔符）；record ~ 记录分隔符

separatrix /ˌsepə'reitriks/ 分界线，分隔号

sequence /'siːkwəns/ 系统，顺序，次序；序列，数列，指令序列，数贯；程序；~ control tape 程序带；~ of function 函数序列；~ of number 数列，数的序列；arbitrary ~ 任意顺序，任意序列；build-up ~ 组合程序；coded ~ 编码序列；cohomology ~ 上同调序列；command ~ 指令序列；control ~ 控制程序；decimal ~ 十进序列；difference ~ 差数序列；fundamental ~ 基本序列；homology ~ 同调序列（下同调序列）；in ~ 顺序，挨次，逐一；irregular ~ 按次序，有条不紊地；minimal ~ 最小序列；operation ~ 工序；plain ~ 顺序；positive ~ 顺序，正序；pulse ~ 脉冲序列；starting ~ 起动程序

sequencer /'siːkwənsə/ 程序装置

sequencing /'siːkwənsiŋ/ 程序化，程序设计；automatic ~ 自动排序

ser. (series) 系列

serial /'siəriəl/ 连续的，顺次的，串联的

serialization /ˌsiəriəlai'zeiʃn/ 系列化

series /'siəriːz/ 系列，组，系统，连续，级数，串联；串行；丛书，刊物；~-parallel 串并联；dimension ~ * 尺寸系列；in ~ 串联式；outside diameter ~ 外径系列

serno (series number) 编号, 序列号, 顺序号; 串联数

serpentine /'sə:pəntain/ 蛇状的, 螺旋形的

serrate /'serit/ 锯齿形的, 使成锯齿状

serve /sə:v/ 为……服务; 符合, 对……适用

service /'sə:vis/ 服务, 业务, 工作; 检查, 维修; 机关, 部门; ～ charts 服务图

servo /'sə:vəu/ 伺服机构, 伺服系统; pulsed ～ 脉冲随动系统, 脉冲跟踪系统; two-input ～ 双端输入伺服机构

servo-amplifier 伺服放大器

servoanalyzer /sɜːˈvəunəlaɪzə/ 伺服分析器

servocontrol /'sə:vəkən'trəul/ 伺服控制系统, 伺服机构, 伺服补偿机; 伺服调整片

servo-driven 伺服系统驱动的

servodyne /'sə:vəudain/ 伺服系统的动力传动装置

servo-gear 伺服机构, 伺服装置; linear ～ 线性伺服机构; nonlinear ～ 非线性伺服机构; on-off ～ 开关伺服系统

servomechanism /'sə:vəu'mekənizəm/ 伺服机构, 随动系统; digital ～ 数字随动伺服机构; feedback ～ 反馈(式)伺服机构

servosimulator /sɜːˈvəusimjuleitə/ 模拟伺服机构

servosystem /'sə:vəu‚sistəm/ 伺服系统, 随动系统; digital ～ 数字跟踪系统, 数字随动系统; sampling ～ 脉冲随动系统, 脉冲跟踪系统

servounit /sɜːˈvəunɪt/ 伺服机构, 助力补偿器

set /set/ 放置, 使, 令; 装置, 仪器, 设备; 组, 套, 台; 集(合); ～ of drawing instruments 制图仪器; ～ off 截取, 分派; ～ out 开始, 布置, 装备; ～ square 三角板; ～ the belt on the wheel 把皮带装在轮子上; adjustable ～ square 可调三角规; character ～ 字符集, 符号集; closed ～ 闭集; empty ～ 空集; finite ～ 有限集; infinite ～ 无限集; non-empty ～ 非空集合; universal ～ 全集

set-on 定位, 安置, 确定, 调节

setover /'set‚əuvə/ 超过位置, 偏置, 偏位; ～ method 跨距法

setscrew /'setskru:/ 固定螺钉, 定位螺钉, 止动螺钉

set-square 三角板, 斜角规

setter /'setə/ 架

setting /'setiŋ/ 安置, 定位, 划线

set-up 固定, 装置, 机构, 组织; 准备; 设立, 建立; 装配, 安装; 配置, 布局; ～ diagram (计算系统)准备(工作框)图

sever /'sevə/ 切断, 割断; 把……分开; 断, 裂开

sewerage /'suəridʒ/ 排水

sexangle /'se‚ksægl/ 六角形, 六边形

sexangular /sek'sæŋjulə/ 六角(形)的

sextant /'sekstənt/ 六分仪

sextic /'sekstik/ 六次曲线, 六级; space ～ 空间六次曲线

SF (semifinish) 半光制; (spot face) 鱼眼坑

SFT (shaft) 轴

SG (standard gauge) 标准量规

SH (scleroscope hardness) 回跳硬度; (sheet) 图表; 张, 页; 薄钢板; (surface-hardness) 表面硬度

shade /ʃeid/ 阴影, 阴暗面, 阴线; 画阴线, 描阴, 着色; 画断面线; (遮)罩; ～ line 阴线; ～s and shadow 投影画, 阴影

shaded /'ʃeidid/ 画上阴影线的, 阴影之下的, 修色的; ～ area 阴线区域; line of ～ 阴界线

shading /'ʃeidiŋ/ 描阴, 润饰; 描阴法(画阴影线法); 遮光; brush-stipple ～ 洒点阴影法; continuous tone ～ 连续色调描阴,

深淡阴影法; hill ~ （地形图上的）山面描阴法; hill ~ by hachures 细线阴影法（地图）; line ~ 直线阴影，线条描阴法; line tone ~ 线条阴影法; smudge ~ 涂黑描阴（法）; surface ~ 表面描阴法

shadow /'ʃædəu/ 影子; 静区（雷达），遮蔽，阴暗（处）; ~ angle 影锥角; ~ area 投影面积; ~ graph 投影画，描影，阴影画; 阴影照相，逆光摄影; ~ line 阴面粗线; ~ lining 加阴影线; ~ moire 阴影纹; line of ~ 影界线

shadowgraph /'ʃædəugra:f/ 影图; X光照片; 放映检查仪器; 描影，影象图

shaft /ʃa:ft/ * 轴，轴柱，杆; 手柄，烟囱，通风管; ~ angle 轴间角; ~ -basis system of fits 基轴制; ~ line 轴线，传动装置; ~ neck 轴颈; basic ~ * 基准轴，主轴; castellated ~ 花键轴; crank ~ 曲轴; crose ~ 横轴; drive ~ 主动轴; driven ~ 从动轴; driving ~ 传动轴，驱动轴; eccentric ~ 偏心轴; gas ~ 通气管; gyro ~ 回转轴; hollow ~ 空心轴; keyed ~ 键轴; main ~ 主轴; revolving ~ 回转轴; solid ~ 实心轴; spline ~ 多键轴，花键轴; splined ~ 花键轴; transmission ~ 传动轴; worm ~ 蜗杆轴

shaftless /ʃa:ftles/ 无轴的

shank /ʃæŋk/ 柄，杆，末梢，后部

shape /ʃeip/ 形，形状，样子，轮廓，外形; 模型; ~ of pass 孔型断面轮廓; ~ -breakdown system 形体分解法; characteristic ~ 特性形状; complicated ~ 复杂形状，复杂零件; conic ~ 圆锥形体; geometric ~ 几何形状; interior ~ 内部形状; structural ~ 构造钢型（型钢剖面形状）; true ~ 实形，真形; true ~ property 真形性; undefined ~ 未定形状; unsymmetrical ~ 不对称形; ‖ *Example sentence:* The plate has a hole of the same shape and size as the workpiece. 例句: 板上有一个形状和大小都和工件相同的孔。

sharp /ʃa:p/ 尖锐的; 明显的; 轮廓鲜明的; 陡的; 准确的

sharpen /'ʃa:pən/ 刃磨，削; ~ pencil or compass lead 削尖铅笔或圆规的铅蕊

sharpener /'ʃa:pənə/ 削刀，锐化器; 磨床，刀具; mechanical ~ 机动铅笔刀; pencil ~ 铅笔刀

sharp-pointed 削尖的，尖锐的

shattering /'ʃætəriŋ/ 断裂

Shaw- /ʃɔ:/ -hardness 肖氏硬度

shear /ʃiə/ 切断，剪切，剪切器

shearing /'ʃiəriŋ/ 错切，剪切，剪断

shearing-off 切断，切面，切口，切片

shear-out 切开，截断; 切口，切痕; 剖面，断面

sheave /ʃi:v/ 滑（槽）轮，三角皮带轮; V ~ 三角皮带轮

sheer /ʃiə/ 全然的，绝对的，极薄的，峻峭的，垂直的; 舷弧; ~ plan 侧视图，侧面图

sheet /ʃi:t/ 图表，张，片，板; 薄钢板（片）; 单（据）; a ~ of paper 一张纸; assembly ~ 装配图，组合图; bill ~ 材料附单; colored ~ 色片; continuous production operation ~ 流水作业图表; contour ~ 等高线图; detail ~ 零件图; field ~ 现场测绘图; flow ~ 程序表，流程表; general location ~ 地盘（位置）图; job ~ 说明图; layout ~ 布置图; layout ~ of exploratory engineering 勘探工程分布图; operation ~ 工序卡，运算卡; overlay ~ 剪贴图; preprinted grid ~ 有格纸; surveying ~ 测量图

sheet-metal 金属片; ~ gage 金属片规，塞规

shell /ʃel/ 外壳，壳体，套（管），罩，轴衬; ~ casting 壳形铸件; ~ structure 薄壳结构; outside ~ 外壳

sherardise (sherardize) /'ʃerədaiz/ 镀锌
shield /ʃi:ld/ 防护装置，掩蔽盾；crasing ～ 擦图片，擦图孔板
shift /ʃift/ 变换，改变，移动，位移；移位，移数，调动；变速，换挡；～ key 变位键；back ～ 二次变换；第二班；gear ～ 变速箱
shift-in (SI) 移入
shift-out (SO) 移出
shim /ʃim/ 薄垫片，(楔形)填隙片
shine /ʃain/ 发光，发亮，光泽
shop /ʃɔp/ 车间，工场，商店；～ drawing 制造图
shore /ʃɔ:/ -hardness 回跳(肖氏)硬度
short /ʃɔ:t/ 短，缺，不足；～ break 短断裂线
short-addendum 短齿
shortcutting /'ʃɔ:tkʌtiŋ/ 简化
shorten /'ʃɔ:tn/ 缩短，减少；～ing coefficient 缩短系数
shorthand /'ʃɔ:t,hænd/ 速记法；～ notation 简化符号；略号
shot /ʃɔt/ -blasting (shot-peening) 喷丸清理
shot-welding 点焊
shoulder /'ʃəuldə/ 轴肩，底，凸缘
show /ʃəu/ 展示，显示，指出，示出，表示，说明，证明；～ construction 以图表示；～ dimensions 示出尺寸；～ give or required data 示出给定或待求的数据；～ us the blue print 把兰图给我们看看；as ～ in fig. 如图所示；‖ *Example sentence: This picture shows China's first 12.000-ton hydraulic free forging press.* 例句：图示为中国制造的第一台12.000吨自由锻水压机。
shrink /ʃrink/ (shrank. shrunk, shrunken)收缩，缩减；～ cavities 缩孔；～ hole 缩孔；～ rule 收缩尺；～ to nothing 渐渐缩小到没有
Sh. s (sheet steel) 钢板
shut /ʃʌt/ 折叠，关闭
shutter /'ʃʌtə/ 活动百页窗；闸板，阀

SI (法文)(systeme Internationald' Unit'es) (=International system of Units) 国际单位制
SI. (shift-in) 移入
side /said/ 边，旁边；端；面，侧面；方面；侧面的，旁边的；枝节的；～ elevation 侧(立)面图；～ elevation view 侧(立面)视图；～ glance 斜视，侧视，暗示；～ view 侧视图；adjacent ～ 邻边；front ～ 前面，正面，正面图；hair ～ 毛面；not-go ～ 上端，不通过端(量规)；opposite ～ 对边
sideline /'saidlain/ 边缘线，边缘区
sidewise /'saidwaiz/ 横向地，侧向地
sight /sait/ 视力，视觉，视界，视野，观测；in ～ (of) 看得见，在视线范围内；line of ～ 视线；to lose ～ of 看不见......，忽略，忘记
sign /sain/ 记号，符号，标志；签署，加符号于；combined plus and minus ～ 正负组合符号；conventional ～ 习用符号；qraphical ～ 图形标志
signature /'signətʃə/ 签名；特征；标记图，图像；add (put) one's ～ to 签名于......
sill /sil/ 窗台
similar /'similə/ 同样的，相似的，类似的；相似；～ views 相似的视图
similarity /simi'læriti/ 相似(性)，相像，类似(性)；～ transformation 相似变换；center of ～ 相似中心
similar-shape 类似形
simple /'simpl/ 简单的，简易的
simplification /simplifi'keiʃn/ 简化
simplify /'simplifai/ 简化
simulate /'simjuleit/ 模拟，仿真
single /'siŋgl/ 单一的，个别的，专一的；～ auxiliary view 单辅视图；～-bevel scale 单面比例尺；～-curved surface 单曲面；～-screen 单屏幕；～-stroke lettering 单笔字
sink /siŋk/ 凹陷，沉下，缩孔，挖掘；counter ～ 锥坑

sinusoid /ˈsainəsɔid/ 正弦曲线, 窦状小管

site /sait/ 地点, 位置, 场所, 现场, 工地; 部位; ～ plan 总平面图, 平面布置图, 总设计图, 地盘图

siting /ˈsaitiŋ/ 设计图, 图案, 平面布置, 建筑工地选择

situate /ˈsitjueit/ (situated) 使位于, 使处于; 位于……的

size /saiz/ 大小, 尺寸; 体积, 度量; 规格; ～ description 绘图尺寸; ～ dimension 大小尺寸; ～ up 估计大小, 估量, 测量大小; absolute ～ 绝对尺寸; actual ～ * 实际尺寸; aperture ～ 孔径尺寸; basic ～ * 基本尺寸; common trimmed ～ 修正尺寸; dead ～ 净尺寸; design ～ 设计尺寸; deviations of ～ 尺寸偏差; drawing sheet ～ 图纸大小; drawing ～ 图的大小; double ～ 双倍尺寸; double elephant ～ 英式最大图纸大小 (40″×27″); eighth ～ 1又1/2″=1又1/2″ 比例尺; enlarged ～ 放大尺寸; feature ～ 特征尺寸, 形体尺寸; final ～ 成品尺寸; free ～ 自由尺寸; full ～ 实足尺寸, 原尺寸, 1:1的图形尺寸; given ～ 已知大小; half the ～ of 尺寸为……的一半; half ～ 1:2缩尺; intended ～ 公称尺寸, 给定尺寸, 额定尺寸; least material ～ * 最小实体尺寸; ledger ～ 美国11″×17″标准纸; legal ～ 美国8又1/2″×14″标准纸; letter ～ 美国8又1/2″×11″标准纸; limiting ～ 极限尺寸, 界限尺寸; limits of ～ * 极限尺寸; mating ～ 作用尺寸; maximum ～ 最大尺寸; maximum limit of ～ * 最大极限尺寸; maximum material ～ * 最大实体尺寸; measured ～ 实测尺寸; minimum ～ 最小尺寸; minimum limit of ～ * 最小极限尺寸; much of the ～ 差不多尺寸; nominal ～ 标称尺寸, 公称尺寸, 名义尺寸; one-half ～ 半尺寸; over ～ 尺寸过大; overall ～ 总尺寸, 轮廓尺寸, 外廓尺寸; quarter ～ 1/4尺寸; real ～ 实际尺寸; regardless of feature ～ (RFS) 不考虑特征尺寸; sixteenth ～ 1:16缩尺; specified ～ 公称尺寸, 名义尺寸; standard ～ 标准尺寸; standard drawing sheet ～s 标准图纸尺寸; statement ～ 美国5又1/2″×8又1/2″标料纸; stock ～ 备料尺寸, 标准尺寸; 库存 (货) 量; take the ～ of 量……的尺寸; theoretical ～ 理论尺寸; toleranceof ～ * 尺寸公差; true ～ 实大, 真实尺寸; twice full ～ 2:1放尺; under ～ 过小, 尺寸不足

sizing /ˈsaiziŋ/ 量尺寸, 定大小; 填料; 校正

skeleton /ˈskelitn/ 轮廓, 构架, 结构; 梗概; 草图, 略图; ～ diagram 单线图, 原理图, 概略图; ～ drawing (lay-out)草图, 简图, 原理图, 结构图, 轮廓图; ～ sketch 轮廓草图, 构架图

skeletonise (skeletonize /ˈskelitənaiz/) 绘草图, 记梗概

skeletonizing /ˈskelitənaiziŋ/ 绘制草图

sketch /sketʃ/ * 草图, 略图; 画草图, 画示意图, 速写; 草稿; ～ book 草稿本; ～ for marking off 号料草图 (船舶); ～ map 草图, 略图; 画草图, 设计图; ～ of brick types 砖型图; aerophotographic ～ 航空摄象略图; architect's ～ 建筑草图; assemble-equilibrium ～ of the rotor of an aero-engine 航空发动机弹性轴装配平衡图; assembly ～ 装配草图, 组合草图; computation ～ 计算草图; design ～ 设计草图; detail ～ 详细草图; detail ～ 零件草图; diagrammatic ～ 示意图, 草图; executive ～ 执行草图; eye ～ 目测图, 草图; field ～ing 现场草图; field ～ 现场测绘; free hand ～ 徒手草图; ground ～ 地形略图; idea ～ 构思草图; maked ～ of 画出……的草图; orthographic ～ 正投影草图; one-point perspective ～ 一点透视草图; outline ～ 轮廓草图, 轮廓图; rough ～

草图; traverse ~ 导线草图(测量)

sketching /'sketʃiŋ/ 划线; 画草图; 画示意图; ~ materials 画草图的材料; ~ technique 画草图的技巧; architectural freehand ~ 建筑徒手草图; axonometric ~ 轴测投影草图; cartographical ~ 手制草图; technical ~ 技术草图

sketchy /'sketʃi/ 大体的,草图的,粗略的

skew /skju:/ 斜的,歪的,偏的,弯曲的; 斜交的, 非对称的; ~ curve 挠曲线,空间(不对称)曲线; ~ gear 交错轴(双曲面)齿轮; ~ line 斜线; ~ ruled surface 不可展直纹曲面; ~ symmetric 斜对称的,反号对称的

skewlines /skju:laɪnz/ 交叉线

skiagraph /'skaɪəgra:f/ 纵断面图,房屋立面图,投影图,X 光照片

skill /skil/ 技巧,技能;熟练,技术,技艺; have ~ in engineering drawing 擅长工程画

skin /skin/ 表皮,外壳; the ~ of an ultrasanic plane 超音速飞机的外壳

skip /skip/ 跳,跳过

skirt /skə:t/ 边缘,边界

skylight /'skaɪlaɪt/ 天窗

skyline /'skaɪlaɪn/ 地平线

S. L. (straight line) 直线

slab /slæb/ 板

slant /sla:nt/ 斜,倾斜,弄斜; ~ from M toN 从 M 向 N 倾斜; ~ wise 倾斜的,歪斜的

SLAR (slant range) 斜距

slate /sleit/ 石板

slew /slu:/ 旋转

slice /slais/ 薄片,切成薄片;切下,切开; ~ section 薄切剖面(即个别剖面)

slide /slaid/ 滑板,幻灯片;滑,滑去;(地质的)断层; ~ rule 计算尺,滑尺

slide-rule 计算尺,滑尺; ~ dial 游标刻度; ~ nomogram 计算尺型列线图

slimline /'slimlaɪn/ 细线,细(长)管

slit /slit/ 切开,纵割;缝,槽,切口

slitting /'slitiŋ/ 切口,切缝,纵切,纵向切分

slitting-up 全切开; ~ method 全切开法

slope /sləup/ 斜度,坡度,斜率;倾斜,坡道,斜面,斜线;滚动; ~ gradient 坡度; ~ indication lines 示坡度; ~ intercept form 斜截式(解析几何); ~ sat 60° 倾斜60°; at a 1 to 1.5 ~ 按 1:1.5的斜度

slot /slɔt/ 缝,槽;开缝;狭槽,小孔,切口;长(方形)孔; cam ~ 凸轮槽,曲线槽; dove-tail ~ 燕尾槽; key ~ 键槽; rim ~ 轮缘槽; side ~ 边坡; T- ~ T 形槽

small /smɔ:l/ 小的,细的,不重要的,少的; ~ end 小端,小头; ~ in size 尺寸小; ~ letters 小写字母; ~ size 小尺寸

small-scale 小型的,小尺寸的,小比例尺的

small-sized 小型的

smooth /smu:ð/ 光滑,光滑的,平滑的,平坦的;校平

smoothness /'smu:ðnis/ 平滑度,光滑

smudge /smʌdʒ/ 污迹,污点

snap /snæp/ 卡规,铆头模,窝模; ~ head 半球头; internal and external ~ 内外径卡规

So. (shift-out) 移去

socket /'sɔkit/ 插座,承窝,套接; ~ head 内孔(六角)头; ~ head cap screw 内六角螺钉

soffit /'sɔfit/ (建筑物)底面

soft /sɔft/ 软的,坡度小的

softening /'sɔfniŋ/ 软化,退火

software /'sɔftwɛə/ (计算机)软件;程序设备;设计计算方法; ~ for display 显示软件; ~ package 软件包; application ~ 应用软件; graphics ~ 绘图软件

SOH (start of heading) 字头,内容开始,标题

开始

solid /'sɔlid/ 立体,固体;立体的,固体的,实心的,坚固;～ angle 立体角;～ axes 空间坐标(轴);～ bounded by plane surface 由平面构成的立体;～ geometry 立体几何;～ line 实线;～-line curve 实线曲线,连续曲线; geometric ～ 几何形体; geometric plane ～ 平面组成之几何体

solidness /'sɔlidnis/ 硬度

solution /sə'lu:ʃən/ 解决,解题;解释,说明,分解,解答;胶水; graphical ～s 图解法

solve /sɔlv/ 解决,解答,溶解,说明;～ing process 解法,解题步骤

sort /sɔ:t/ 排序,分类

so-so 平常的,一般的,马马虎虎,还过得去

south /sauθ/ 南,南方;～ by east 南偏东

SP (series parallel)复联,串并联; (space) 空间,地方,距离,间隔; (special) 特殊的,专门的; (specific) 特殊的,专门的;特性,详细说明书; (standard pitch) 标准线距

space /speis/ * 间距;空间,宇宙,太空;地方,场地;间隔,距离,空白,区;～ analysis 空间分析;～ diagram 空间图,立体图,位置图,矢量图;～ height 空间高度;～ of three dimension 三维空间;～ problems 空间问题; available ～ 可用范围; blank ～ 空白范围; clearance ～ 余隙; higher ～ 高维空间,高度空间; intertooth ～ 齿间; quartuple ～ 四维空间,四度空间; quintuple ～ 五维空间,五度空间; septuple ～ 七维空间,七度空间; tooth ～ 齿间隔; topological ～ 拓扑空间; topologically complete ～ 拓扑完全空间; vector ～ 矢量空间,矢性空间; virtual ～ 虚拟空间; visual ～ 视空间

spacer /'speisə/ 隔片,隔离器,分离器,分规;～ point 分规针头

spacers /'speisəs/ 弓形小分规

spaces /speisiz/ 间开,拉开

spacewise /speiswaiz/ 空间坐标

spacing /'speisiŋ/ 间隔,距离,空白,空隙,位置,布置,留间隔; center-to-center ～ 中心距

span /spæn/ 跨距,翼展(飞机两翼端之距); central ～ 中心跨距; extreme ～ 最大跨距; total ～ 总跨度

spanner /'spænə/ 扳手; bent ～ 弯头扳手; crocodile ～ 鳄头扳手; hexagon ring ～ 六角孔扳手

spare /spɛə/ 多余的,备用的,备件

spat /spæt/ 流线形的轮盖

spatial /'speiʃəl/ 空间的,存在于空间的;篇幅的;立体的

spec. (specification) 规范,细则,说明书,一览表

special /'speʃəl/ 特殊的,特别的,附加的;主要的;～ celluloid tools 专门的赛璐珞模板(在赛璐珞薄片上刻的图案模板);～ drafting tool 特制绘图模板;～ position of view 视图的特殊位置,特殊位置视图

specific /spə'sifik/ 特殊的,确定的;比,比较的; unless otherwise ～ed 除另有规定外

specification /spesifi'keiʃn/ 详细说明,说明书;明细表,清单;规格,规范;技术要求,技术条件,列举,逐一指明; meet the ～ 符合规格,满足技术要求; standard ～ 标准规格

specify /'spesifai/ 规定,指定,确定;列举,详细说明; unless otherwise ～ied 除非另有规定

specimen /'spesimin/ 样品,样机;试料

speed /spi:d/ 快,迅速;速度,速率; drawing ～ 绘图速度

SPG (spring) 弹簧

sphere /sfiə/ 球,球体,范围;球面;～ tangent 切球面; oblate ～ 扁圆球; stlll ～ 钢球

spheric(al) /'sferik(əl)/ 球形的

sphericity /sfeˈrisiti/ 球状圆体，球形度，圆球度，球表面积对体积的关系

spherics /ˈsferɪks/ 球面几何学，球面三角学

spheroid /ˈsfɪərɔɪd/ 球状体，回转扁(椭)球(体); oblate ～ 扁球体，椭球体

spheroidal /sfɪəˈrɔɪdəl/ 球状的，球体的，椭球体的

spherometer /ˌsfɪəˈrɒmɪtə/ 球径计，球面曲率测量仪

spigot /ˈspɪgət/ (plug) 塞，孔堵，插头

spin /spin/ 旋转

spindle /ˈspindl/ 轴，心轴; sleeve ～ 套轴; splined ～ 花键轴; tail ～ 顶尖轴; tapering ～ 锥形轴

spiral /ˈspaɪərəl/ 螺旋的，螺纹的；螺旋，螺旋线，螺旋状，涡线; ～ angle 螺旋角; ～ burr 螺纹; ～ gear 斜齿轮，螺旋齿轮; ～ of Archimedes 阿基米德涡线; Archimedean ～ 阿基米德涡线

spiral-rise 螺旋上升

splendent /ˈsplendənt/ 发亮的，有光泽的；显著的

spline /splain/ * 花键；曲线软片，挠性曲线尺，活动曲线规；缝隙片；方栓，齿槽，齿条；开槽；嵌键于；用花键连接；仿样，样条，样条函数; ～ function 样条函数; ～ shaft 花键轴; ～ weight 曲线条尺(压铅); B-～ B样条函数; external ～ 外花键; internal ～ 内花键; involute ～ * 渐开线花键; rectangle ～ * 矩形花键; spring ～ 弹簧键

split /split/ 劈裂，分开，裂缝，部分的，对开的，开口的，开尾的; ～ bolt 开尾螺栓; ～ nut 开缝螺母; ～ pin 开尾销，开口销; ～ ring 开口环

SPNR (spanner) 扳手

spoke /spəuk/ 辐，舵轮把柄; wheel ～ 轮辐

spot /spɒt/ 点，圆点，光点，斑点；地点，部位；发现，看出；局部的，少量，少许；在……上用点子作记号；打点，钻定心孔; ～ annealing 局部退火; ～ face 鱼眼坑，螺钉孔上刮的平面，切鱼眼; ～ map 点示图; center ～ 钻孔导向点; high ～ 突出部分; on the ～ teaching 现场教学

spot-face 鱼眼坑，切鱼眼

spotting /ˈspɒtɪŋ/ 测定点位，钻定心孔

spr (spacer) 垫圈，垫片

spread /spred/ 展开，伸展，伸开，散布，范围；加宽，扩充，拉长；涂(漆)；距; ～ of axles 轴间距

spring /sprɪŋ/ 弹簧，弹性；发生，跳起；泉水，起点; ～ assembly 簧片组; ～ bow instrument 弹簧弓型仪器; ～ cotter 开口销，开尾销; ～ of curve 曲线起点; ～ pitch 弹簧节距; belleville ～ * 碟形弹簧; blade ～ 板簧; coil ～ 螺旋弹簧，盘簧; compression ～ 压缩弹簧，压力弹簧; conical ～ * 截锥螺旋弹簧，锥形弹簧; cup ～ 盘形弹簧; cylindrical ～ 圆柱弹簧; cylindrical helical compression ～ * 圆柱螺旋压缩弹簧; cylindrical helical extension ～ * 圆柱螺旋拉伸弹簧; cylindrical helical torsion ～ * 圆柱螺旋扭转弹簧; cylindrical helical ～ * 圆柱螺旋弹簧; diameter of ～ wire 簧丝直径，电线直径; disk ～ 盘形弹簧; draught ～ 拉力弹簧; extension ～ 拉伸弹簧，拉力弹簧; flat ～ * 片弹簧; helical ～ * 螺旋弹簧; inside diameter of ～ 弹簧内径; leaf ～ * 板弹簧; outside diameter of ～ 弹簧外径; semi-elliptic leaf ～ 弓形板弹簧; spiral ～ * 平面涡卷弹簧，螺旋弹簧; tension ～ 拉力弹簧; torsion ～ 扭力弹簧; volute ～ * 截锥涡卷弹簧，涡线弹簧，锥形弹簧; wire ～ * 异形弹簧，细(线)弹簧

spring-style 草图设计
sprite /sprait/ 鬼怪图形
sprocket /ˈsprɔkit/ 链轮,星形轮
spur /spəː/ 迹,痕迹;齿,齿轮;排出口,孔;推动;～ bevel gear 正伞齿轮,直齿圆锥齿轮;～ gear 正齿轮,圆柱直齿齿轮;～ pinion 直齿小齿轮;～ rack 齿条
spurnormal /spəːˈnɔːməl/ 最大斜度线(迹线垂直线);the ～ to H 对 H 面的最大斜度线;the ～ to V 对 V 面的最大斜度线;the ～ to W 对 W 面的最大斜度线
spur-parallel 迹线平行线
spur-wheel 正齿轮
sq (square) 方形的,平方的,二次的,正方形
sq. ft 平方英尺,□′
sq. in 平方英寸,□″
square /skwɛə/ 平方,正方形,方场,方格,直角尺,矩尺;～ed paper 方格纸,坐标纸;～ to (with) 与……成直角,垂直于;adjustable angle ～ 活动角尺;bevel ～ 斜角规;caliber ～ 测径尺;center ～ 求圆心规;combination ～ 组合角尺;on the ～ 成直角;out of ～ 不成直角,不规则,歪斜,不正确;rootmean ～ 均方根;set ～ 三角板(英);steel ～ 直角钢尺;T- ～ 丁字尺;tee ～ 丁字尺;triangular set ～ 三角板;try ～ 矩尺,直角尺
squared /ˈskwɛəd/ 闭合
square-edged 方边的,成90° 角的
squarely /ˈskwɛəli/ 成方的,方方正正的;笔直,对准;正面地
square-neck 方颈
squareness /ˈskwɛənis/ 垂直度,(面)垂直度;方(形),方形度,正方度
squint /skwint/ 倾斜,偏移,倾向于,斜视的,斜视角,斜孔小窗
SR (split ring) 开口环;(study requirement) 学习要求

S/R (slant range) 斜距
SS (sections) 截面,区域,部门;(set screw) 定位螺钉
stack /stæk/ 堆积,叠加;组套;竖管,烟囱
stage /steidʒ/ 级,阶,层;分级;程度,阶段,步骤;～ pump 多级泵
stagger /ˈstægə/ 交错,错开;参差;交错的,错开的,错开排列;～ed section view 阶梯剖(视图);～ing dimension 交错排列尺寸
stair /stɛə/ 楼梯,阶梯;～ case 楼梯;～ chart 阶形图(靠拢的铅直式统计图表,象台阶形)
stair (-) way 楼梯,阶梯,梯子,楼梯间
stamp /stæmp/ 标出,表明;痕迹,印模;冲压,压花;特征;种类;模具;商标
stand /stænd/ 座;box ～ 箱座
standard /ˈstændəd/ 标准,规格,规范,样品,标准的;落地式的;机架;～ drawing 标准图;～ drawing sheet sizes 标准图纸尺寸;～ parts 标准零件;～ specification 标准说明书;administration ～ 管理标准;American National ～ (ANS) 美国国家标准;basic ～ 基础标准;be up to ～ 合乎标准;Chinese National ～s 中国国家标准;company ～ 企业标准;drafting-room ～ 制图室标准;engineering construction ～ 工程建设标准;environmental protection ～ 环境保护标准;internal ～ 内部标准;international ～ 国际标准;international metric ～ 国际公制标准;Japanese Industrial ～ (JIS) 日本工业标准;line ～ 线标准;method ～ 方法标准;ministerial ～ 部标准;national ～ 国家标准;product ～ 产品标准;SAE ～s SAE 标准;safety ～ 安全标准;specialized ～ 专业标准;system of ～ 标准体系;technical ～ 技术标准;technique ～ 工艺标准;tentative ～ 试用标准,暂行标准,标准

standardization 138

草案; terminological ～ 术语标准; unified ～ 统一标准; United States ～ 美国标准; working ～ 通用标准, 技术单位

standardization /ˌstændədaiˈzeiʃn/ 标准化

standard-sized 标准尺寸(大小)的

standing /ˈstændiŋ/ 直立的, 固定的

stand-off 远距离的; 投射的; ～ distance 投射距离

start /staːt/ 开始, 出发; 引起, 启动, 开动; 创建; ～ing plane 起画面(画轴测投影时, 最初画的物体上的面); ～ing point 起点; graphics ～ 绘图开始; plotter ～ 绘图机起动

stated /ˈsteitid/ 规定的, 用符号表示的; 陈述; unless otherwise ～ 除另有说明外

statement /ˈsteitmənt/ 语句(计算机), 陈述; control ～ 控制语句; drawing ～ 图注; executable ～ 可执行语句; external ～ 外部语句; general ～ 概述; input/output ～ 输入/输出语句; non-executable ～ 非执行语句; specification ～ 说明语句; subprogram ～ 子程序语句; type ～ 类型语句

state-of-(the)-art(s) 技术发展水平, 科学发展动态, 工艺现状

state-specified 国家规定的

station /ˈsteiʃn/ 测量点; 位置, 站, 台; ～ map 火车站图; ～ orainates 站线(船舶); ～ point 驻点(透视图中的观察点)

st. c. (steel casting) 铸钢件

STD (standard) 标准, 标准的

stdn (standardization) 标准化

steel /stiːl/ 钢, 钢制的, 钢筋; ～ casting 铸钢; all ～ 全钢; alloy ～ 合金钢; angle ～ 角钢; carbon ～ 碳素钢; cast ～ 铸钢; channel ～ 槽型钢; chrome ～ 铬钢; cold drawn ～ 冷拉钢; cold-rolled ～ (CRS) 冷轧钢, 冷乳钢; corrosion-resistant ～ 耐蚀钢; hot-rolled ～ 热轧钢; joist ～ 工字钢, 梁钢; manganese ～ 锰钢; mild ～ 低碳钢; nickel ～ 镍钢; profile ～ 型钢; round ～ 圆钢; SAE ～s SAE 规格钢; shape ～ 型钢; spring ～ 弹簧钢; stainless (non- rust) ～ 不锈钢; tee ～ 丁字钢; tool ～ 工具钢

steering /ˈstiəriŋ/ 转向

stem /stem/ 字根, 植物之杆, 杆, 柄

step /step/ 步, 级, 梯阶, 台阶; 阶段; 踏步; 步骤; 踏板, 节距, 跨距, 行程; 轴瓦; ～ off 量出, 量开, 量离; compilation ～s 编译步骤; stair ～ 楼梯踏步; the ～ of construction of drawing 作图步骤

STEP (standard for the exchange of product model data) 产品模型数据交换标准

step-like 阶梯形的(指曲线)

stepped /stept/ 分级的, 阶梯的, 有台阶的

stereo /ˈstiəriəu/ 立体, 立体照片; ～ comparator 体视比较仪, 立体坐标量测仪

stereoautograph /ˌstiəriəˈɔːtəgraːf/ 体视绘图仪

stereogram /ˈsteriəgræm/ 实体图, 立体图, 体视图, 极射图, 赤面投影图; 立体照片; 多边形

stereograph /ˈstiəriəgraːf/ 立体平面图, 立体照片

stereographic(al) /ˌstiəriəˈgræfik(əl)/ 立体几何学的, 立体画法的, 立体照相的

stereography /ˌsteriˈɔgrəfi/ 体视法, 立体平画法, 立体几何学

stereometry /ˌstiəriˈɔmitri/ 立体几何(学), 测体积术

stereopair /ˈstiəriəˌpɛə/ 体视对

stereo-perception 立体感

stereopicture /ˈstiəˈriːpiktʃə/ 立体图像

stereoplanigraph /ˌstiəriəuˈplenigraːf/ 精密立体测图仪

stereoplotter /ˈsteriəplɔtə/ 立体测图仪, 立体绘图仪

stereoscopic(al) /ˌstiəriəˈskɔpik(əl)/ 立体的,体视(镜)的,立体镜

stereotome /ˈsteriːəutəm/ 立体图片

stick /stik/ 杆,棒,棍;粘贴,伸出;手柄;～ out 伸出长度

stiffener /ˈstifənə/ 支肋,加强板,弯曲肋力板

stigmatic /stigˈmætik/ 共点的

stigmatize /ˈstigmətaiz/ 描绘成

stimulate /ˈstimjuleit/ 促进,激发

stipple /ˈstipl/ 点画,点描(不用线条)

stippling /ˈstipliŋ/ 洒点法(描阴影),洒点

stl (steel) 钢

stlc (steel cast) 钢铸件

STN (stainless) 不锈钢

stock /stɔk/ 托盘;后把;钻柄;材料;平凡的,普通的;～ key 普通(平)键;～ size 备料尺寸,标准尺寸

stone /stəun/ 石,砂轮;oil ～ 油石,油磨石

storage /ˈstɔːridʒ/ 存贮;存贮器

store /stɔː/ 储存,存入,库,店;～ closet 储藏室

storey /ˈstɔːri/ 室(楼),层;the first ～ (英)二楼;(美)底层,一楼

STR. (straight) 直线,直的

straight /streit/ 直,直线;直接;直的,马上,连续的;～ angle 平角;～ between two curves 两弧间直线;～ cut 纵向切割;～-cut gear 直齿齿轮

straightaway /ˈstreitəˌwei/ 直线段,笔直的,直线行进的

straightedge /ˈstreitˌedʒ/ 直尺,标尺,用直尺检验;把……一边弄直

straight-face 直线面

straight-line 直线(的)

straightness /ˈstreitnis/ 直线度,直线性;正直度,平直度

stratum /ˈstreitəm/ (地)层

streamline /ˈstriːmlain/ 流线型;使现代化;使合理化;革新;to ～ (fairing) 使流线型化,使曲线或曲面光滑

streamliner /ˈstriːmlainə/ 流线形物

street /striːt/ 街,道,马路

strength /streŋθ/ 强度,力(量);bending ～ 抗弯强度;compressive ～ 抗压强度;shear (ing) ～ 抗剪强度;tensile ～ 抗拉强度

stress /stres/ 应力,受力状态;压力;着重点,强调;～ concentration 应力集中;～ diagram 应力图

stretch /stretʃ/ 伸展,伸张,伸长;拉长;加宽,铺开,展开;～ force 张(拉)力;～-out (小延伸或延伸的)展开法

stretching /ˈstretʃiŋ/ 延伸,伸展,拉长,伸缩变换

stretchout /ˈstretʃaut/ 伸长线;～ line 把……拉直的线,把……拉长的线,周围线

strict /strikt/ 笔直的;严格的,精确的

strike /straik/ 刺透,穿透;组成;落在……上,照到;敲,击;～ off 清楚正确地描绘;～ out 做成,设计成

striking /ˈstraikiŋ/ ～ out 展平(皮革)

string /striŋ/ 绳,带,线,索,弦;伸直,拉直;排成一串,成串地展开;行,列,串;字符串;alphabetic ～ 阿拉伯数字串;binary-digit ～ 二进制数串;character ～ 字符串;symbol ～ 符号串

strip /strip/ 拆卸,拆开,剥去

stroke /strəuk/ * 螺纹行程(写字、绘画的)一笔,一划,笔划;～ edge 笔划边缘;～ centerline 笔划中线;～ filtering 筛选;～ function 笔划功能;～ input 笔划输入;open-style arrowhead ～ 开尾式箭头画法;～ width 笔划宽度

structure /ˈstrʌktʃə/ 构造,结构,机构;assembling ～ 装配结构;casting ～ 铸件结

构; data ～ 数据结构; drilling ～ 钻孔结构; forging ～ 锻件结构; machining ～ 机械加工件结构; technical ～ 工艺结构

stub /stʌb/ 短, 残片, 存根; 短(截)线; ～ line 短截线, 线段

stud /stʌd/ * 双头螺柱, 螺柱; 轴, 销子, 接合器; 柱头螺栓; 墙筋; ～ bolt* 全螺纹螺柱, 双头螺柱; ～ end 柱栓端, 旋入端; ～ nut 柱栓螺母; ～ with undercut (groove)* 带退刀槽的螺柱; waisted ～ * 腰状杆螺柱; welded ～ * 焊接螺柱

stuffing /'stʌfiŋ/ 填料, 填充; ～ box 填料盒, 填料函; ～ gland 填料压盖

S-turn S形转弯; split ～ 拼合S形转弯

STWY (stairway) 楼梯, 阶梯

style /stail/ 名称, 称呼; 型, 款式, 态度, 样式, 格式

stylographic /stailə'græfik/ 尖头铁笔的, 尖头铁笔书写的; ～ pen 管形笔

stylus /'stailəs/ 记录针, 笔尖, 笔头, 感应笔, 触笔; curve following ～ 描绘曲线的笔尖

SU (servo-unit) 伺服机构, 随动机构, 助力机构

subassembly /sʌbə'sembli/ 部件, 组合件, 组件, 机组; 局部装配; ～ drawing 局部组合图, 部件装配图; ～ of pump 水泵装配图

subdivide /sʌbdi'vaid/ 细分, 再分, 复分

subdivision /'sʌbdiviʒən/ （城市）分区图

subduple /sʌb'dju:pl/ 二分之一的

subgrade /'sʌbgreid/ 路基, 地基; high-way ～ 公路路基

subgraph /'sʌbgra:f/ 子图; complete ～ 完全子图; fuzzy ～ 模糊子图; null ～ 空子图

subject /'sʌbdʒikt/ 目的物; 主题, 题目, 学科, 科目; 使服从, 隶属的

subpoint /'sʌb,point/ 投影点, 下点

subprogram /sʌb'prəugræm/ 子程序; ～ statement 子程序语句

subquadrate /sʌb'kwɔdrit/ 近正方形的; 正方形带圆角的

subquadruple /sʌb'kwɔdrupl/ 1:4的, 四分之一的

sub-quality 不合格

subquintuple /sʌb'kwintjupl/ 1:5的, 五分之一的

subrogate /'sʌbrəgeit/ 代替, 取代

subroutine /'sʌbru:'ti:n/ 子程序; ～ level （电脑）子程序级; axes ～ 画坐标轴的子程序; basic ～ 基本子程序; function ～ 功能子程序

subscriber /səb'skraibə/ 用户

subscript /'sʌbskript/ 下标, 角注, 写在字母右下角的; ～ed variable 下标变量（数组元素）

subsequence /'sʌbsikwəns/ 顺序, 子序列

subset /'sʌbset/ 子集; character ～ 字符子集, 符号子集

substandard /sʌb'stændəd/ 不标准的, 次级标准

substitute /'sʌbstitju:t/ 代替, 替换, 取代; 代替人, 代替物

subtend /səb'tend/ 对; 弦对弧, 边对角; a hypotenuse ～s a right angle 斜边对直角

subtitle /'sʌb,taitl/ 副标题, 小标题

subtriangular /sʌb,trai'æŋgjulə/ 近似三角形的

subtriple /'sʌb'tripl/ 1:3的, 三分之一的

sussession /sək'seʃn/ 连续, 接续; 次序

successively /sək'sesivli/ 依次, 陆续, 接连, 相继

sufficient /sə'fiʃənt/ 充分的, 足够的

suggestion /sə'dʒestʃən/ 示意, 暗示; 建议; 意见

suggestive /sə'dʒestiv/ 暗示的, 示意的, 可作参考的

suit /sju:t/ 使配合, 适合

suitable /ˈsju:təbl/ 合适的, 适当的; 相对的

summarization /ˌsʌməraiˈzeiʃən/ 归纳, 总结

summit /ˈsʌmit/ 顶点, 最高峰

sunk /sʌk/ 埋头的, 沉头的, 凹下去的

superdimensioned /sju:pədiˈmenʃənd/ 超尺寸的

supererogatory /sju:pəreˈrɔɡətəri/ 多余的, 不必要的

superficies /sju(:) pəˈfiʃi:z/ 表面, 外表, 外观, 外貌

superfluity /sju:pəˈflu(:)iti/ 多余, 太多, 多余之量, 不必要的东西

superfluous /sju(:)ˈpəˈflu:əs/ 多余的, 过多的, 不必要的

superscript /ˈsju:pəskript/ 写在某一字上面的, 标在某一字的左(或右)上角的; 标在某一字上面(或下面)的字(或符号); 上标

superscription /sju(:) pəˈskripʃən/ 题字, 标题, 题目

supersede /sju(:) pəˈsi:d/ 代替, 取而代之, 作废; ～d by X 本图由 X 取代; ～s X 本图取代 X

supervacaneous /sju:pəvəˈkeiniəs/ 多余的, 不需要的

supplementary /ˌsʌpliˈmentəri/ 补充的, 辅助的

supreme /sju(:)ˈpri:m/ 最高的, 最上的, 极度的, 最主要的

surface /ˈsə:fis/ 面, 表面, 外观; 表面加工, 表面处理; ～ coordinates 曲面坐标; ～ defect 表面缺陷; ～ gemetry 表面几何形状; ～ limit 面之极限, 转向素线; ～ of contace 接触面; ～ of revolution 回转面; ～ of rotation 旋转面; ～ of second order 二阶曲面; ～ quality 表面质量; ～ roughness 表面粗糙度, 表面不平度; ～ structure 表面结构; ～ structure code 表面结构代号; ～ structure symbols 表面结构符号; ～ texture 表面纹理; ～ waviness 表面波纹度; angular ～ 斜面; aspheric ～ 非球面; auxiliary ～ 辅助面, 辅面; base ～ 基面, 底面; bottom ～ 底面; boundary ～ 界面; circular conical ～ 圆锥面; circular cylindrical ～ 圆柱面; classification of ～s 面的分类; close-fitting ～s 密吻合面; conical ～ 圆锥面, 锥面; conic curved ～ 锥曲面; contact (ing) ～ 接触面; convolute ～ 切线面; curved ～ 曲面; cylindrical ～ 圆柱面, 柱面; datum ～ 基(准)面; definite curved ～ 定线面; developable ～ 可展曲面, 可展表面; double curved ～ 双曲面, 二次曲面, 复曲面; double curved ～ of revolution 曲线回转; double curved ～ of transposition 曲线柱面; double ruled ～ 双直纹面; double ruled warped ～ 双直线翘曲面; extended ～ 展开面; fair ～ 流线型面; faying ～ 搭接面, 重叠面; fitting ～ 配合面; flat ～ 平面; flat lower ～ 平底面; front ～ 前面; glassy ～ (玻璃状)光泽面; helical ～ 螺旋面; helicoidal ～ 螺旋面; horn ～ 牛角面; illuminated part of a ～ 阳面; indefinite curved ～ 变线面; inclined ～ 单斜面; inner ～ 内表面; irregular 不规则曲面; junction ～ 接合面; level ～ 水平面; machined ～ 已加工面, 加工面; mating ～ 配合面; median ～ 中界面; modified single curved ～ 变形单曲面; molded hull ～ 船体型表面; non-developable ～ 不可展曲面, 不可展表面; nonmatings ～ 非接触面; non-regular curved ～ 不规则曲面; normal ～ 正垂面; oblique ～ 斜面; outer ～ 外表面; pitch ～ (齿)节面; plain bearing ～ 平承面; plane ～ 平面; quadric ～ 二次曲面; receding ～ 退隐面(垂直面); reference ～

基面; regular curved ～ 规则曲面; ruled ～ 直线面,直纹曲面,规则表面; running ～ 波状表面,跑合面; sheep horn curved ～ 羊角曲面; single curved ～ 一次曲面,单曲面; single curved ～ of revolution 旋转单曲面; single cutting ～ 单一剖切面; skew ruled ～ 不可展直纹面; spheric ～ 球面; spiral ～ 螺旋面; tagent ～ 切线面; turned ～ 旋转(曲)面; under ～ 下面; undevelopable ～ 不能展开曲面; unfinished ～ 非加工面; unmachined ～ 不(未)机械加工表面; unruled quadric ～ 非直纹二次曲面; upper ～ 上面; warped ～ 翘曲面,扭曲面; washer faced bearing ～ 垫圈承压面; working ～ 工作面,加工面

survey /sə'vei/ 测量图,测量,调查,摄影测绘; ～ and drawing 测绘; ～ the topography 地形测量; aerial ～ 航空测量; construction ～ 施工测量; grid ～ 方格测量图

surveying /sə'veiŋ/ 测量学

sweat /swet/ 焊,熔化,熔焊

sweep /swi:p/ (swept, sweeping) 曲线,弯曲; 范围,扫除

SWG (standard wire gauge) 标准线规

switch /switʃ/ -hook 钩键; contact ～ 接触开关; double pole double throw ～ 双极双投开关;

swivel /'swivl/ 转体,旋转,转环; ～ pen 转笔(曲线笔)

Sym (symbol) 符号;/symmetric(al)/ 对称的

symbol /'simbəl/ 符号,记号,型号; ～ classification 符号分类; ～ classification code 符号分类代码; combined ～ 组合符号; contour ～ 熔接道表面形状符号; ～ for machining 切削加工符号; ～ definition 符号定义; ～ description 符号描述; ～ detail 符号细节; ～ elements 符号要素; ～ for machining 切削加工符号; ～ function 符号

功能; ～ name 符号名称; ～ of symmetry 对称符号; ～s for instructur's corrections 供教师批改作业用的符号; ～s for use on drawings 供图上用的符号; ～ significant detail 符号重要细节; ～ string 符号串; ～ type 符号类型; aiming ～ 目标符号; all-around ～ 围焊符号; ANSI regular ～ 美国标准正规符号(螺纹); ANSI simplified ～ 美国标准简化符号(螺纹); architectural ～ 建筑符号; ASA ～ 美国标准符号; ASA thread ～ 美国标准螺纹符号; basic ～ 基本符号; block ～s 方框符号; building ～s 建筑符号; compound ～s 组合符号; concave ～ 凹陷符号; control ～ 控制符; convex ～ 凸起符号; cutting-plane ～ 切割面符号; datum ～ 基准符号; documentation ～ (产品技术)文件用图形符号; electrical ～ 电工符号; feature control ～ 几何公差符号; field weld ～ 现场熔接符号; finish ～ 表面加工方法符号(熔接); flowchart ～ 流程图符号,程序符号; flush ～ 嵌平符号; foot ～ 英尺符号('); function ～ 功能符号; graphical ～ occurrence 图形符号扩展; graphical ～s 图形符号; graphical ～s for agricultural machinery 农业机械图形符号; graphical ～s for coal mine machinery 煤矿机械图形符号; graphical ～s for use on equipment 设备用图形符号; graphical electrical ～ 电工符号; hole ～ 孔的标号; inch ～ 英寸符号("); ISO projection ～ ISO 投影符号; letter ～ 字母符号; limit ～s 限定符号; machining ～ 切削符号; marine ～ 航海符号; ordinary ～ 一般符号; pictographic ～ 象形符号; plane ～ 平面符号; product ～ 产品符号; public ～ 公用符号; public information ～s 公共信息图形符号; reference ～ 基准符号; schematic ～ 图表符号; simplified external-thread ～ 外螺纹

简化符号; simplified internal-thread ~ 内螺纹简化符号; supplementary weld ~ 熔接道辅助符号; surface roughness ~ 表面光洁度符号; symbolic ~ 象征符号; tolerance ~ 公差代号; topographic ~ 地形符号; tpd (technical product documentation) ~ 产品技术文件用图形符号; weld all around ~ 周围焊符号; welding ~ 焊缝符号; window and door ~s 门窗符号; wiring ~ 线路符号(电)

symbolize /ˈsimbəlaiz/ 象征, 用符号表示

symmetric(al) /siˈmetrik(əl)/ 对称的, 匀称的

symmetry /ˈsimitri/ 对称, 匀称, 调和; 对称性, 对称度; the axis of ~ 对称轴

synop. (synopsis) 对照表, 一览

sys. (syst. system) 系统, 装置, 制度

system /ˈsistəm/ 体系, 系统, 制度; 次序, 规律; 分类法; 方式, 方法; 类; ~ of axis 坐标轴系; ~ of basic hole 基孔制; ~ of basic shaft 基轴制; ~ of corrdinates 直角坐标系; ~ of fit 配合制; advanced cartographic ~ 高级绘(制)图系统; aligned ~ （写尺寸数字）对齐制; artwork process ~ 原图处理系统, 版图处理系统; basic hole ~ 基孔制; basic angle tolerance ~ 基准角度公差制; basic shaft ~ 基轴制; British ~ 英制; complete decimal ~ 完全小数制; computer-assisted mapping ~ 电脑绘图系统; file ~ 文件系统; fit ~ 配合制度; grid ~ 图之分区法; grid reference ~ 图幅分区; hole-basic ~ of fits* 基孔制; horizontal ~ 水平制（写尺寸数字的单向制）; limit ~ 极限制度, 公差制度; metric ~ 公制; multiple ~ 复图制; off-line ~ 脱机系统; online ~ 联机系统; open ~ 开放系统; parts designation ~ 零件编号法; scaling ~ 计算图, 设计图, 换算电路; shaft-basic ~ of fits* 基轴制; single-drawing ~ 单图制; third angle ~ 第三角法; two datum ~ 双基准制; two-projection-plane ~ 两投影面体系; unidirectional dimensions ~ 单向注尺寸制; unidirectional ~ 单向制; unified building modular ~ 建筑统一模数制; unilateral tolerance ~ 单向公差制

T

T (Tee) T 形接头; T 型钢

tab. (table) 表格, 工作台, 制表符

table /ˈteibl/ 桌, 台, 板; 表（格）, 目录; 写入板(计); ~ look up 一览表; ~ of allowance 公差表; ~ of off sets 型值表; adjustable drawing ~ 活动制图桌; cable collocation ~ 电缆配置表; chart ~ 图桌, 图表架; connection ~ 接线表; drafting ~ 制图桌, 制图台; drawing ~ 绘图桌, 绘图台; interconnection ~ 互联接线表; looking up ~ 查表; plotting ~ 绘图桌, 图形显示幕; unit connection ~ 单元接线表

tablet /ˈtæblit/ 图形输入板, 写入板; data ~ 数据图形输入板

tabulate /ˈtæbjuleit/ 制表

tack /tæk/ 图钉, 小钉, 平头钉; 方针, 方法; center ~ 圆心板; thumb ~ 图钉, 揿钉

tack-weld 搭焊

tad (top assembly drawing) 顶部装配图
tag /tæg/ 特征, 标志, 标记, 卡片; 标签; garment ～ 外表特征; parts ～ 零件标签
tail /teil/ 尾部, 末端; ～ of thread 螺纹收尾
takedown /ˈteikˈdaun/ 移出, 移去; 取出, 取下; 卸下
tall /tɔ:l/ 高的
tallness /ˈtɔ:lnis/ 高度
tally /ˈtæli/ 标签, 名牌; 骑缝号
tan. (tanget) 切线, 正切
tandem /ˈtændəm/ 串联, 串列; 双轴; 前后排直的, 纵列的
tangent /ˈtændʒənt/ 切线, 正切, 切线的; ～ method 正切法; common ～ 公切线; exterior common ～ 外公切线; ‖ *Example sentence: Drawing a line tangent to a circle from a point outside the circle.* 例句: 过圆外一点作圆的切线。
tangential /tænˈdʒenʃəl/ 切线的; ～ circles 相切圆; ～ point 切点
tank /tæŋk/ 槽, 箱, 罐, 桶, 柜; 坦克车; pen 槽笔, 储水笔
tap /tæp/ 塞子, 龙头; 丝锥; 攻丝; ～ drill 螺孔钻头; ～-drill hole 内螺纹预钻孔; ～ wrench 铰杠, 丝锥扳手; bottom ～ 三攻螺丝攻; bottoming ～ 三攻螺丝攻; taper ～ 头道螺丝攻
tape /teip/ 带, 纸带, 卷尺, 皮尺; 磁带, 录音带; 用卷尺量; 用胶布把……粘牢; 用磁带为……录音; adhesive ～ 胶布带, 胶皮带; celluloid ～ 透明胶水纸; drafting ～ 制图胶带; gummed ～ 胶纸带; line ～ 卷尺; magnetic ～ 磁带; masking ～ 胶纸带; paper ～ 纸带; punched ～ 穿孔纸; scotch ～ 胶黏带, 透明胶带; sequence control ～ 程序带; steel (steel-band) ～ 钢卷尺
taper /ˈteipə/ 锥形, 圆锥, 锥度, 斜度, 坡度; 渐缩的, 微光; ～ cone 圆锥; ～ fit 锥度配合; ～ hole 锥形孔; ～ 1 in 100 dia 锥度 1:100直径; ～ perfoot 每呎(英寸)斜度; ～ pin 锥销; ～ socket 锥孔; ～ tolerance 锥度公差; ～ tolerance zone 锥度公差域; back ～ 倒锥; conic ～ 圆锥锥度; dimensioning ～ 锥度标注; direction of ～ 锥度方向; flat ～ 斜; Morse ～ 莫氏锥度; pattern ～ 拔模斜度
task /ta:sk/ 任务(计)
TB (tee-bend) T形弯头
T-bar 丁字(形)铁
T-beam 丁字(形)梁
tbg (tubing) 管道, 管路
T-bolt T形螺栓
TC (Technical Committee) 技术委员会
TC 10 (ISO/TC 10) ISO 下设的第十技术(制图)委员会
teardown /ˈteədaun/ 拆卸
technic /ˈteknik/ 技巧, 技术, 技艺, 工艺; 工艺的, 工艺学; assembling ～ 装配工艺
technical /ˈteknikəl/ 技术的, 工艺的; 学术的; 专门的, 专业的; ～ certificate 技术证书; ～ conditions 技术条件, 技术规范; ～ institute 工业学院; ～ manual 技术规范, 技术手册; ～ provisions 技术条件; ～ order 技术指标(规程); ～ regulations 技术规程; ～ requirements 技术要求; ～ sketching 技术草图; ～ specifications 技术说明书, 技术规范, 技术条件; ～ standard 技术规格, 技术标准
technique /tekˈni:k/ 技术, 技巧; ～ of drafting 绘图技巧; power media ～ 多媒体技术
technologic(al) /teknəˈlɔdʒik(əl)/ 工艺的, 工艺学上的
technology /tekˈnɔlədʒi/ 工艺学, 工艺, (工业)技术; 术语学
tee /ti:/ T形、T形物; ～ square 丁字尺; to

a ～ 恰好地, 丝毫不差地
television /ˈtelivɪʒən/ 电视, 电视机
tell /tel/ -tale 信号装置, 计算器, 寄存器, 指示器
temper /ˈtempə/ 回火, 淬硬, 淬火; ～ hardening 回火硬化, 二次硬化
tempering /ˈtempəriŋ/ 回火; high ～ 高温回火; low ～ 低温回火; surface ～ 表面回火
template /ˈtemplit/ (templet) 样板, 模板; circle ～ 圆模板; dimensioning ～ 注尺寸用模板; drafting ～ 绘图模板; drawing ～ 绘图模板, 制图技术; electro symbol ～ 电器符号模板; ellipse ～ 椭圆模板; hole ～ 孔形模板; isometric ellipse ～ 等角椭圆模板; lettering ～ 字体样板; radius ～ 圆角模板; radius tangent ～ 圆弧切线模板; screw & nut ～ 螺钉螺帽模板; surface symbol ～ 表面符号模板; tilted hex drafting ～ 斜六角形模板; tooling ～ 工具模板; welding symbol ～ 熔接符号模板
tensile /ˈtensail/ 拉力的, 抗拉的; ～ strength 抗拉强度
tension /ˈtenʃn/ 拉力, 张力, 压力; 拉紧
term /tə:m/ 名称, 术语, 地位, 相互关系; 期间, 期限, 学期
term. (terminal) 终点, 终端
terminal /ˈtə:minl/ 末端的; 接头; 终端, 终点; 引线; 中断; 极限的; earth ～ 接地端子(电); interactive ～ 交互式终端; video ～ 显示终端
terminology /ˌtə:miˈnɔlədʒi/ 术语, 专门名词; 术语学, 词汇; view ～ 视图术语
terraced /ˈterəst/ 阶梯形的, 台阶状的; ～ fields 梯田; ～ roof 平台屋顶
terrazzo /teˈrɑ:tsəu/ (意大利语)水磨石
test /test/ 检验, 测验, 试验, 考查, 小考; ～ condition 试验条件; ～ data 试验数据; ～ number (～ No.) 试验号; ～ piece 试样; ～ result 试验结果
tetragon /ˈtetrəgən/ 四边形
tetrahedral /ˈtetrəˈhedrəl/ (tetrahedron) 四面形, 四面体; dimensional ～ 尺度四面体; local ～ 局部四面形; pyramidal ～ 锥形四面体; unit ～ 单位四面体
tetrahedroid /ˈtetrəˈhi:drɔid/ 四面体
tetrahedron /ˈtetrəˈhedrən/ 四面体, 三棱锥; regular ～ 正四面体
text /tekst/ 原文, 正文, 课文; 题目, 标题; 教科书, 文本; ～ hand 粗体正楷; ～ path 正文路径
texture /ˈtekstʃə/ 结构, 特征; 文理, 网纹; ～ of surface 表面特征
tfr. (transfer) 过渡, 转送
THD (thread) 螺纹
T-head bolt T 形头螺栓
theorem /ˈθiərəm/ 定理, 原理, 原则, 法则; rightangle's projection ～ 直角投影原理
theory /ˈθiəri/ 学说, 理论, 原理; ～ of axonometry 轴测投影原理; ～ of machines 机械原理; graphical ～ 图学理论
thermolize /ˈθə:məlaiz/ 表面热处理
thick /θik/ 厚的, 粗的; 厚度, 浓度; 稠
thickness /ˈθiknis/ 厚度, 厚薄, 最厚部分, 稠, 粗; arc ～ 弧线齿厚; chordal ～ 弦线齿厚; circular ～ 弧线齿厚; length width and ～ 长, 宽, 高; tooth ～ 齿厚; wall ～ 壁厚
thin /θin/ 薄, 细, 瘦, 稀; ～ sections 细薄剖面; ～ webs 薄壁, 薄肋
thing /θiŋ/ 物体
third /θə:d/ 第三(个), 三分之一(的); ～ angle 第三角, 第三象限; ～ dimension 第三度, 第三尺寸; a ～ (one) 三分之一; two ～s 三分之二
thk. (thickness) 厚度
thole /θəul/ 圆屋, 圆屋顶, 桨座
tholobate /ˈθɔləbeit/ 圆屋顶座

thousandth /ˈθauzənθ/ 千分之一，第一千

thread /θred/ 螺纹，螺旋；索；螺线，螺距；～ angle* 牙型角，螺纹角；～ clearance 螺隙；～ connection 螺纹连接；～ contact height* 螺纹接触高度；～ depth 螺纹深度；～ed hole 螺孔；～ed rod 螺杆；～ form 螺纹牙型；～ gage 螺纹规；～ groove width* 螺纹槽宽；～ height* 牙型高度；～ of screw 螺纹；～ pitch 螺距；～ ridge thickness* 螺纹牙厚；～s perinch 每吋（英寸）螺纹数；～ symbol 螺纹型号；Acme ～ 爱克米螺纹（英制梯形螺纹，夹角29°）；aero ～ 航空螺纹；American standard ～ 美国标准螺纹；American National screw ～ 美国标准螺纹；angular ～ 三角螺纹；angular depth of ～ 螺纹理论深度；axis of ～* 螺纹轴线；basic form of ～ 螺纹基本牙型；basic ～ form 螺纹基本牙型；brass ～ 英国铜匠螺纹；breechblock ～ 枪镗螺纹；Brigg's standard pipe ～ 布立格标准管螺纹；British Association (B.A.) ～ 英国标准学会 (B.A.) 螺纹；British standard ～ 英国标准螺纹；British Standard Fine ～ 英国标准（韦氏）细螺纹；British Standard pipe ～ 英国标准管螺纹；Brown and sharpe worm ～ 布朗沙普蜗杆螺纹；buttress ～ 锯齿形螺纹，直三角螺纹；casing ～ 壳螺纹；coarse ～ 粗牙螺纹；coarse pitch ～ 粗螺纹；common ～ 普通螺纹，公制螺纹；complete ～* 完整螺纹；conical ～ 圆锥螺纹；cutted ～ 车削螺纹；Dardelet ～ 是一种自锁螺纹，与 Acme 螺纹相似，为法国一位军官设计；depth of ～ 螺纹深度；depth of ～ spacing 螺纹大小径间隔；double ～ 双线螺纹，双头螺纹；double start ～ 双头螺纹；dryseal straight pipe ～ 气密直管螺纹；dryseal taper pipe ～ 气密斜管螺纹；external ～* 外螺纹，阳螺纹；fastening ～ 连接螺纹；female ～ 内螺纹，阴螺纹；fine ～ 细牙螺纹；fine ～ series 细螺纹级；flat ～ 平螺纹，方螺纹；form of ～* 螺纹牙型；form line of ～ 螺线；free end of the ～ 螺纹伸出端，螺纹不配合部分；French standard ～ 法国标准螺纹；full (form) ～ 完全螺纹，全牙；full diameter of ～ 螺纹大径；gas (pipe) ～ 管螺纹；half of ～ angle* 牙型半角；inch ～ 英制螺纹；incomplete ～* 不完整螺纹；inside ～ 内螺纹；internal ～* 内螺纹，阴螺纹；international standard ～ 国际标准螺纹；ISO metric ～ 公制螺纹；knuckle (screw) ～ 圆螺纹；leaning ～ 梯形螺纹（一面坡螺纹）left-hand ～* 左旋螺纹，反螺纹；length of ～ engagement* 螺纹旋合长度；male ～ 外（阳）螺纹；metric ～ 米制螺纹，公制螺纹，普通螺纹；minus ～ 负螺纹（螺纹外径等于未切螺纹部分的直径）；multiple ～ 多线螺纹，多头螺纹；multi-start ～* 多线螺纹；negative ～ 阴螺纹；non-standard ～ 非标准螺纹；number of ～s 螺纹扣数；nut ～ 内螺纹；oil drive back ～ 回油线；outside ～ 外螺纹；perfect ～ 完全螺纹；pipe ～ 管螺纹；pitch ～ 径节螺纹；plus ～ 正螺纹（螺纹内径等于未切螺纹部分的直径）；power-transmitting ～ 传动螺纹；quadruple ～ 四线螺纹；quick (-pitch) ～ 粗扣螺纹，大螺距螺纹；regular screw ～ 普通螺纹，基本螺纹；revolving direction of ～ 螺纹旋向；right-hand ～* 右旋螺纹，正丝扣；rolled ～ 滚压螺纹；round ～ 圆螺纹；SAE Standard ～ SAE 标准螺纹；screw ～* 螺纹；screw ～ pair* 螺纹制；seller's screw ～ 塞勒（美制60°）螺纹；semiconventional representaion of ～ 螺纹半写实画法；sharp V ～ 三角螺纹（非截顶）；simplified representation of ～ 螺纹简化画法（美式）；single start ～* 单线螺纹，单头螺纹；

special ～ 特殊螺纹; square ～ 方牙螺纹; standard ～ 标准螺纹; standard full V ～ 正三角形标准螺纹; steep-lead ～ 大导程螺纹; step-pitch ～ 大螺距螺纹; stop line of ～ 螺纹终止线; straight ～ 圆柱形螺纹; taper pipe ～ 锥形管螺纹; taper screw ～ * 圆锥螺纹; tapping screw ～ * 自攻螺纹; three-start ～ 三头螺纹; transmission ～ 传动螺纹; trapezoidal ～ 梯形螺纹; tri-angular ～ 三角螺纹; triple ～ 三线螺纹; two-start ～ 二头螺纹; unified screw ～ （英、美、加）统一（标准）螺纹; United States ～ 美国螺纹; useful ～ * 有效螺纹; useful length of screw ～ 有效螺纹长度; U.S.S. ～ 美国标准螺纹（牙形角60°）; V ～ V形螺纹; vanishing of ～ 退刀纹, 螺(纹)尾(扣); washout ～ * 螺尾, 不完整的螺纹; Whitworth ～ 惠氏螺纹（英国的标准螺纹）; wood screw ～ * 木螺钉螺纹; worm ～ 蜗杆螺纹

threaded /ˈθredid/ 有螺纹的; ～ hole 螺孔; ～ rod 螺纹杆; internally ～ part 内螺纹件

threading /ˈθrediŋ/ 车螺纹, 攻丝

three-dimensional 三度的, 空间的, 三维的, 立体的; ～ charts 三向图表

three-space 立体的, 空间的

three-way 三通的

thrice /θrais/ 三次, 三倍的

throat /θrəut/ 咽喉, 焊(缝)喉, 喉部; ～ diameter 喉圆直径

through /θru:/ 穿过, 通过

through-hardening 全部硬化

throughout /θru:ˈaut/ 贯穿, 通, 全部, 遍及

thru (through) 穿过, 通过

thumb /θʌm/ 拇指, 翻阅, 查; 手拧(螺丝); ～ nail 拇指甲, 略图, 短文; ～ nut 蝶形螺母, 元宝螺母; ～ pin 图钉; ～ screw 手拧螺丝; ～ tack 图钉; ～ through 翻查; as a (by) rule of ～ 根据经验, 凭经验

thumbnail /ˈθʌmˌneil/ 拇指甲, 草图, 略图; 短文

thumbscrew /ˈθʌmˌskru:/ 蝶形螺母

thumbsketch /θʌmsketʃ/ 草图, 略图

thumbtack /ˈθʌmˌtæk/ 图钉, 揿钉

TIB (Technical Information Bureau) 技术情报局

tick /tik/ 点, 小记号, 滴嗒声; time ～ 记时记号, 记时器

tide /taid/ -staff 标尺

tie /tai/ 联系, 关系; 联络线, 连接件, 拉杆

tie-line 直接连接线

tile /tail/ 瓦, 砖片; cored ～ 空心砖; grey ～ 青砖, 青瓦; single-lap ～ 平瓦

tilt /tilt/ 倾斜, 高低角, 仰角

timber /ˈtimbə/ 木料, 原木; ～ back 顶梁

tin /tin/ 锡, 镀锡

tint /tint/ 阴影, 阴影线; flat ～ 均匀阴线; graded ～ 渐变阴影线

tip /tip/ 末梢, 尖端; 触点, 接头; 顶

tit /tit/ (title) 标题, 题目, 书名; 名称, 称号

title /ˈtaitl/ 标题, 题目; 名称; 学位, 官衔; 字幕, 图标符号; 小点, 微量, 一点点

title-block * 标题栏, 明细表

titles /ˈtaitls/ 极微之物, 字母上的区别符号; 标题, 书名

T-joint T字形接头

T-L (true length) 实长, 真长

TM (technical manual) 技术手册, 技术指南

to /tu:, tə/ 比; 使; 朝, 向; 到; 对于; 按照; the ratio of 6 ～ 4 6与4之比(6:4) ～ fairing 使流线型化, 使曲线或曲面光滑; draw…～ scale 按一定比例把……画出来; ‖ *Example sentence:* As two is to three, so is four to six. *例句:* 四比六等于二比三。

toe /təu/ 足尖, 趾; 底, 下端; 钻孔底, 齿顶

（高），柄梢；凸轮的从动部工具

token /'təukən/ 标记, 记号, 特征

tol. (tolerable) 可容许的; (tolerance) 公差, 容限

tolerance /'tɔlərəns/ （尺寸）公差, 容限; ～ grade* 公差等级; ～ in form 形状公差; ～ in position 位置公差; ～ of concentricity 同心度公差; ～ of form 形状公差; ～ of position 位置公差; ～ of size* 尺寸公差; ～ on fit 配合公差; ～ symbols 公差代号; ～ unit 公差单位; ～ zone* 公差带; zone of size* 尺寸公差带; angle ～ 角度公差; basic ～ 基本公差; bilateral ～ 双向公差; bilateral ～ system 双向公差制; casting ～ 铸造公差; close ～ 精密公差; concentricity ～ 同心度公差; crest diameter ～ 螺纹顶径公差; crest diameter ～ symbol 螺纹顶径公差符号; critial ～ 临界公差; cumulative ～ 累积公差; cylindrical ～ zone 圆柱形公差带; dimensional ～ 尺寸公差; fit ～ * 配合公差; fit ～ zone* 配合公差带; flatness ～ 平直度公差; form ～ * 形状公差; fundamental ～ * 标准公差; general ～ 通用公差; general ～ note 通用公差说明; geometric ～ 表面形状和位置公差, 几何公差; grade of ～ 公差等级; location ～ * 位置公差, 装置公差, 安装公差; manufacturing ～ 制造公差; margin ～ 公差范围; maximum ～ 最大公差; minimum ～ 最小公差; orientation ～ * 定向公差; position ～ * 位置公差; profile ～ 轮廓公差; projection ～ zone* 延伸公差带; run-out ～ * 跳动公差; standard ～ * 标准公差; standard ～ unit* 公差单位; unilateral ～ 单向公差; zero ～ 零公差; zone ～ for a contour 轮廓的带式公差

toning /'təuniŋ/ 着色

tool /tu:l/ 工具, 仪器; measuring ～ 测量工具

tooth /tu:θ/ 齿, 牙; 切齿; 啮合, 咬合; 凸轮, 粗糙面; ～ space 齿间距; ～ thickness 齿厚; angular ～ 三角形齿; cycloid gear ～ 摆线轮齿; cycloidal ～ 摆线齿; double helical ～ 人字齿; helical ～ 螺旋齿; involute ～ 渐开线形齿; knuckle ～ 圆顶齿; rectangular ～ 矩形齿; spiral ～ 螺旋（齿轮的）齿; straight ～ 直齿; stub ～ 短齿; trapezoidal ～ 梯形齿

toothed /tu:θt/ 有齿的, 齿形的; ～ gearing 啮合齿轮, 齿轮传动装置

top /tɔp/ 顶, 上部, 头, 顶头; 盖上, 给……涂上保护层; ～ view 顶视图, 上视图, 俯视图

topmost /'tɔpməust/ 最高的, 顶高的

topnotch /tɔp'nɔtʃ/ 顶点

topographic(al) /tɔpə'græfik(əl)/ 地形的, 地形学, 地形测量的; ～ map 地形图

topography /tə'pɔgrəfi/ 地形, 地志, 地形学

topology /tə'pɔlədʒi/ 拓扑学, 拓扑（结构）, 地表学, 局部解剖学; combinatorial ～ 组合拓扑学

toroid /'tɔurɔid/ 环面, 圆环, 超环面的; inner-surface of ～ 内环面; outer-surface of ～ 外环面

torse /tɔ:s/ 扭曲面, 可展曲面, 残缺（未完成）的东西; transition ～ 渐变曲面

torsion /'tɔ:ʃn/ 扭转, 扭曲, 挠曲; 扭力

torsional /'tɔ:ʃənl/ 扭转的

torus /'tɔ:rəs/ (ring) 环面, 环形体

total /'təutl/ 总的, 全部的; ～ depth 总深

touch /tʌtʃ/ 接触, 碰到, 达到

TP (T-piece) T 形接头; (T-pipe) T 形管

tpd (technical product documentation) 产品技术文件

T. P. I (teeth per inch) 每英寸齿数; (threads

per inch) 每英寸螺纹数

tr (technical report) 技术报告; (traced) 描绘的; (tracer) 描图员

trace /treis/ 迹点, 迹线; 交点, 交线, 轨迹; 扫描, 描绘, 描图, 画曲线; 跟踪, 寻找, 微量; ～ of line 线的迹点; ～ of plane 面的迹线; collective point of ～ 迹线集合点; determination of ～ 迹点的求法; frontal ～ 正面迹点(线); horizontal ～ 水平迹点(线); profilo ～ 侧面迹点(线); the ～ of line 直线的迹点; the ～ of plane 平面的迹线; to ～ an outline on paper 在纸上画一略图; to ～ out 画出, 计划

tracer /ˈtreisə/ 描图员, 追踪者; 描图器, 仿形器; form ～ 定型靠模

tracing /ˈtreisiŋ/ 描图, 摹图, 画曲线; 追踪; 迹线, 底图; ～ cloth 描图布; ～ machine 描图机(电子轨迹描绘器); ～ paper 描图纸; ～ sheet 描图纸; automatic ～ 自动描绘扫迹; computer-aided ～ 计算机辅助绘图; duplicating ～ 复制描图; inked ～ 墨水描图; order of ～ 描图顺序; pencil ～ 用铅笔描图; ray ～ 电子描绘扫迹, 光线追迹, 声线描迹

track /træk/ 轨迹, 轨道; 跨距; 跟踪; 磁道

trammel /ˈtræməl/ 椭圆规, 量规, 规, 束缚物; 游标卡尺; 束缚; ellipse by the ～ method 用束缚法画椭圆, 用椭圆规画椭圆; elliptic ～ 椭圆规; steel beam ～ 钢杆规

TRANS. (transactiona) 学报, 学会会刊; 论文集; (transpose) 变换

transect /træn'sekt/ 横切, 横断

transection /træn'sekʃən/ 横切, 横断, 横断面

transfer /ˈtrænsfə:/ 转换, 变换, 转移, 传递; ～ by rubbing 拓印法; ～ instruction 转移指令, 传递指令; ～ protocol 传输协议; outside ～ caliper 移动外径规

transfigure /trænsˈfigə/ (使)变形

transform /trænsˈfɔ:m/ 转变, 变换, 变化, 变形, 反式, 变换式; direct ～ 直接变换; integral ～ 积分变换; inverse ～ 反变换, 逆变换; object-image ～ 物象变换

transformation /trænsfəˈmeiʃn/ 变换, 换算; 变形, 变化; 转化, 转变; ～ by reciprocal direction 倒方向变换, 逆向变换; ～ of projections 投影变换; ～ of similitude 相似变换, 类似变换; abrupt ～ 不连续变换; affine ～ 仿射变换; circle ～ 圆变换; colineation ～ 角素变换; 直射变换; 共线变换; coordinate ～ 坐标变换; conformal ～ 保角变换; 保形变换; 共形变换; 共形映射; congruent ～ 全等变换, 合合变换; conjugate ～ 共轭变换; correlation ～ 异素变换; 对射变换; cyclic ～ 循环变换; dualistic ～ 对偶变换; elementary ～ 初等变换; elliptic ～ 椭圆变换; equiform ～ 相似变换; equilong ～ 等距变换; geometric ～ 几何变换; homographic ～ 等画变换; homology ～ 透射变换; homothetic ～ 同位相似变换; 位似变换; identity ～ 恒等变换; integral ～ 积分变换; inverse (inversion) ～ 反变换; 逆变换; 反演变换; involution (involutory) ～ 对合变换; isogonal ～ 等角变换, 保角变换; linear ～ 线性变换; 一次变换; non-linear ～ 非线性变换; normalization ～ 规格化变换; point ～ 点变换; point-curve ～ 点线变换; point-surface ～ 点面变换; polarity ～ 配极变换; projective ～ of conic 二次曲线的射影变换; quadratic ～ 二次变换; reflection ～ 反射变换; rotation ～ 旋转变换; similarity ～ 相似变换; singlevalued ～ 单值变换; stereographic proiection ～ 体视投影变换, 球面投影变换; symmetry ～ 对称变换; topological ～ 拓扑变换; topologicale quivalent ～ 拓扑等价变换; viewing ～ 视

口变换

transformer /træns'fɔːmə/ 变形接头；变压器

transit /'trænsit/ 变换，转变，通过；过渡；飞越，经过；~ fit 过渡配合

transition /træn'ziʃən/ 变换，转变，过度，换接，通过；跃过；~ curve 过渡曲线，介曲线；~ fit 过渡配合；~ piece 变形接头；~ point 转折点；convolute ~ 盘旋变口体

transitional /træn'ziʃənl/ 过渡的；变过的，转移的，推移的

transitive /'trænsitiv/ 过渡的，传递的

translate /træns'leit/ 翻译，转化，使平移

translation /træns'leiʃən/ 平移，平动，直线运动；译文，翻译；变换，调换，换置，移动，位移，位移，转移；logic ~ 逻辑变换；~ rectilinear ~ 直线平移

translator /træns'leitə/ 翻译程序，自动编码器，移码器；转换器，变换器；card ~ 卡片译码器；image ~ 图像变换器

transmission /trænz'miʃn/ 透射，发射，传播，传动，装置，变速器；~ box 传动箱，变速箱，变速箱；bevelgear ~ 圆锥齿轮传动；cylindrical gear ~ 圆柱齿轮传动；worm and worm gear ~ 蜗杆、蜗轮传动

transmit /trænz'mit/ 透射，发射，传导，传送，传动，透过

transparency /træns'pɛərənsi/ 幻灯片，印有图像或图案的透明玻璃

transparent /træns'pɛərənt/ 透明的，明显的，清楚的

transpose /træns'pəuz/ 移，位置，换位，交叉，变调

transposition /trænspə'ziʃən/ 移位，换位；对换

transversal /trænz'vəːsəl/ 横向的；横切的；横断的；贯线的；贯线，截线，截断线，横断线

transverse /'trænzvəːs/ 横向的，横截的，横断的，横切的；横放的；（椭圆）长轴；横梁；~ axis 横向轴；~ circular thickness 端面弧线齿厚；~ section (cross section) 横截面；~ section of underground drainage piping 管道纵断面图；~ shaft 横轴；~ strength 横向（抗弯）强度

trapezium /trə'piːziəm/ 梯形（英国）；不规则四边形（美国）

trapezoid /'træpizɔid/ 梯形（美国）；不规则四边形（英国）；isosceles ~ 等腰梯形；regular ~ 正梯形

travel /'trævl/ 移动；运动；行程；冲程；~ of piston 活塞冲程；~ of valve 阀行程

traverse /'trævəːs/ 横过，横断；横切；横截；切断；相交，交叉

tread /tred/ 距；踏板；梯面；级宽；踩；踏

treatment /'triːtmənt/ 处理；加工；处治；计算；antiseptic ~ 防腐处理；chromate ~ 镀铬；cold ~ 冷处理；graphical ~ 图解（法）；heat ~ of steel 钢的热处理；mechanical ~ 机械加工；passivating ~ 钝化处理；superficial ~ 表面处理；thermal ~ 热处理

trefoil /'trefɔil/ 三叶饰，三叶花样，三叶形

trench /trentʃ/ 管沟；pipe ~ 管沟

trend /trend/ 方向，走向，倾向，趋向；方位，趋势

trestle /'tresl/ -board 大图板

triangle /'traiæŋgl/ 三角形；三角板（美），三角铁；45° and 30°—60°—~s 45° 及 30°—60° 三角板；~ of vectors 矢量三角形；acute ~ 锐角三角形；acute-angled ~ 锐角三角形；adjustable ~ 可调三角板；Braddock-Rowe ~ 字格三角板；circular ~ 圆弧三角形；congruent ~s 全等三角形，叠合三角形；equiangular ~ 等角三角形；equilateral ~ 等边三角形；fixed-angle ~ 固定

角三角板(45°, 30°—60° 三角板) funda-mental ~ * 原始三角形; homothetic ~ 同位相似三角形, 位似三角形; inscribed ~ 内接三角形; isosceles ~ 等腰三角形; obtuse ~ 钝角三角形; plain ~ 无刻度三角板; right ~ 正三角形, 直角三角形, 勾股形; right-angled ~ 直角三角形; scalene ~ 不等边三角形, 不规则三角形; scalene obtuse ~ 不等边钝角三角形; scalene right ~ 直角三角形; similar ~s 相似三角形; solid ~ 黑三角形(基准); solution by a ~ 三角形解法; spherical ~ 球面三角形; spheroidal ~ 椭球面三角形; three arc ~ 弧三角形

triangular /traiˈæŋɡjulə/ 三角(形)的; ~ prism 三棱柱; ~ pyramid 三棱锥; ~ scale 三棱尺; ~ thread 三角螺纹

triangulate /traiˈæŋɡjuleit/ 使成三角形, 把……分成三角形; 对……作三角测量; 三角形的; 有三角形花样的

triangulation /trai,æŋɡjuˈleiʃn/ 三角测量, 三角剖分, 三角形法, 分成三角形, 三角形展开法; graphical ~ 图解三角测量; plane table ~ 平板三角测量

triaxial /traiˈæksiəl/ 三度的, 三维的, 空间的; 三元的, 三轴的; ~ diagram 三轴线图

trick /trik/ 技巧, 窍门, 诡计, 奸计

tridimensional /ˌtraidiˈmenʃənəl/ 三度的, 三维的; 立体的

trigon /ˈtraiɡən/ 三角形, 三角板; 三角规(测时三角日晷), 三角(形)的; ~al 三角形的

trigonometric(al) /ˌtriɡənəˈmetrik(əl)/ 三角法的, 三角学的

trigonous /ˈtriɡənəs/ 三角形的, 有三角的

trihedral /traiˈhedrəl/ 有三面的, 三面角的; 三面形

trihedron /traiˈhedrən/ 三面体

trilateral /ˈtraiˈlætərəl/ 三边的; 三角形, 三边形

trilinear /traiˈliniə/ 三线的; ~ chart 三线图表, 三线坐标图

trim /trim/ 整理, 修理, 修整; 切边, 去毛刺, 使整齐; ~ size 实际尺寸

trimetric /traiˈmetrik/ 三度(投影), 不等角; 斜方(晶)的; ~ projection 三度(测)投影, 正三测投影

triple /ˈtripl/ 三重的, 三倍的

triplet /ˈtriplit/ T形接头; 三重线; 三件一套

tripod /ˈtraipɔd/ 三脚架; axes ~ 轴三脚架 (轴测投影的三个轴)

trisecant /traiˈsiːkənt/ 三度割线, 三重割线

trisect /traiˈsekt/ 三分, 三截, 三等分; 把……分成三份

trisection /traiˈsekʃən/ 三等分, 三重割; ~ of an angle 角的三等分, 三等分角

trisectrix /traiˈsektriks/ 三等分角线

trisquare 曲尺, 矩

trmt. (treatment) 处理, 加工

trochoid /ˈtroukɔid/ 次摆线, 余摆线; 长短辐(辐点)旋轮线; 滑车形的, 圆锥形的; ~ pump 余摆线泵

trop /trou/ (法语)多余的, 无用的, 碍事的

troublesome /ˈtrʌblsʌm/ 困难的, 麻烦的; ~ problem 难题

trough /trɔf/ 槽

trough-line 槽线

tr. pt. (transition point) 转折点

true /truː/ 真实; ~ length 真实长度, 真长; ~ length of an inclined line 倾斜线的实长; ~ length of an oblique line 倾斜直线的实长; ~ projection 真实投影; ~ shape 实形, 真形; ~ shape of plane face 平面的实形; ~ to size 真实尺寸, 尺寸正确的

trueness /ˈtruːnes/ 精确度, 准确度, 真实性, 正确

true-up 校准

truncate /'trʌŋkeit/ 截断, 截尾, 切头; 截法, 截头的, 平头的, 斜截头的; 切除, 削去, 修剪, 缩短

truncated /'trʌŋkeitid/ 截头 (棱锥体), 斜截的, 被截的, 截断的, 截面的, 平切的; ~ cone 截锥, 斜截圆锥体; ~ octahedron 平截八面体; ~ pyramid 截头棱锥; ~ sphere 截球

truncation /trʌŋ'keiʃən/ 截, 削, 切断; 使尖端钝化; height of ~ 切除高度 (螺纹)

truss /trʌs/ 架, 构架; 一束, 一捆; 支柱, 支架, 支带; ~ member 桁架构件; roof ~ 屋架

T. S. (tensile strength) 抗拉强度 (极限); (true size) 实大, 实际尺寸; (tool steel) 工具钢

T-section T 形剖面

T-slot T 形槽; inversed ~ 倒 T 形槽

T-square 丁字尺, T 尺; adjustable head ~ 活头丁字尺; fixed-head ~ 头部固定的丁字尺

T-steel T 型钢

tube /tju:b/ 管子, 管道, 管材; angular ~ 折角管; flare ~ 喇叭管

tubing /'tju:biŋ/ 管, 管材, 管系, 管装置; disehargе ~ 排气管系

tubular /'tju:bjulə/ 管系, 管形的, 空心的; ~ beam compass 管形量规

tumbling /'tʌmbliŋ/ 翻滚

tunnel /'tʌnl/ 地道, 烟囱

turn /tə:n/ 回转, 旋转, 圈, 转; 车削, 变向, 弯曲, 转弯处, 转折点, 转数; ~ out 制造

turning /'tə:niŋ/ -point 转折点

tutorial /tju:'tɔ:riəl/ 指导 (计)

TV (televison) 电视

twice /twais/ 两次, 再次; 两倍于; ~ actual size 比实物大一倍, 实际大小的两倍

twin /twin/ 双的, 成对的, 孪生的; ~ axis 共轭轴

twist /twist/ 扭, 扭转, 扭弯; 歪曲, 曲解; 缠绕, 盘旋; 搓, 捻, 扭曲, 绞; 螺旋状, 缠绕物

twisty /'twisti/ 扭曲的, 盘旋的; ~ed curve 空间曲线, 挠曲线

two /tu:/ -dimensional 二维的, 二度空间的, 平面的; ~ charts 两向图表

type /'taip/ 式, 型; 型号; 式样, 类型, 典型, 标准的, 模范; 打印, 显示; ~ of bearings 轴承类型; ~ of name 名字的类型; ~ of variable 变量的类型; abstract graphical ~ 抽象图形类型, 抽象图示类型; bracket ~ 支架类; chest ~ 箱体类; data ~ 数据类型; disk and cover ~ 盘盖类; mixed ~ 混合式; shaft and sleeve ~ 轴套类; split ~ 拼合式; tandem ~ 串联式; topological ~ 拓扑型

typic (al) /'tipik(l)/ 典型的, 标准的

U-bar 槽钢

U-bend 马蹄弯头, U 形弯, U 形弯头

U-bolt U 形螺栓

UDC (Universal Decimal Classification) 国

际十进制分类法

ultimate /ˈʌltəmət/ 极限, 最后; 极端的, 最后的, 基本的

umbra /ˈʌmbrə/ 空间影区, 阴暗空间

umbriferous /ʌmˈbrifərəs/ 投影的, 有阴影的, 成荫的

unbend /ʌnˈbend/ 展平, 弄直, 伸直

uncoil /ʌnˈkɔil/ 展开, 解开

UNC thread (Unified Coarse thread) 统一标准粗牙螺纹; (Unified National Coarse thread) 国家统一的粗牙螺纹

undercoat /ˈʌndəkəut/ 底漆, 内涂层

underga(u)ge /ˈʌndəˌgeidʒ/ 尺寸不足, 短尺

underline /ʌndəˈlain/ 在……下划线, 在下边加横线; 底线, 划在下边的线, 插图下的说明; ～ a sentence 在句子下划线; to ～ 在……下划线

underlying /ʌndəˈlaiŋ/ 下边的, 基本的, 基础的

undermost /ˈʌndəməust/ 最下的, 最低的

underneath /ʌndəˈni:θ/ 在下面, 在底下; 下面的, 底层的

underproof /ˈʌndəˈpru:f/ 不合格的, 不合标准的

underream /ˈʌndəri:m/ 扩孔, 扩眼

underside /ˈʌndəsaid/ 下侧, 下面

undersize /ˈʌndəˈsaiz/ 尺寸不足, 尺寸不够; 负公差尺寸, 减少尺寸; 小型的

undersized /ˌʌndəˈsaizd/ 尺寸不足的, 不够大的

undo /ʌnˈdu:/ 拆开, 解开, 拆卸, 作废; 复原

undulant /ˈʌndjulənt/ 波浪线的, 起伏的

undulate /ˈʌndjuleit/ 波浪形的, 起伏的; 使波动, 使起伏

undulation /ʌndjuˈleiʃən/ 波浪形, 不平度

unequal /ʌnˈi:kwəl/ 不等的

uneven /ʌnˈi:vən/ 不平的, 不规则的; 不整齐, 不光滑; ～ surface 粗糙表面

UNF thread (Unified National Fine thread) 国家统一的细牙螺纹

unfavo(u)rable /ʌnˈfeivərəbl/ 不适宜的, 相反的

unfin. (unfinished) 未完工的

unfinished /ʌnˈfiniʃt/ 未精加工的

unfold /ʌnˈfəuld/ 展开, 推开, 打开; 伸展, 发展; 开花

unidimensional /ˈju:nidiˈmenʃənl/ 线性的; 一维的, 一度(空间)的, 直线型; 同尺寸的

unidirectional /ˈju:nidiˈrekʃənl/ 单(方)向的; ～ system (写尺寸数字的) 单向制

unification /ˌju:nifiˈkeiʃən/ 划一

uniform /ˈju:nifɔ:m/ 均匀的, 一致的; 始终如一的, 不变的, 相等的

unify /ˈju:nifai/ 统一, 联合, 使成一体, 使一致

unilateral /ˈju:niˈlætərəl/ 单向的, 单侧的; 一方的, 片面的; ～ surface 单侧曲面; ～ system 单向(尺寸)系统(制); ～ tolerance 单向公差

union /ˈju:njən/ 联接, 管接头, 管套节

unit /ˈju:nit/ 装置机组, 器械; 单元, 单位, 组; 零(部)件; ～ assembly drawing 部分组合图, 单元组合图; ～ construction 部件, 组件, 单元结构; ～ of fit 配合单位; central processing ～ 中央处理单元; fundamental tolerance ～ 基本公差单位; imperial ～ 英制单位; serve ～ 伺服机件

unite /ju(:)ˈnait/ 结合, 联合, 合并

unitization /ˌju:nitaiˈzeiʃən/ 统一化

unity /ˈju:niti/ 统一性, 统一体, 联结

United States thread 联邦螺纹

universal /ˌju:niˈvə:səl/ 通用的, 万能的; 普遍的; 普通的, 一般的; 宇宙的, 全世界的; ～ compass 万能两脚规

unlimited /ʌnˈlimitid/ 无限的, 无边际的

unload /ʌn'ləud/ 转储，卸下

UNO (United Nations Organization) 联合国组织

unpolished /ʌn'pɔliʃt/ 没有磨光的，粗糙的

unprocessed /'ʌn'prəusest/ 未加工的，未处理的

unruled /ʌn'ru:ld/ 不规则的

uns. (unsymmetrical) 不对称的

unseen /ʌn'si:n/ 不可见的，看不见的，未看见的

unshaded /ʌn'ʃeidid/ 无阴影的，无遮蔽的

unslotted /ʌn'slɔtid/ 无槽

unsupported /ʌnsə'pɔ:tid/ 无支撑的；未经证实的，自由的；～ height 自由高度

unsymmetric(al) /ʌnsi'metrik(əl)/ 不对称的，不匀称的，不平衡的

upbuilding /ʌp'bildiŋ/ 堆积

update /ʌp'deit/ 更新

upmost /'ʌpməust/ 最上的，最高的，最主要的

upper /'ʌpə/ 上面，上头，上部，上限；上面的，较高的；～ base 上底；～ bound 上限；～ case (字母的) 大写体；～ limit 上限尺寸

upper-case 大写字母，大写体；大写的；用大写字母排印

upper-frame 顶架，顶框

uppermost /'ʌpəməust/ 最上的，最高的，最主要的；最上，最高，最初，最先

upright /'ʌprait/ 垂直，竖立；垂直的，直立的；笔直，柱

ups-and-downs 高低起伏

upside /'ʌpsaid/ 上面，上部；～ down 颠倒，倒转，混乱的

upstroke /'ʌpstrəuk/ 上划 (笔划)

upthrow /'ʌpθrəu/ 隆起

up-to-date 现代（化）的，最新（式）的

up-to-size 具有标称尺寸的

upward /'ʌpwəd/ 向上，……以上；在上面

u. s. (undersize) 尺寸过小，小型的；减少尺寸；(ut supra)（拉丁语）如上所示，如上所述

U. S. A. (United States America) 美利坚合众国

usage /'ju:zidʒ/ 惯例，习俗；习惯法；用法；使用，运用；语法；by ～ 习惯上，老是

usance /'ju:zəns/ 惯例，习惯，使用

USBS (United States Bureau of Standards) 美国标准局

use /ju:z/ 用，使用，利用；用法，用途；使用价值；～ to 习惯于；～ing the pencil 铅笔的用法；to be ～d 使用；to be of ～ 有用；to make ～ of 利用，使用

useful /'ju:sful/ 有用的，实用的

useless /'ju:slis/ 没用的，无益的，无效的

user /'ju:zə/ 用户

USG (United States Standard Gauge) 美国标准量规

USS (United States Standard) 美国（工业）标准（规格）

USSG 同 USG

USST (United States Standard thread) 美国标准螺纹

usual /'ju:ʒuəl/ 惯例的，正规的，正常的；～ position 正常位置

utmost /'ʌtməust/ 极度（的），极端（的），最高（的），最大（的），最远（的）；～ limits 极限

UTS (ultimate tensile strength) 抗拉强度极限

utter /'ʌtə/ 全部的；说明，表达

vacancy /'veikənsi/ 空格, 空白; 空虚, 空闲
vacant /'veikənt/ 空的, 空着的, 空白的, 未被占用的
val. (value) 数值
valid /'vælid/ 正确的, 有根据的; 有效的
valley /'væli/ 波谷, 山谷, 沟底; 流域; 曲线上的凹部; 屋谷, 屋面天谷（屋顶两斜面间连接小于180°之角）; 谷值
value /'vælju:/ 价值, 数值, 值; 有效的; approximate ～ 近似值; average ～ 平均值; limiting ～ 极限值; maximum ～ 最大值; mean ～ 平均值; minimun ～ 最小值; root mean aquare ～ 均方根值; tolerance ～ 公差值
valve /vælv/ 电子管, 阀, 活门; ～ body 阀体; angle ～ 折角阀; back-pressure ～ 止回阀; ballcheck ～ 球止回阀; blow-off ～ 安全阀, 排泄阀; butterfly ～ 蝶形阀; check ～ 止回阀; cross ～ 三通阀; gate ～ 闸阀; globe ～ 球阀; plug ～ 塞阀; relief ～ 安全阀; safety ～ 安全阀; three way ～ 三通阀
vandyke /væn'daik/ 凡代克（纸名, 为薄质感光纸）
vane /vein/ 舵, 叶片, 翼, 轮叶, 风标
vanish /'væniʃ/ 消失, 变为零; 消失不见, ～ing line 没影线, 灭线; ～ing plane 没影面; ～ing point 没影点, 灭点（透视图中平行线条的会集点）; ～ing point perspective of building 灭点法建筑透视图
var. (variance) 偏差, 变动
variable /'vɛəriəbl/ 变量; dummy ～ 虚拟变量; integer ～ 整型变量; name of ～ 变量名; real ～ 实型变量
variation /vɛəri'eiʃn/ 偏差, 公差; ～ of fit* 配合公差; allowable ～ 容许变量
variograph /'vɛəriəgra:f/ 变异书写器
various /'vɛəriəs/ 不同的, 各种的; ～ oblique position 不同的倾斜位置
variplotter /'vɛəriplɔtə/ 自动曲线绘制器, 自动作图仪
varisized /'vɛərisaizd/ 各种大小的, 不同尺寸的
vary /'vɛəri/ 变化, 变换, 变更, 改变, 偏离; 不同, 不一致; 修改; 增量; to ～ (directly) as 与……成正比; to ～ from... to...从……到……不等; to ～ from unit to unit……各不相同; to ～ in size 大小不同
vault /vɔ:lt/ 拱顶, 拱顶室, 地下室, 地窖; 使作成穹形
V-belt V 形带, 三角带
V. D. P. (vertical datum plane) 正立基准面
VDU (Visual Display Unit) 荧幕显示单元
vec. (vector) 向量
vectogram /'vektəgræm/ 矢量图, 向量图
vectograph /'vektəgra:f/ 矢量图, 向量图; 立体电影
vector /'vektə/ 矢量, 向量; 航向; 引导, 动力; 矢径; ～ diagram 矢量图, 结构线图; ～ polygon 向量多边形; ～ product 矢量积; concurrent ～ 共点矢量; coplanar ～ 共面矢量; non-coplanar ～ 非共面矢量; radius ～ 矢径, 动径; velocity ～s 速度矢量
vee /vi:/ V 字形物, V 字形的; double ～ 双

V 形, X 形的
vee-cut V 形掏槽, 楔形掏槽
veed /vi:d/ V 形的
vee-trough V 形槽
vein /vein/ 脉, 矿脉, 水脉; 木纹, 纹理; ore ～ 矿脉
velocity /vi'lɔsiti/ 速度
ventage /'ventidʒ/ 孔隙, 出口
ventilation /venti'leiʃn/ 通风; ～ map 通风系统图; ～ network 通风网络图
verify /'verifai/ 核实
vernier /'və:njə/ 游标, 游标尺, 微调的, 微动的; ～ bevel protractor 活动游标量角器; ～ callipers (～ calipers) 游标卡尺; ～ control 微调; ～ gauge 游标尺; ～ scale 游标尺; a ～ depth gauge 精细深浅尺; angular ～ 角度游标; direct ～ 顺游标; linear ～ 长度游标, 线性游标
versatile /'və:sətail/ 通过的, 万能的; 多用途的, 活动的; 易变的, 反复无常的; ～ digital computer 通用数字计算机
version /'və:ʃn/ 解释, 方案, 译文, 译本; 翻译
verso /'və:səu/ 纸张的背面, 书的左页 (双数页, 与 recto 相对)
vert. (vertical) 垂线, 垂直的, 立式的
vertex /'və:teks/ 顶, 顶点, 至高点; ～ of a cone 锥顶; ～ of the angle 角的顶点; edit ～ 编辑顶点
vertical /'və:tikəl/ 垂直的, 垂线, 垂面, 竖向, 铅垂方向; 直立的; ～ angle 垂直角, 上仰角, 对顶角; ～ axis 铅垂轴; ～ line 垂直线; ～ plane 垂直面; ～ projection 垂直投影; ～ view 俯视图; out of the ～ 不垂直的
verticality /'və:tikəl/ 垂直度, 垂直性, 垂直状态; 竖直
V-gutter 三角形槽, V 形槽
VHN (Vickers hardness number) 维氏硬度值

VI (vertical interval) 垂直间距, 等高线的垂直距离
vice versa /,vaisi'və:sə/ (拉丁语) 反之亦然, 反过来也是这样
Vickers-hardness /'vikəz'ha:dnis/ 维氏硬度
vid 或 **vi'de** /'vaidi/ (拉丁语) 见, 参看; ～ ante 见前; ～ infra 见下, 见后; ～ p.15 见第15页; ～ post 见后; ～ supra 见上, 参看上文
view /vju:/ * 视图, 视形, 外观, 外形, 式样; 图; 视, 查看; ～ port 视见区, 视口; ～s and sizes 外表和尺寸; ～ to be changed 改正视图; ～ to be supplied 补足视图; additional ～ 附加视图, 辅助视图; adjacent ～ 相邻视图; aerial ～ 鸟瞰图; aero ～ 鸟瞰图; aeroplane ～ 空瞰图; aligned ～ 旋转视图, 展开画法; aligned section ～ 旋转剖视图; alternate position ～ 变换位置视图; auxiliary ～ 辅助视图, 辅视图; ‖ *Example sentence: An auxiliary view is a view projected on any plane other than one of the three principal planes of projection (frontal, horizontal, or profile).* 例句: 辅助视图是投影在三个主要投影面 (正立投影面、水平投影面、侧立投影面) 以外的任一个平面上的视图。auxiliary-adjacent auxiliary ～ 复辅助视图; back ～ 后 (背) 视图; basic ～ 基本视图; bird ～ 鸟瞰图; bird's-eye ～ 鸟瞰图; bottom ～ * 仰视图, 底画视图; broken ～ 中断视图; broken-out section ～ 局部剖视 (图); choice of ～s 视图选择; circular ～ 圆形视图; close-up ～ 近景图, 特景图; 近视图, 近观图; 全貌图, 特写镜头; combination of ～s 视图的排列; complete ～ 全视图; complete auxiliary ～ 全部辅视图 (不仅表示局部斜面, 其他部分也画出); conceptual ～ 概念视图; contour ～ 外形视图;

coupling of opposite half ～s 两半（视图）合一; cross-sectional ～ 横剖视图; cut-away ～ 剖视图; cut-open ～ 剖视图; developed ～ 展开图; diagrammatic(al) ～ 简图, 图表, 图示; distant ～ 远距离观察; dorsal ～ 背视图; double auxiliary ～ 复辅视图, 第二辅视图; echelon section ～ 阶梯剖视图; edge ～ 边视图, 重影视图; ‖ *Example sentence: The true size of the dihedral angle is observed in a view in which each of the given planes appears in edge view.* 例句: 二面角的真实大小反映在给定两平面都是重影的视图上。edge ～ of a plane 平面的重影视图(平面的边视图); edge ～ of plane 平面的边视图, 平面的重影视图; elevation ～ 立面图, 垂直投影; end ～ 端视图, 侧视图; end auxiliary ～ 侧辅视图; enlarged ～ 放大视图; exchange of opposite ～s 视图换位; expanded ～ 展开图; expansion ～ 展开视图; exploded ～ 分解图, 解剖图, 展示图; 立体影象; exterior ～ 表面图, 外观图, 外形图; external ～ 外视图, 外观图; false ～ 虚拟视图; field of ～ 视界; finite logical ～ 有限逻辑视图; front ～ * 主视图, 前视图, 正视图; front elevation ～ 正视图, 主视图, 正立面图; full ～ 全视图; fullsection ～ 全剖视图; general ～ 全视图, 总图; 大纲, 概要; grouped ～ 分组视图; half-section ～ 半剖视图, 半截面; half-sectional ～ 半剖视图; head-on ～ 正视图; hydraulic engineeringbird's-eye ～ 水利工程鸟瞰图; interrupted ～ 折断画法; isometric ～ 等角视图; lateral ～ 侧视图; left ～ * 左视图; left-side ～ 左侧视图; local ～ 局部视图; longitudinal ～ 纵向视图; master ～ 主要特征视图; mulit-revolved ～ 多次旋转视图(化工); normal ～ 法面视图, 真形视图; normal ～ of a line or plane 一条线或一个面的法面视图(反映线的真长或面的真形); normal ～ of a plane 平面的真形视图; oblique ～ 斜视图, 辅视图; offset section ～ 阶梯剖的剖视图; orthographic ～ 正投影图; outside ～ 表面视图, 外观视图; part ～ 局部视图; partial ～ 局部视图, 部分视图; partial auxiliary ～ 局部辅视图; partial section ～ 局部剖视图; part sectioned ～ 局部剖视图; perspective ～ 透视图, 远景图, 鸟瞰图; phantom ～ 假想视图, 经过透明壁的内视图; pictorial ～ 写生图, 插图; plan ～ 平面图; plan auxiliary ～ 平面辅视图, 上辅视图; point ～ 点视图, 线的端视图; point ～ of a line 直线的点视图(直线的端视图); primary auxiliary ～ 第一辅助视图, 单辅视图; ‖ *Example sentence*①: *A primary auxiliary view is an auxiliary view obtained by projection from one of the six basic views.* 例句①: 第一辅视图是从六个基本视图之一投影得到的。*Example sentence*②: *A primary auxiliary view is obtained by projection on a plane that is perpendicular to one of the three principal planes of projection and is inclined to the remaining two.* 例句②: 第一辅助视图是投影在与三个主要投影面之一垂直, 与另两个投影面倾斜的投影面上的视图。principal ～ 主要视图, 主视图, 前视图; rear ～ * 后视图; reduced ～ 缩小视图; reference ～ 参考视图; relative position of ～s 各视图的排列位置; relevant ～ 相关视图; removed ～ 移出视图; revolving ～ 旋转视图; revolving section ～ 旋转剖的剖视图; right ～ * 右视图; right end ～ 右侧视图; right-hand ～ 右侧视图; right-side ～ 右侧视图; secondary auxiliary ～ 第二辅视图, 复辅视图; section ～ * 剖视图(剖视), 剖面图(剖面); sectional auxiliary ～ 剖面辅视图; selection of

~ 视图选择; semisectional ~ 半剖面图, 半剖视图; side ~ 侧视图; side auxiliary ~ 侧辅视图; single auxiliary ~ 单辅视图, 第一辅视图; single section ~ 单一剖视; small-scale ~ 缩尺视图; special ~ 特殊视图(非基本视图); special position of ~ 特殊位置视图; staggered section ~ 阶梯剖的剖视图; stretch-out ~ 展开图; successive auxiliary ~ 连续辅助视图; supplementary ~ 补充视图; three dimensional ~ 立体图; top ~ * 俯视图, 顶视图, 上视图, 平面图; top auxiliary ~ 上辅视图, 平面辅视图; un-dimensioned ~ 无尺寸的视图; unnecessary ~ 不必要视图; up-ward ~ 仰视图; vertical ~ 俯视图

viewer /'vjuə/ 观察者, 观众; 指示器
view-factor 视觉因素
viewing /'vju:iŋ/ 观察
viewless /'vju:lis/ 不可见的, 看不见的
viewpoint /'vju:pɔint/ 视点, 观察点; 看点, 看法, 观点
viewport /'vju:pɔ:t/ 视口; ~ input 视口输入; ~ transformation 视口变换; overlapping ~ 视口重叠; set ~ 设置视口
vignette /vi'njet/ 小插图, 装饰图案; 简介
virgin /'və:dʒin/ 洁白的, 纯粹的; 首次的
virtual /'və:tjuəl/ 有效的, 实际上的
virus /'vaiərəs/ 病毒
viscose /'viskəus/ 粘胶; ~ glue 胶水; ~ paper 粘胶纸
viscosity /vis'kɔsiti/ 黏度
vise /vais/ (vice) 虎钳, 夹紧
visibility /vizi'biliti/ 可见性, 视界, 可见的, 能见距离, 能见度, 显著, 明显; ~ of lines 线的可见性; ‖ Example sentence: Complete the visibility of the pyramid and add a right-side view. 例句: 完成棱锥的可见性并增加右视图。
visible /'vizəbl/ 可见的, 明显的; ~ line 可见线, 外形线; ~ outline 可见轮廓线; ~ portion 可见部分
vision /'viʒən/ 视线, 视力; 影象, 象; 想象力; 视觉; axis of ~ 视轴, 视觉; center of ~ 视觉中心; have ~s of 想象到
vista /'vistə/ 透视图; 远景, 长而狭之景
visual /'vizjuəl/ 直观的, 可见的; 视力的, 凭视力的; 看的; 形象化的; ~ angle 视角; ~ line 视线; ~ ray 可见射线
visualization /vizjuəlai'zeiʃn/ 目测(方法), 用肉眼(检查); 想象; 直观化, 形象化; 形象
visualize /'vizjuəlaiz/ (visualise) 目测, 观察, 显现; 设想, 想象, 形象, 形象化; 使可见
visualized /'vizjuəlaizd/ 直观的, 具体的, 形象化的
vocabulary /və'kæbjuləri/ 词汇, 词汇表, 指令表, 字符表
volume /'vɔlju:m/ 书, 册, 卷, 部, 体积, 容积; ~ diagram 体积图
volute /və'lju:t/ 螺旋形, 盘蜗形; (泵的)渐伸套
VP (vertical plane) 竖直面, 直立面
V-parallel 平行于 V 面的面(正平面)
VPH (Vickers Pyramid hardness) 维氏硬度
VPN (Vickers Pyramid number) 维氏(钻石锥体)硬度值
V-projection V 面投影
V-projector V 面投射线
VRL (vertical reference line) 垂直基准线
V-shaped V 型的
V-thread 三角牙形螺纹
V-trace V 面迹点(线)
V-type thread 三角牙形螺纹
vulcanite /'vʌlkənait/ 硬橡皮; 硬橡胶; 胶木
v. v (vice versa) (拉丁语)反之亦然, 反过来也一样

W (west) 西; (whitworth) 惠氏螺纹; (width) 宽度; (wire) 电线, 金属线

wadding /ˈwɔdiŋ/ 填料, 填塞物, 衬料

wall /wɔːl/ 壁, 墙; 工作面; ～ of partition 隔墙; ～ section 墙壁剖面; partition ～ 隔断, 间壁; retaining ～ 挡土墙; thin ～ 薄壁

wane /wein/ 变小, 衰退; (月) 亏; 缺; wax and ～ 盈亏, 盛衰

want /wɔnt/ 需要, 应该; 必须; 缺, 欠, 不足, 差; it ～s 3cm of 2m 两米差三厘米; it ～s 1cm of the regulation length 比规定长度少一厘米

ware /wɛə/ 制品, 仪器; hard ～ 硬件; soft ～ 软件

warning /ˈwɔːniŋ/ 警告

warp /wɔːp/ 翘曲, 拖船索; 扭曲, 弯曲; ～ed helicoid 翘曲螺旋面; ～ed surface 翘曲面

warpage /ˈwɔːpeidʒ/ 翘曲, 变形

washer /ˈwɔʃə/ * 垫圈; 洗净器; ～ faced 垫圈承压面; bright ～ 光垫圈, 精制垫圈; conicalspring ～ * 锥形弹性垫圈; curved spring ～ 鞍形弹性垫圈; double coil spring lock ～ 双圈弹簧垫圈; elastic ～ 弹性垫圈; external tab ～ * 外舌止动垫圈; face 垫圈承压面; finishing ～ 光制垫圈; half-bright ～ 半光垫圈; half-finished ～ 半光垫圈; internal tab ～ * 内舌止动垫圈; plain ～ * 平垫圈; retaining ～ 弹簧垫圈; rough ～ 粗制垫圈; round ～ with square hole* 方孔圆垫圈; single chamfer plain ～ * 单面倒角平垫圈; slot (ted) ～ 开口(开缝) 垫圈; snap ～ 开口垫圈; split ～ 开口垫圈; spring ～ * 弹性垫圈, 弹簧垫圈; square taper ～ * 方斜垫圈; square ～ with round hole* 方垫圈; tab ～ with long tab* 单耳止动垫圈; tab ～ with long tab and wing* 双耳止动垫圈

washout /ˈwɔʃaut/ 冲溃, 水冲蚀; ～ thread 不完整螺纹

water /ˈwɔːtə/ 水

waterlines /ˈwɔːtəlainz/ 水线(船舶)

water-lining 水纹线

water-quenching 水淬火

watersupply /ˈwɔːtəsʌpˈlai/ 给水

wave /weiv/ 波(浪); 波浪形

waveform /ˈweivfɔːm/ 波形

waviness /ˈweivinis/ 波度; 波动性; 波浪, 波纹, 波状起伏

wavy /ˈweivi/ 起浪的, 多浪的, 波状的; ～ line 波浪线

way /wei/ 路线, 道路, 通路; 方法; key ～ 键槽

WB (wheel base) 轴距

WD (whole depth) 齿全深; (wiring diagram) 布线图, 线路图, 安装图

Wdf (wdrf) (woodruff) (键名) 半圆键

web /web/ 缩颈, 撑杆, 轮辐, 辐板; 坯料; 连接板

wedge /wedʒ/ 楔, 楔状物; 劈开嵌入

weight /weit/ 加重, 重量, 重; 重要; 用铅笔加粗; (线型的)重度; 载荷, 负荷; ～s of line 线之粗度; bare ～ 空重, 皮重; dead ～ 净重; full ～ 全重; gross ～ 总重; lead ～

压铅, 文镇, 镇纸

weld /weld/ 焊, 焊接; ~ bead 焊缝; ~ drawing 焊接图; acetylene ~ 乙炔焊; alternating current arc ~ 交流电弧焊; arc ~ 电弧焊; autogeous ~ 气焊, 乙炔焊; automatic ~ 自动焊接; basic ~ symbol 焊接基本符号; bead ~ 珠焊, 圆缘焊接; bevel ~ 斜角焊; built-up ~ 堆焊; butt ~ 对接焊; button-spot ~ 点焊; corner ~ 角焊; effective length of ~ 焊缝有效长度; end lap ~ 搭头焊; field ~ 现场焊接; field peripheral ~ 现场全周焊接; fillet ~ 角焊, 条焊; flare-V ~ 喇叭型焊; flash ~ 闪光焊; field peripheral ~ 现场全周焊接; groove ~ 起槽焊接; intermittent ~ 间断焊; included angle of ~ 焊道夹角; increment lenth of ~ 熔接焊道每段长度; jam ~ 对接焊; lap ~ 搭接焊; leg of fillet ~ 角焊宽度, 角焊脚长; loudspeaker ~ 喇叭型焊; metal-arc ~ing 金属极电弧焊; oxy-acetylene ~ 氧乙炔焊; pad ~ 堆焊, 垫块焊; peripheral ~ 全周熔接; plug ~ 塞焊; point ~ 点焊; projection ~ 突焊; site ~ 现场焊接, 工地焊接; spot ~ 点焊; square ~ I型焊; Tee ~ T 形焊; V ~ V 形焊

west /west/ 西, 西方

WF (wide flange shape) H 型钢

WG (wire gauge) 线规; (Working Group) 工作组

wheel /hwi:l/ 轮, 齿轮; 旋转; 滚动; ~ and rack 齿轮与齿条; angle ~ 斜齿轮; angular ~ 锥形齿轮; bevel (gear) ~ 伞齿轮, 圆锥齿轮; blade ~ 叶轮; cam ~ 凸轮; chain ~ 链轮; chevron ~ 人字齿轮; double helical spur ~ (同 double helical gear) 人字齿轮; drive ~ 主动轮; driven ~ 从动轮; follower ~ 从动轮; hand ~ 手轮; helicoidal ~ 斜齿轮, 螺旋齿轮; herringbone ~ 人字齿轮; lobed ~ 叶轮; main driving ~ 主动轮; ratchet ~ 棘轮; sprocket ~ 链轮; spur ~ 正齿轮; thumb ~ (图形显示终端上的定位)指轮; turbine ~ 涡轮; vane ~ 叶轮; worm ~ 蜗轮

wheelbox /'hwi:lbɔks/ 齿轮(变速)箱

wheel-web 轮辐

whirl /hwə:l/ 旋转

white /hwait/ 白色; 白色的; ~ metal 巴氏合金; Chinese ~ (zinc white) 锌白

whiteprint /'wait'print/ 白印, 晒图; ~ machine 白印机, 晒图机

WHIT THD (Whitworth thread) 惠氏螺纹

whole /həul/ 全部, 一体; 完整的; 整个的; considered as a ~ 视为一体

whole-foot 全尺, 整尺

WI (wrought iron) 熟铁

wide /waid/ 宽的, 阔的, 宽广的, 宽度; 幅(宽); ~ flange beam 宽缘工字钢; be two metres ~ 宽两米; tooth ~ 齿宽

widen /'waidn/ 加宽, 放宽; 扩大, 扩展; ~ing on curve 曲线加宽

width /widθ/ 宽(度), 宽, 幅宽, 阔度, 广度; ~ across flats 对面宽度; ~ of row (interrow) 行距; ~ of tooth 齿宽; 12 feet in ~ 宽12英尺; inner ~ 空隙

wild /waild/ -card 任意卡片

wind /wind/ 风, 气流; 方向; ~ rose 风玫瑰, 风向频率玫瑰, 风向徽

winding /'waindiŋ/ 围绕, 缠绕的, 绕组, 绕圈; 绕法

window /'windəu/ 窗子, 窗户; 窗孔, 窗口, 窗玻璃; 橱窗; 窗状开口, 给……开窗; ~ frame 窗框; ~ pane 玻璃窗; ~ screen 纱窗; ~ sill 窗槛, 窗盘, 窗台; ~ stool 窗台; blank ~ 假窗, 盲窗; casement ~ 竖铰链窗; dormer ~ 屋顶窗; non-opening ~ 固定窗; revolving ~ 旋转窗; sliding ~ 拉(移)窗

windshield /ˈwindˌʃiːld/ 风挡

wing /wiŋ/ 翼, 飞机翼; 边, 叶片（轮）

wing-nose （飞机）翼前缘

wipe /waip/ 擦, 拂; 擦去, 抹去

wire /ˈwaiə/ 金属线, 金属丝; 电缆, 电线; ～ gage 线规, 金属丝规; ～ hemming 包边线, 用线画边; ～ spring 线弹簧; earth ～ 地线（电）; spring ～ 簧丝

wiring /ˈwaiəriŋ/ 线路, 导线; 布线; ～ diagram 配线图, 布线图, 线路图; ～ outlet 电线出口; ～ pattern 布线图案; printed ～ 印制线路

witness /ˈwitnis/ 证明, 引证, 证据; 表明, 表示, 说明, 作证; 目睹, 目测; ～ line 作图线, 引证线（细实线）

WL (water line) 水平面线

WMG (Washburn & Moen Wire Gage) W&M 线号规

wont /wəunt/ 习惯, 惯常作法

wood /wud/ 木头, 木材; 木制的东西; 树木, 树林; ～ cut 木刻, 木版画; ～ engraving 木刻术, 木版画; ～ screw 木螺钉; along grain of ～ 纵剖木纹; balsa ～ 轻木; elm ～ 榆木; pine ～ 松木

woodruff /ˈwudˌrʌf/ (woodroof, woodrow) 香车叶草; ～ drill 半圆钻; ～ key 半圆键, 月形键; 半圆键; 月牙销, 半月销

word /wəːd/ 字, 单词, 字码; ～ composition 字的组合; ～ spacing 字的间隔

work /wəːk/ 工作; 功; 作品; 操作; 工件; 成果; ～ piece 工件; construction ～ 建筑工程

workblank /ˈwɜːkblæŋk/ 毛坯

working /ˈwəːkiŋ/ 工人的, 工作的; 操作的, 运转的; 加工, 制造, 处理; ～ drawing 工作图, 施工图; ～ line 工作线（即构件之重力线, 用以计算构件之应力与大小）; ～ point 工作点; ～ position 工作位置; ～ surface 工作面, 加工面; close ～ fit 紧滑配合

workmanship /ˈwəːkmənʃip/ 手艺, 工艺, 工作质量, 做工

workpiece /ˈwəːkpiːs/ 工件

works /wəːks/ 工厂

workshop /ˈwəːkʃɔp/ 工厂, 车间

workstation /ˈwəːkˌsteiʃən/ 工作站; open ～ 打开工作站

world /wəːld/ ～-wide 世界范围的

worm /wəːm/ 蜗杆, 螺杆, 蛇管, 旋管; 蚯蚓, 蠕虫; ～ and gear 蜗杆与蜗轮; ～ gear; 轮（蜗杆）付; ～ gearing 蜗轮传动装置; ～ lead angle 蜗杆螺纹导角（螺旋角）; ～ reduction gear 蜗轮减速器; ～ screw 蜗杆; ～ wheel 蜗轮; ～ conveying 螺旋传送装置; ～ twin 双头蜗杆

W-projector W 面投射线

wrench /rentʃ/ 扳手, 钳; end ～ 平扳手

wrinkle /ˈriŋkl/ 折叠, 折皱

write /rait/ 写出

writer /ˈraitə/ 打字机

writing /ˈraitiŋ/ 写法

wrong /rɔŋ/ 错误, 错的, 不正当的, 有毛病的; 反的; the ～ side of paper 纸的反面

w-r-t (w. r. t) (with respect to) 关于, 相对于, 对应于

wt. (weight) 重量

wye /wai/ Y 字形物, Y 形连接, 三通

X-axis X 轴线, X 轴, 横坐标轴
X-axle X 轴
X-coordinate X 坐标
X-direction X 轴方向
xerographic /ziərəuˈgræfik/ 静电印刷的
xerography /ziˈrɔgrəfi/ 静电印刷术, 干印术
xerox /ˈziərɔks/ 用静电印刷术复制, 静电复印机; 静电复印件
X-line X 轴线
xpln. (explanation) 解释, 说明
X-Y plotter X-Y 绘图仪

yard /jɑːd/ 场, 场地; 堆置场, 仓库; 工作场; 码［英制长度单位, 1码=3英尺=91.44厘米］
yarn /jɑːn/ 绳, 线; asbestos ～ 石棉绳; spun ～ 线绳
yawn /jɔːn/ 间隙, 缝隙
Y-axis Y 轴线, Y 轴, 纵坐标轴
Y-coordinate Y 坐标
Y-curve 叉形曲线, Y 形曲线
yd. (yard) 码（等于91.44厘米）
Y-direction Y 轴方向, 沿纵轴
yield /jiːld/ 生, 生产, 出产; 得出
Y-joint 叉形接头, Y 形接头
Y-line Y 轴线
yoke /jəuk/ 轭, 套, 座
Y-piece 叉形肘管, Y 形肘管
Y-pipe 斜角支管, 叉形管, 三通管
YR (year) 年
Y-section 三通管接头

Z

Z-axis Z 轴, Z 轴线
Z-bar Z 字形铁
Z-chart Z 形算图, Z 字图
Z-coordinate Z 坐标

Z-direction Z 轴方向，沿 Z 坐标
zee /ziː/ Z 字形，Z 形钢
zero /ˈziərəu/ 零，零度，冰点，零位，零点；～ bevelgear 零位圆锥齿轮；～ curve 零线；～ line 零位线，基准线；～ offset 零点偏置；～ point 零点；原点；relative ～ 相对零
zigzag /ˈzigzæg/ "之"字形，锯齿形的，曲折的，交错的；～ line 锯齿线，错纵线；～ nomograph 曲折算图
zinc /zik/ 锌；镀锌
zip /zip/ ～-a-tone screen 捷伯通片（为描阴影之纤维素）
ZL (zero line) 零位线，基准线

zone /zəun/ 地带，区域，范围；将……分区；区，层，带；～ curve 距限曲线；carburization ～ 炭化层；hard ～ 硬化区；tolerance ～ 尺寸公差带，公差范围；wave ～ 波段；wedge-shaped tolerance ～ 扇形公差带
zoning /ˈzəuniŋ/ 图面分区
zoning map 土地分区图
zoning ordinance 分开规划
zoom /zuːm/ 缩放，图形变比
zooming /ˈzuːmiŋ/ 图像电子放大，缩放（计）
Z-section Z 形截面，Z 形剖面
Z-steel Z 字钢

B 部

汉–英

(CHINESE-ENGLISH)

B部汉语词条音节索引

1. 按每一音节右侧的汉字,可查阅同一音节字头的词条。
2. 汉字右侧的数字,指该词条字头所在的页码。

音节	词条字头	页码	音节	词条字头	页码
	A			笔	180
a	阿	179	bian	边	180
an	安	179		编	180
	鞍	179		扁	180
ao	凹	179		变	180
	B		biao	标	181
	B	179		表	181
ba	八	179	bing	并	182
	巴	179		病	182
	拔	179	bo	拨	182
bai	百	179		波	182
	摆	179		玻	182
ban	板	179	bu	补	182
	版	179		不	182
	半	179		布	182
bang	棒	180		部	182
bao	包	180		**C**	
	保	180		C	183
	报	180	ca	擦	183
	刨	180	cai	材	183
bei	北	180		采	183
	备	180		彩	183
	背	180		菜	183
ben	本	180	can	参	183
bi	比	180	cang	舱	183
			cao	操	183

续表

音节	词条字头	页码	音节	词条字头	页码
	草	183		次	186
ce	侧	184	cu	粗	186
	测	184	cui	淬	186
ceng	层	184	cun	存	186
cha	插	184	cuo	错	186
	查	184		**D**	
chai	拆	184			
chan	产	184	da	打	187
chang	长	184		大	187
	常	184	dai	代	187
chao	超	184		带	187
che	车	184	dan	单	187
cheng	成	184	dang	当	188
	承	184		挡	188
	程	184	dao	导	188
chi	尺	184		倒	188
	齿	185		道	188
	赤	185	deng	等	188
chong	虫	185	di	笛	188
	重	185		底	188
chou	抽	185		地	189
chu	出	185		第	189
	初	185	dian	典	189
	除	185		点	189
	处	186		电	190
	储	186		垫	190
chuan	穿	186	diao	吊	190
	传	186	die	碟	190
	船	186	ding	丁	190
	橡	186		钉	190
	串	186		顶	190
chuang	窗	186		定	190
chui	垂	186	dong	东	190
ci	磁	186		动	191

续表

音节	词条字头	页码	音节	词条字头	页码
du	独	191	fou	否	195
	读	191	fu	服	195
	度	191		浮	195
	镀	191		符	195
duan	端	191		辐	195
	短	191		俯	195
	断	191		辅	195
	锻	191		复	196
dui	对	191		覆	196
dun	钝	191	**G**		
duo	多	191		G	196
E			gai	改	196
e	轭	192		概	196
er	耳	192	gan	干	196
	二	192		甘	196
F				感	196
fa	发	193	gang	钢	196
	法	193		港	197
fan	翻	193	gao	高	197
	反	193	ge	格	197
	泛	193		隔	197
fang	方	193		个	197
	防	193	gen	跟	197
	房	193	geng	更	197
	仿	194	gong	工	197
	访	194		弓	198
	放	194		公	198
fei	飞	194		功	198
	非	194		共	198
fen	分	194		供	198
feng	风	195	gou	勾	198
	封	195		沟	198
				钩	198

续表

音节	词条字头	页码	音节	词条字头	页码
	构	198		横	201
gu	估	198	hou	喉	201
	鼓	198		后	201
	固	198		厚	201
	故	198	hu	弧	201
gua	刮	198		互	201
	卦	198	hua	花	201
	挂	198		滑	201
guan	关	198		化	201
	管	198		划	201
	贯	199		画	201
	灌	199	huan	环	202
guang	光	199		缓	202
gui	规	199		幻	202
	轨	199		换	202
gun	滚	199	huang	黄	202
guo	过	199		簧	202
	锅	200	hui	灰	202
	国	200		回	202
	H			汇	202
				绘	202
hai	海	200	hun	混	202
han	函	200	huo	活	202
	罕	200		**J**	
	汉	200			
	焊	200	ji	机	203
hang	行	200		奇	203
	航	200		积	203
hao	号	200		基	203
he	合	201		激	204
	核	201		极	204
	赫	201		集	204
hei	黑	201		几	204
heng	恒	201		给	204

续表

音节	词条字头	页码	音节	词条字头	页码
	计	204		解	209
	记	205		界	209
	技	205	jin	金	209
	迹	205		紧	209
	既	205		进	209
jia	加	205		近	209
	夹	205	jing	经	209
	家	205		晶	209
	甲	205		精	209
	假	206		井	209
	架	206		颈	209
jian	尖	206		警	209
	监	206		径	209
	兼	206		净	209
	检	206		静	209
	减	206		镜	210
	剪	206	jiu	纠	210
	简	206		九	210
	件	206		酒	210
	间	206		救	210
	建	206	ju	局	210
	渐	207		矩	210
	键	207		句	210
	箭	207		距	210
jiao	交	207		锯	210
	胶	207		聚	210
	角	207	juan	卷	210
	铰	207	jue	绝	210
	脚	207	jun	军	210
	校	207		均	210
jie	阶	208		竣	210
	接	208			
	节	208		**K**	
	结	208	ka	卡	210
	截	208	kai	开	211

续表

音节	词条字头	页码	音节	词条字头	页码
kan	勘	211		亮	214
ke	可	211	lie	列	214
	刻	211	lin	邻	214
kong	空	211		临	214
	孔	211		檩	214
	控	212	ling	菱	214
kou	口	212		零	214
	扣	212		另	215
ku	库	212	liu	流	215
kua	夸	212		柳	215
kuai	块	212		六	215
kuan	宽	212	lou	楼	215
kuang	矿	212	lu	炉	215
	框	212		路	215
kuo	括	212		露	215
			lü	铝	215
L				滤	215
la	拉	212	lun	轮	215
	喇	212	luo	罗	216
lan	蓝	212		逻	216
	缆	212		螺	216
lei	累	212		洛	217
	肋	212		落	217
	类	212			
leng	棱	212	**M**		
	冷	213	ma	麻	217
li	理	213		马	217
	立	213		码	217
lian	连	213	mai	埋	217
	联	213	man	漫	217
	链	213	mao	毛	217
liang	量	214		锚	217
	梁	214		铆	217
	两	214	mei	玫	217

续表

音节	词条字头	页码	音节	词条字头	页码
	煤	217	nü	女	220
	每	218	nuan	暖	220
	美	218	nuo	诺	220
men	门	218			
meng	蒙	218	**O**		
mi	密	218	ou	偶	220
mian	面	218	**P**		
miao	描	218			
	秒	218	pai	排	221
mie	灭	218		派	221
ming	名	219	pan	盘	221
	明	219		判	221
	命	219	pao	抛	221
mo	摹	219	pei	配	221
	模	219	pen	喷	221
	磨	219	pi	批	221
	莫	219		皮	221
	墨	219	pian	偏	222
mu	母	219	pin	频	222
	木	219	ping	平	222
	目	219		屏	223
N			po	坡	223
			pou	剖	223
nai	耐	219	pu	普	223
nan	南	219	**Q**		
nao	挠	219			
nei	内	219	qi	七	224
ni	泥	220		齐	224
	拟	220		其	224
	逆	220		企	224
niao	鸟	220		起	224
niu	牛	220		气	224
	扭	220		汽	224
nong	农	220		契	224

音节	词条字头	页码	音节	词条字头	页码
	砌	224		散	228
qian	千	224	sao	扫	228
	铅	224	se	色	228
	签	224	sha	沙	228
	前	224		砂	228
qiao	桥	224	shai	晒	228
	翘	224	shan	山	228
qie	切	224		删	228
qing	青	225		闪	228
	轻	225		扇	228
	倾	225	shang	商	228
qiu	求	225		上	228
	球	225	she	舍	228
qu	区	225		设	228
	曲	225		射	229
	取	226	shen	伸	229
	去	226		深	229
quan	圈	226		渗	229
	全	226	sheng	升	229
que	缺	226		生	229
R				省	229
rao	绕	226	shi	施	229
re	热	226		十	230
ren	人	226		石	230
	任	226		识	230
ri	日	227		实	230
rong	容	227		蚀	230
	熔	227		矢	230
ruan	软	227		示	230
rui	锐	227		世	231
run	润	227		市	231
S				视	231
san	三	227		拭	231
				室	231
				释	231

续表

音节	词条字头	页码	音节	词条字头	页码
shou	手	231		塔	234
	首	231	tai	胎	234
	受	231		太	234
shu	书	231	tan	弹	234
	枢	231		碳	235
	输	231	tao	套	235
	熟	231	te	特	235
	属	231	ti	梯	235
	鼠	231		提	235
	术	231		体	235
	竖	231	tian	天	235
	数	231		添	235
shuan	栓	232		填	235
shuang	双	232	tiao	挑	235
shui	水	232		条	236
shun	顺	233		调	236
shuo	说	233		跳	236
si	司	233	tie	铁	236
	斯	233	ting	停	236
	四	233	tong	通	236
song	松	233		同	236
su	素	233		铜	236
	塑	234		统	236
suan	算	234	tou	头	236
sui	随	234		投	236
sun	损	234		透	237
	榫	234	tu	凸	237
suo	缩	234		涂	237
	索	234		图	237
	锁	234		徒	239
T				土	239
	T	234	tui	推	239
ta	塌	234		退	240
			tuo	拖	240

续表

音节	词条字头	页码	音节	词条字头	页码
	脱	240		舾	243
	椭	240		吸	243
	拓	240		习	243
W				细	243
				系	243
wa	挖	240	xia	狭	243
	蛙	240		下	243
	瓦	240	xian	掀	243
wai	歪	240		舷	243
	外	240		弦	243
wan	弯	241		显	243
	完	241		线	244
	万	241		现	244
wang	网	241		限	244
wei	微	241	xiang	相	244
	韦	241		详	245
	维	241		向	245
	尾	241		项	245
	纬	241		象	245
	未	241		橡	245
	卫	242	xiao	肖	246
	位	242		削	246
wen	文	242		消	246
wo	涡	242		销	246
	蜗	242		小	246
wu	圬	242	xie	楔	246
	屋	242		斜	246
	无	242		写	247
	五	242	xin	心	247
	物	242		芯	247
	误	242		锌	247
X				信	247
xi	西	243	xing	星	248
				形	248

续表

音节	词条字头	页码	音节	词条字头	页码
	型	248		仪	250
	醒	248		遗	250
	性	248		以	250
xiu	修	248		已	250
	袖	248		艺	250
	锈	248		翼	250
xu	虚	248		异	250
	序	248	yin	因	250
xuan	旋	248		阴	250
	选	248		引	251
	渲	249		隐	251
xun	循	249		印	251
	Y		ying	英	251
				影	251
ya	压	249		应	251
	鸭	249		映	251
	牙	249		硬	251
	亚	249	yong	用	251
yan	延	249	you	优	251
	沿	249		油	251
	研	249		游	252
	颜	249		有	252
	檐	249		右	252
	眼	249	yu	隅	252
	燕	250		余	252
yang	阳	250		鱼	252
	仰	250		语	252
	氧	250		与	252
	样	250		雨	252
ye	野	250		裕	252
	业	250		预	252
	页	250	yuan	原	252
yi	一	250		源	253
	移	250		元	253

续表

音节	词条字头	页码	音节	词条字头	页码
	圆	253	zhou	周	259
	缘	254		轴	259
yun	允	254	zhu	逐	260
	运	254		主	260
Z				注	260
				驻	260
	乙	254		柱	260
zai	再	254		铸	260
zan	暂	254	zhuan	专	261
zeng	增	254		砖	261
zha	轧	255		转	261
	闸	255	zhuang	装	261
zhai	窄	255		桩	261
zhan	展	255	zhui	追	261
	栈	255		锥	261
	站	255	zhun	准	261
zhao	照	255	zhuo	着	261
zhe	折	255	zi	资	261
zhen	针	255		子	261
	真	255		字	262
zheng	整	255		自	262
	正	255	zong	综	263
zhi	支	256		棕	263
	直	256		总	263
	执	258		纵	263
	纸	258	zu	足	263
	只	258		组	263
	指	258	zuan	钻	263
	质	258	zui	最	263
	制	258	zuo	左	264
	终	258		作	264
zhong	中	259		坐	264
	重	259			

阿基米德涡线 Archimedean spiral
安全标准 safety standard
安全符号 safety symbol
安全性 security
安装尺寸① fixing dimension, installation dimension
安装的安全性 installation security
安装简图 installation diagram

*安装图② *installation drawing, setting drawing, installation diagram, set-up diagram
鞍形键 saddle key
鞍形面 saddle-shape surface
凹凸板 buckle plate
凹圆锥 female cone
凹坑 concave pit
凹面 concave face

B

B样条函数 B-spline
八边形 octagon
八点法 eight-point method
*八角螺母 *octagon nut
*八角头螺栓 *octagon bolt
八面体 octahedron
巴氏合金 Babbit metal, white metal
拔模斜度 pattern draft
百分比半圆形图 percentage semicircle
百分比棒（条）形图 percentage bar chart
百分比图 percent chart
百分比圆形图 percentage pie chart
百叶窗 shutter, jalousie, louver
摆线 cycloid

*板弹簧③ *leaf spring
版权 copyright
半导体集成电路芯片图 layout of semiconductor IC chip
半对数标值（图尺） semilogarithmic scale
半集中表示法 semi-assembled representation （电）
半径 radius
半径标注（曲线）尺寸法 dimensioning by radii
半径规 radius gage
半宽水线图 half breadth plan
半立方抛物线 half-cubic parabola
半六角形槽 semihexagonal slot
半剖视 half section

注：① 安装尺寸：将部件安装到其他零、部件上，或将机器、设备安装在基础上所需要的尺寸。
② *安装图：表示设备、构件等安装要求的图样。
③ *板弹簧：单片或多片板材（簧板）制成的弹簧。

半剖视图① half section view
半斜投影 cabinet projection
半透明纸 semi-transparent paper
半箱法 semi-box construction
半斜图 cabinet drawing
半写实画法 semiconventional representation
*半圆键 *Woodruff key, half round key
*半圆头铆钉 *semi-round head rivet, *button head rivet
半圆周 semi-circumference
半展开图 half development
半自动铅笔 semiautomatic pencil
棒（条）形图 bar chart
包容原则 envelope principle
包线边 wire hemming, wired edge（板金）
保安矿柱图 safety pillars plan
保存，存 save
保护层 protective coating, armour coat
报警符 bell character
刨花板 shaving board, chip board
北立面图 north elevation
备份 backup
备件图② spare parts drawing
背景图像 backgroung image
背景显示图像 background display image
背立面图 rear elevation, back elevation
背视图 dorsal view
（本图）取代 X supersedes X
（本图）由 X 取代 superseded by X
*比例③ *scale
比例变换 scaling
比例尺 scale
比例尺分格法 division by using scale

比例规 proportional divider
笔划 stroke
笔划边缘 stroke edge
笔划宽度 stroke width
笔划设备 stroke device（计）
笔划顺序 sequence of stroke
笔划中线 stroke centerline
笔划字符发生器 stroke character generator
笔杆 penholder
笔式绘图机 pen plotter
笔头 nib, pen nib
笔头清洁剂 pen cleaner
边 side
边界表示法 boundary representation
边界面 boundary surface
边界线 boundary line, margin line
边距 edge distance
边框 border, enclosure
边坡 side slope
边视图④ edgewise view, edge view
边视图法 edge view method
编辑行 edit line
编码方案 coding scheme
编码图像 coded image
编码图形 coded graphics
编译 compile
*扁环螺母 *flat nut
变换 transformation
变换法 alternate method
变换矩阵 transformation matrix
变换位置 alternate position
变换位置法 change-of-position method
变换位置视图 alternate position view

注：① 半剖视图：当机件具有对称平面时，在垂直于对称平面的投影面上，以对称中心线为界，由一半剖视图和一半外形视图组合而成的图形。
② 备件图：指机器、设备内易损件（备用件）的零件图。
③ *比例：图中图形与实物相应要素的线性尺寸之比。
④ 边视图：沿平行某个表面的方向投影所做出的视图。

变换位置线　alternate position line
变口体　transition piece
变量　variable
变量图表　amount of change chart
变率图表　rate-of-change chart
变位键　shift key
变位复曲面　double curved surface of transposition
变形　deformation
变形单曲面　modified single curved surface
变形接头　transition piece
变址，下标，索引　index
标尺　rod
标点符号　punctuation mark
标高　elevation, ordinance datum
*标高投影① *indexed projection
标记　marker
标记，特征位② flag
标记符号图　legend drawing（电）
标识符　identifier
标题，题目　heading
*标题栏③ *title block
标题栏格式　form of title block
标题形式　form of title
标值（图尺）　scale
标值（图尺）方程　scale equation
标值（图尺）系数　scale modulus
标志④ mark, denotation
标志用图形符号　graphical symbols for use on sign
标注用符号　symbols for indicating

标准　standard
标准比例　standard scale
标准大小，标准型号　standard size
*标准公差⑤ *standard tolerance
标准公差单位　standard tolerance unit
标准化　standardization
标准化组件　standard module
标准件⑥ standard parts
标准螺纹　standard thread
标准体系　system of standard
标准图　standard drawing' typical drawing, standard design
标准图纸　standard sheet
标准 U 形钢　standard U channel
标准文献　document of standard
标准详图　standard details
表　list
表格　table
表格零件图　*tabulated detail drawing
表格图⑦ *tabular drawing
表面　surface
表面波纹度　surface waviness
表面粗糙度　surface roughness, surface rating, surface finish
表面符号　surface quality symbol
表面符号模板　surface symbol template
表面几何形状　surface geometry
表面结构　surface structure
表面结构代号　surface structure code
表面结构符号　surface structure symbols
表面模型　surface model

注：①*标高投影：在物体的水平投影上，加注其某些特征面、线、控制点等的高度数值的正投影。
②标记，特征位：为了标识而使用的位、字或符号。
③*标题栏：技术图样上，由名称、代号区、签字区、更改区和其他区组成的栏目。
④标志（mark）：用图形、文字、符号、颜色等，在产品、包装上，或在某些场所，表示其特性或要求的记号。标志（efnotation）：表示计算机语言语义的一种数学函数。
⑤*标准公差：标准规定的，用以确定公差带大小的任一公差。
⑥标准件：按国家标准、行业标准或企业标准制造的零部件。
⑦*表格图：用图形和表格表示机构相同而参数、尺寸、技术要求不尽相同的产品的图样。

表面描阴法 surface shading
表面缺陷 surface defect
表面视图，外观视图 outside view
表面特征 texture of surface
表面纹理 surface texture
表面质量 surface finish quality, surface quality
表面硬度 skin hardness
*表图① *chart
并行接口 parallel interface
病毒 virus
拨号 dial
波度 wavine
波浪线，断裂线 freehand continuous line, reak line, wavy line
玻璃 glass
玻璃图板 glass drawing board
玻璃制件图 unit drawing of glass products
补充视图 supplementary view
补图 complement of a graph, complementary graph
不必按比例 unnecessary scale
不必要尺寸 unnecessary dimension
不必要视图 unnecessary view
不变式，不变量 invariant
不垂直 out of square
不带公差尺寸 untoleranced dimension
不等边三角形 scalene triangle
不等角投影 trimetric projection
不等角 anisometric drawing, trimetric drawing
不对称 asymmetry

不对称形 unsymmetrical shape
不够粗 not heavy enough
不够黑 not dark enough
不规则曲面 irregular surface
不规则曲线板 irregular curve
不合比例尺寸 out-of-scale dimension
不加工螺栓 unfaced bolt
不考虑形态尺寸 regardless of feature size (RFS)
不可展开的面 nondevelopable surface
不可展曲面 nondevelopable curved surface
不可展直纹面 skew ruled surface
不(未)机械加工表面 unmachined surface
不(未)机械加工部分 unmachined part
不锈钢 stainless steel
不用箭头注尺寸 arrowless dimensioning
不圆 out of roundness
布尔运算 Boolean operation
布局 layout
布氏硬度 Brinell hardness, ball hardness
布图 block out
布线图 wiring diagram, wiring scheme
布置图 layout drawing, layout sheet
部标准 ministerial standard
部分配管图 spool drawing
部分视图 partial views
部分图 partial graph
部件 subassembly
部件测绘② survey and drawing of parts
部件装配图 part assembly drawing, partial assembly

注：① *表图：用点、线图形和必要的变量数值，表示事物状态或过程的图。
② 部件测绘：根据现有部件实物画其部件图和零件图的过程。

C

C 形垫圈　C-washer
C 形夹　C-clamp
擦图片　erasing shield
材料　material
材料表　material list
材料单　bill of material
采剥工程断面图　cross-section of stripping work
采剥工程综合平面图　stripping work map
采掘工程层面图　excavation engineering bedding drawing
采掘工程立面图　excavation engineering elevation
采掘工程平面图　excavation engineering plan, mine map
采掘工程图　excavation engineering drawing
采掘计划图　excavation plan
采掘竣工图　excavation finish plan
采掘设计图　excavation design
采暖管道系统图　heating piping system drawing
采暖平面图 ①　heating layout
采暖通风空调　heating vertilation & airconditioning (HVAC)
采区开采设计图　design of stope
彩色编码　colour coding
彩色打印机　colour printer
彩色喷墨绘图机　colour jet ink plot

彩色图示系统　color graphic system
彩色图形（学）　color graphics
彩色图形打印机　color graphics printer
彩色位图　color bitmap
彩色显示　colour display
彩色显示器　colour displayer
菜单　menu
参考尺寸　reference dimension
参考视图　reference view
参考位图　bitmaps by reference
舱壁（结构）图　bulkhead plan
舱底图　hold plan
舱容图　capacity plan
操纵系统图 ②　controlling system drawing
操作　operation
操作安全　operating security
操作空间　operating space
操作系统　operating system
草测图　reconnaissance map, reconnaissance drawing
草稿纸　scratch paper
草体字母　script
*草图 ③，草稿图 *　sketch, scheme, draft, rough draft, rough sketch, scratch board drawing, sketch drawing, cartoon
草图板　scratch board, sketch board, sketch plate
草图簿　sketch book
草图纸　sketch paper

注：① 采暖平面图：表达采暖管道和设备的平面布置图。
　　② 操纵系统图：表示机器、设备内操纵功能系统中各零（元）件间连接程序的示意图。
　　③ *草图：以目测估计图形与实物的比例，按一定画法要求，徒手（或部分使用绘图仪器）绘制的图。

侧垂线 frontal horizontal line
侧滚法 rolling method
侧棱 lateral edge
侧立面图 side elevation
侧立投影面 profile projection plane
侧邻辅助图 side-adjacent auxiliary
侧面 lateral face, side face
侧面螺钉弹簧规 side-screw bow instrument
侧面投影 profile projection
侧面投影法 profile projection method
侧面图 profile, sheer plan, profile plan
侧平面 profile plane
侧平线 profile line
侧视图 side view
测角仪 bevel protractor
测量图 survey map, surveying sheet
测量坐标网 surveying coordinate drawing
测图器 chartometer
层 layer（计）
层辨别 layer discrimination（计）
层次化 layering（计）
插补 interpolation
插头 plug
插座 socket
查看 view
查询 inquiry
拆卸画法[①] disassembling representation, dismounted presentation, dismantlement representation
产品标准 product standard
长（度） length
长臂（杆）圆规 beam compass
长的短划 long dash

长度尺寸 linear dimension
长方体 cuboid
长方形 rectangle
长划点线 long dashed dotted line
长划短划线 long dashed short dashed line
长划及短划 long and short dash
长划三点线 long dashed triplicate-dotted line
长划双点线 long dashed double-dotted line
长划双短划线 long dashed double-short dashed line
长链线 long chain line
长体字 compressed letter
长折断面 long break
长折断线 long break line
长轴 major axis
常用比例 regular proportion
常用字法（建筑） office lettering
超平面 hyperplane
超图[②] hyper-graphics
超图的覆盖 covering of a hypergraph
超图的链 chain of a hypergraph
车站（线路房屋）布置图 layout of sta-tion
成角透视 angular perspective
成套曲线板 curves
承影面 shadow surface
承招图，核准图样 approved plan
程序图 process drawing, routing diagram
程序 program, programme
程序设计 programming
程序纸 codingsheet
*尺寸[③] *dimension, size
尺寸安置 placement of dimensions

注：① 拆卸画法：绘制机械装配图或土建结构图时，假想拆去部件或构件上某些零件，以便清晰地画出其余部分视图的图示方法。
② 超图：表达复杂的多维关系、哲理、构思等超越传统的图形概念的图。它是计算机图学、心理学及现代几何结合在一起而发展成表达复杂的多维关系的新方法。
③ *尺寸：在技术图样上，用特定长度或角度单位表示的数值。

*尺寸公差① *tolerance of size, dimensional tolerance
*尺寸公差带② *tolerance zone of size
尺寸更改 revision of dimension
尺寸过多 over dimensioning
尺寸弧长注法 dimensioning around the arc
尺寸基准 datum
尺寸界线③ extension line
尺寸链 dimensional chain
*尺寸偏差④ *deviation, deviations of size, dimension deviation
尺寸数字 dimension figure
尺寸线⑤ dimension line
尺寸线之延长线 extension of dimension line
尺寸形式 dimension forms
尺寸注法 dimensioning, inscription of dimensions
尺度四面体 dimensional tetrahedra
齿顶高 height of addendum
齿顶圆 addendum circle, outside circle
齿高 tooth height
齿根高 height of dedendum
齿根圆 dedendum circle, root circle
齿厚 tooth thickness, circular thickness
齿间 tooth space
齿宽 tooth width
*齿轮 gear
齿数 number of teeth
齿条 rack, gear rack
赤平极射投影 stereographic projection
虫瞻图 worm's eye view
重叠画法 lapped representation

重叠角隅 identified corner
重叠面 superimposed surface
重叠投影法 method of lapped projections
重叠图 over lapping drawing (矿)
重复尺寸 duplicate dimension
重复形态 repeated detail, repetative feature, repetative information
重合法 coincidence method
重合断面⑥ coincidental section, revolved section
重合截面 interpolated section
重新启动 restart
重影(性) coincidence of projections
重影点⑦ coincidence points
重影视图 edge view
抽象图形类型 abstract graphical type
出版和发行阶段(计) issue and distribution
初步设计建筑立面图 preliminary design architectural elevation
初步设计建筑平面图 preliminary design architectural plan
初步设计建筑剖面图 preliminary design architectural section
初步设计总平面图 site plan preliminary design
初步设计图 preliminary design drawing
初步图样 preliminary drawing
初次辅助投影 primary auxiliary projection
初始行 initial line
初始化 initializing
初始图形交换规范 initial graphics exchange specification
除另有规定外 unless otherwise specified

注：①*尺寸公差：简称"公差"。允许的尺寸变动量。
②*尺寸公差带：在公差带中上下偏差所限定的区域。
③尺寸界线：限定尺寸范围的线。
④*尺寸偏差：某一尺寸减其基本尺寸所得的代数值。
⑤尺寸线：表示尺寸起止的线，规定用细实线绘制。
⑥重合断面：将切断面图形画在相关视图之内的断面。
⑦重影点：在某一个投影面上投影重合的两个点，称为该投影面的重影点。

除另有说明外 unless otherwise stated
处理 process
储量计算图 calculating drawing of reserves, reserves estimation map
穿点 piercing point
穿孔机 puncher
穿孔带 punched tape
穿孔卡片 punched card
传动螺纹 power-transmitting thread
传动系统图① drive system drawing
传输安全性 communication security
传输协议 transfer protocol
船舶制图员 naval draftsman
船底塞布置图 bottom plug arrangement plan
船体曲线板 ship curve
船体型表面 molded hull surface
船体制图② drawing for hull of ships
船体中横剖面图 midship section
椽条 rafter
串行接口 serial interface
窗口 window
窗框 window frame
窗台 sill, window sill, window stool
垂线间长 length between perpendiculars（船）
垂直 perpendicular
‖ 例句①：如果直线垂直于平面，它必垂直于平面内的每一直线。Example sentence①: If a line is perpendicular to a plane, it is perpendicular to every line in the plane.
‖ 例句②：如果平面内相交两直线都垂直于给定直线，则平面垂直于该直线。Example sentence②: A plane is perpendicular to a line if the plane contains two intersecting lines each of which is perpendicular to the given line.
垂直地质剖面图 vertical geological section
垂直度 perpendicularity
垂直控制 vertical control
垂直面 perpendicular plane
垂直面，直立面 vertical plane
垂直平分线 perpendicular bisector
垂直剖面 normal section
垂直线 perpendicular lines
磁带 magnetic tape
磁带绘图系统 magnetic tape plotting system, tape magnetic system
磁鼓 magnetic drum
磁盘 disk, dise
磁盘操作系统 diskette operation system
次投影 secondary projection
次透视 secondary perspective
粗糙 rough, coarse
粗糙度 roughness
粗糙度截取长度 roughness-width cut-off
粗糙面 rough surface
粗点划线 chain thick line
粗实线 continuous thick visible line
粗线鸦嘴笔 Swedish pen
粗牙 coarse thread
粗牙螺纹 coarse thread
淬火 quenching line, full line,
存储阶段 storage phase
存储块分配图 block allocation map
存储器 storage
存取框图 access block diagram
错切 shearin

注：① 传动系统图：表示机器、设备内传动系统中各零件间连接程序的示意图。
② 船体制图：研究船体的形状、结构、布置和工艺要求等方面的表达和绘制船舶图样的学科。

D

打样 design
打印机 printer
打字机 writer
*大径① *major diameter
大楷字母高度线 cap line, capital line
*大六角螺母 *heavy series hexagon nut
大小尺寸 size dimension
大写字母 capital letter
代码 code
*带孔销 *pin with split pin hole
带头键 gib headed key
带头斜键 gib-head taper key
带式打印机 band printer
*带用螺栓 *belting bolt
带状法(锥面法) zone method (polyconic method)
单笔箭头 one-stroke arrowhead
单笔字 one-stroke letter, single-thickness letter, single-thickness letter
单对数坐标折线图(统) aemilogarithmine line chart
单行数字 numerals in a single line
单件图 separate drawing
单面投影 one-plane projection, single plane projection
单面投影法 one-plane method
单面印制板零件图 one-sided printed board detail drawing
单屏幕 single-screen
单曲率线, 平面曲率 line of single curvature

单曲面 single-curved surfaces
单曲面件 single curvature parts
单曲线 single curved line
单色显示 monochrome display
单式条形图 bar chart
单式象形图 pictorial chart
单式圆形图 pie chart
单条形图表 single bar chart
单头螺纹 single thread
单图 single-part drawing
单图制 single-drawing system
单位四面体 unit tetrahedra
单线表示法 single-line representation
*单线螺纹② *single-start thread
单线电路图 one-line diagram, single-line diagram
单线管系图 diagrammatic piping drawing
单线图 one-line diagram, single line drawing, elementary diagram (电)
单线运行图 single line train graph
单向公差 unilateral tolerance
单向公差制 unilateral tolerance method, unilateral tolerance system
单向制 unidirectional system
单向制(尺寸数字方向) horizontal system
单向注尺寸制 unidirectional dimensions system
单斜法 monoclinic plane
单斜面 angular surface, inclined plane, inclined surface

注:①*大径(螺纹大径):与外螺纹牙顶或内螺纹牙底相切的假想圆柱或圆锥的直径。
②*单线螺纹:沿一条螺旋线所形成的螺纹。

单叶双曲回转面 ① hyperboloid of revolution of one sheet
单一尺寸 single dimension
单元 cell
单元接线表 unit connection table
单元接线图 unit connection diagram
单元组合图 unit assembly drawing
单张零件图 single-part drawing
单折线图 conversion chart
单直纹面 single ruled surface, singly ruled surface
当前 current
*挡圈 ② * (closing) ring
挡土墙 retaining wall
*导程 ③ * lead
导电图形 conductive pattern
导面 ④ director plane
导线 ⑤ directrix
导线换位图 schematic diagram of conductor transposition
导线面 conductor side
导线跳线间隔棒安装示意图 installation diagram for conductor
倒喇叭接合 inverted flare joint
倒棱角 chamfer
倒圆角 fillet
倒转法 method of rabatterment
道路标准横断面图 typical road cross-section
道路断面 roadway section
道路平面图 road plan
等边三角形 equilateral triangle
等粗线 uniform line

等高线，等深线 contour line, contour
等高线笔 contour pen
等高线地图 contour line map
等高线图 contour chart, contour map
等积投影 authalic projection
等角格子纸 isometric sketching paper
等角立方体 isometric cube
等角面 isometric plane
等角剖面 isometric section
等角视图 isometric view
等角投影 isometric projection
等角投影尺 isometric scale
等角图 isometric drawing
等角图法 isometric graphical presentation
等角椭圆模板 isometric ellipse template
等角线 isometric line
等角圆 isometric circle
等角圆弧 isometric circle arc
等角轴 isometric axis
等角轴上线长 isometric length
等角坐标 isometric coordinates
等距离投影 equidistant projection
等距平面 isoplane
等距曲线 equidistant curve
等效电路图 equivalent-circuit diagram
等斜投影　斜等测投影 cavalier projection
等斜图　斜等测图 cavalier drawing
等腰三角形 isosceles triangle
等雨量线 contour rain line
笛卡尔图 Descartes chart
笛卡尔坐标 Cartesian coordinate
*底图 ⑥ * traced drawing, tracing

注：① 单叶双曲回转面：交叉两直线中的一直线绕另一直线旋转所形成的曲面。
②*挡圈：紧固在轴上的圆形机件，可以防止装在轴上的其他零件窜动。
③*导程（螺纹导程）：同一条螺旋线上的相邻两牙在中径线上对应两点间的轴向距离。
④ 导面：约束母线运动的面。
⑤ 导线：约束母线运动的线。
⑥*底图：根据原图制成可供复制的图。

*底径 *root diameter
地板构造平面图 floor construction plan
地沟平面图 utilities trench plan
地籍图 cadastral map
*地脚螺母 *fundation nut
*地脚螺栓 *foundation bolt
地景图 landscape map
地理图 geographical map
地面布置图，地板平面图 floor plan
地盘图 siteplan, location drawing
地平面 ground plane
地平线 ground line
地球定向卫星 earth-orien teel satellite
地球观察卫星 earth observation satellite
地势图 relief map
地图 map
地图比例尺 map scale
地图投影 map projection
地图制图 cartography
地下室平面图 basement floor plan, basement plan
地下系统 under floor system
地下主线管 underground main
地线（电）earth wire, ground neutral
地形符号 topographic symbol
地形高低符号 relief symbol
地形图 ① topographic drawing, topographic map
地形学 topography
地质剖面图 ② geological section
地质图 geological map
第二分角 second angle

‖例句①：第一辅助视图是投影在与三个主要投影面之一垂直，与另两个投影面倾斜的投影面上的视图。Example sentence①: A primary auxiliary view is obtained by projection on a plane that is perpendicular to one of the three principal planes of projection and is inclined to the remaining two.
‖例句②：第一辅助视图是从六个基本视图之一投影得到的。Example sentence②: A primary auxiliary view is an auxiliary view obtained by projection from one of the six basic views.
第二原图 secondary master copy
第三分角 third angle
第三角画法 ③ *third angle method
第三角投影法 third angle (quadrant) projection method
第三投影面，侧投影面 third plane of projection
第四分角 fourth angle
第一分角 first angle
第一辅助视图 a primary auxiliary view
*第一角画法 ④ *first angle method
第一角投影 first angle projection
第一角投影法 first angle (quadrant) projection method
典型零件 typical detail
点 point
点对点式接线图 point to point type of connection diagram
点划线 dashed dotted line, dot-dash-line
点画；点刻 stipple

注：① 地形图：用标高投影法按较大比例尺绘制的，着重表示有限范围内地物、地貌的位置和高程的地图。
② 地质剖面图：假想用平面将大地剖开，绘出剖切面上所呈现地层构造的图样。
③ *第三角画法：将物体置于第三分角内，并使投影面处于观察者与物体之间而得到正投影的方法。
④ 第一角画法：将物体置于第一分角内，并使其处于观察者与投影面之间而得到正投影的方法。

点绘 plotting
点绘交线 plotted intersection
点绘器 plotter
点式打印机 dot printer
点视图 point view
点图表 scatter diagram
点图形① dot pattern
点线 dotted line
点线字母 dotted letter, phantom letter
点圆规 pump compass
点阵绘图仪 dot matrix plotter
点阵图 dot chart
点阵字符发生器 dot matrix character generator
电动擦图器 motor eraser
电工符号 graphical electrical symbol
电缆配置表 cable collocation table
电缆配置图 cable collocation diagram
电力平面图 powerx supply layout
电力系统图 power supply system drawing
电路接线图 electrical hookup
*电路图② *circuit diagram
电路作用描述图 circuit-description diagram
电平图 level diagram
电气符号模板 electro symbol template
电气工程图 electrical engineering drawing
电气原理图 elementary diagram
电气制图③ electrical drawing
电气照明 electric lighting

电原理图 circuit diagram
电子地图 electronic map
电子管底接线图 base diagram
电子管位置图 tube location diagram
电子数据处理 electronic data processing
电子图 electronic drawing
电子系统图 electronic schematic diagram
电子线路图 electronic diagram
*垫圈④ *washer
垫圈承面 washer face
*吊环螺钉⑤ *lifting eye bolt
*吊环螺母 *lifting eye nut
*碟形弹簧⑥ *belleville spring
丁字尺 tee square, T-square
钉线椭圆画法 gardener's ellipse
顶点 vertex, acme
*顶径 *crest diameter
顶棚 ceiling
定比例 scaling
定测图 location drawing
定位尺寸⑦ locating dimension
*定位公差⑧ *location tolerance
定位孔 locating hole
定位器 locator
定位轴线 orientation axis
*定向公差⑨ *orientation tolerance
定形尺寸⑩ form dimension
定型图 typical drawing, typical design
东立面图 east elevation

注：① 点图形：在模式识别中，由点阵构成的图形，它用点的疏密和有无来表示物体。
② *电路图：用图形符号表示电路设备装置的组成和连接关系的简图。
③ 电气制图：研究绘制和阅读电气工程图的一门学科。
④ *垫圈：放在螺母或螺钉头与被连接件之间的薄金属垫。
⑤ *吊环螺钉：头部为环状的螺钉，通常用于超吊。
⑥ *碟形弹簧：外廓呈碟状的弹簧。
⑦ 定位尺寸：确定物体各部分相互位置的度量数值。
⑧ *定位公差：关联实际要素对基准在位置上所允许的变动全量。
⑨ *定向公差：关联实际要素对基准方向上所允许的变动全量。
⑩ 定形尺寸：确定物体及其各部分的形状和大小的度量数值。

动画 animation
动画制作系统 animation system
动画制作语言 animation language
动力配电盘 power panel
动态的 dynamic
动态仿真 kinematics
动态图像 dynamic image
动态运动 dynamic motion
动圆 describing circle
动作顺序图 action sequence diagram
独立分号运行图 independent sectional train graph
独立原则 independent principie
读—写头 read/write head
度 degree
度量几何 [1] metric geometry
度量坐标 metric coordinate
镀涂层 plating, coating, cladding
端点 terminal
端面 end suface
端视图 end view
端子功能图 terminal function diagram
端子接线表 terminal connection table
端子接线图 terminal connection diagram
短轴 minor axis
断路器 breaker
断面图 section, cut
锻件图 [2] forging drawing
对称度 symmetry
对称符号 symbol of symmetry
对称面 plane of symmetry
对称视图 opposite hand view
对称轴 axis of symmetry
对点 opposite point

对合，对合对应 involution
对合变换 involution transformation
对合线束 pencil of involutions
对角分割 diagonal division
对角面 diagonal plane
对角线 diagonal
对偶超图 dual hypergraph
对偶图 dual graph
对偶性 duality
对齐制注尺寸法 aligned dimensioning
对数标值（图尺） logarithmic scale
对数坐标算图 logarithmic chart
对应迹 corresponding trace
对应投影 corresponding projection
对中符号 centring mark
对准 lined up
钝角 obtuse angle
多边形 polygon
多层 layers
多层印制板零件图 multiplayer printed board detail drawing
多次旋转视图 multi-revolved view
多角柱 polygonal column
多媒体 multimedia
多媒体技术 power media technique
‖例句：多面视图是各正投影视图在单一平面（图纸）上有规则地排列。*Example sentence:* A multiview drawing *is a systematic arrangement of orthographic views on a single plane (the drawing paper).*
多面视图 multiview
多面体 polyhedron
多面投影 multiple-plane projection
多条形图表 multiple bar chart

注：① 度量几何：也称欧氏（欧几里得）几何。研究图形度量性质的几何学。图形经过一切等距变换，保持不变的几何性质叫作度量性质。
② 锻件图：锻压加工毛坯时使用的图样。

多维画法几何 [1] n-dimensional descriptive geometry
多线表示法 multiline representation
*多线螺纹 [2] *multi-start thread
多向 multi-direction
多余尺寸 redundant dimension, superfluous dimension
多元棒(条)形图 multiple-bar chart
多圆锥投影 polyconic projection
多折线图 multiline chart

E

轭螺栓 yoke bolt
耳 lug
二笔箭头 two-stroke arrowhead
二层平面图 second floor plan (美国), first floor plan (英国)
二叉(树形)图 binary graph
二重相切，复切 double contact
二重元素 double elements
二次曲面 quadric surface
二次曲线线束 pencil of conics
二底图 duplicated original
二点透视 two-point perspective
二分图 bipartite graph
二进制图像 binary image
二进制图像处理技术 binary image precessing technique
二进制图像分析 binary image analysis
二面角 dihedral, dihedral angle
‖例句①: 两个相交平面所形成的角称为二面角。Example sentence①: The angle formed by two intersecting planes is called a dihedral angle.
‖例句②: 二面角的真实大小反映在给定两平面都是重影的视图上。Example sentence②: The true size of the dihedral angle is observed in a view in which each of the given planes appears in edge view.
二十面体 icosahedron
二视图法 two view method
二维显示 two dimension display
二元全息图 binary hologram

注: ① 多维画法几何: 研究多维空间里各种几何要素和形体的图示图解方法的几何学。
② *多线螺纹: 沿两条或两条以上的螺旋线形成的螺纹，该螺旋线在轴向等距分布。

F

发布阶段 distribution phase
发动机吊挂图 mounting location system of an aero-engine
发动机简图 schematic drawing of an aero-engine
发动机冷却和封严系统图 aircooling system and sealing system of an aero-engine
发动机轮廓图 contour drawing of an aero-engine
发动机弹性轴装配平衡图 assemble-equilibrium sketch of the rotor of an aero-engine
发蓝(煮黑) color-harder
发散光线 diverging rays
法兰(凸缘) flange
翻动图形 flipping drawings
翻滚 tambling
反对称图 anti-symmetric graph
反馈图像 feedback image
反曲点 point of inflection
反射 reflection
反射变换 reflection transformation
反演变换 inversion transformation
泛水 flashing
方案设计 schematic design
方案设计建筑立面图 schematic design architectural elevation
方案设计建筑平面图 schematic design architectural plan
方案设计建筑剖面图 schematic design architectural section

*方案图① * conceptualxdrawing sche-matic design drawing
*方垫圈 *square washer with round hole
方法标准 method standard
方格法 graticulation
方格法(绘制地图用) gratication
方格纸 squar-lined paper, coordinate paper, squared paper, cross-section paper, graph paper, commercial graph paper
*方孔圆垫圈 *round washer with square hole
方框法 enclosing-square method
方框符号 block symbol
方框图(方块图) block diagram, bar chart, block chart
方键 square key
方块图 block chart, block diagram
方块型 block form
*方螺母 *square nut
*方头螺钉 *square head screw
方位标 bearing mark (化)
方位角 azimuth
方位投影 zenith projection
方箱法 box construction
方箱法 boxing method, box-method
方形 square
方牙螺纹 square thread
防潮层 damp-proof course(D. P. C)
防水层 water-proof course, water tight course
防锈 rust-proof
房屋地区图 building plot

注: ①*方案图: 概要表示工程项目或产品设计意图的图样。

房屋建筑制图统一标准 unified standard of architectural drawing
房屋正面 facade
仿射几何 [1] affine geometry
仿射平面 affine plane
仿射群 affine group
仿射坐标 affine coordinate
仿形图 profile drawing
仿真 simulation
访问，存储 access
放尺 enlarged scale
放大 enlargement
放大比例 enlarged scale
放大尺寸 enlarged size
放大机 enlarger
放大视图 enlarged view
放大图 enlarged drawing
放大样 lofting
飞机绘图器，航行绘图尺 aircraft plotter
飞机接线图 air craft wiring diagram
飞机三面图 [2] three-view drawing of an airplane
飞机水平测量图 horizontal survey map of an airplane
飞机专业制图规定 specialized standard for aeroplane
飞机总体布局图 general layout of an aeroplane assembly
非标准螺纹 non-standard thread
非垂直线 nonperpendicular lines
非等角线 non-isometric line
非对称形体 asymmetrical feature
非功能尺寸 non-functional dimension
非固有点，无穷远点 point at infinity
非共面直线 non-coplanar lines
非几何曲线 nongeometric curve
非几何形的多锥形地图 non-geometrical polyconic map
非几何形的极点地图 non-geometrical polar map
非几何形的圆锥形地图 non-geometrical conical map
非迹线平面 non-trace plane
非加工面 unfinished surface
非加工面尺寸 unfinished dimension
非接触面 nonmatings surface
非平行运行线 non-parallel train graph
非球面 aspheric surface
非同平面线 nonplanar lines
非透视立体表示法 pseudo-pictorial representation
非图形 non-graphical
非圆曲线 non-circular curve
非周期图 acyclic graph
分 minute
分辨率 resolution
分部地图 sectional map
分层揭示 representation in layers
分段尺寸 intermediate dimension
分段划分图 hull block division plan
分段结构图 struction plan of section
分格转绘法 craticulation（建）
分隔符 separator
分隔线 demarcation
分规 divider
分号运行图 sectional train graph

注：[1] 仿射几何：研究图形仿射性质的几何学。图形凡经过一切仿射变换，保持不变的几何性质叫仿射性质。仿射变换是一种范围比等距变换更广泛的几何变换。
[2] 飞机三面图：又称全机理论图。将飞机理论外形尺寸按规定比例缩小，以航向自右向左的飞机侧面全图为主视图绘制出的三面视图。

*分角 [1] *quadrant
分解装配图，立体分解系统图 exploded assembly drawing
分开表示法 detached representation
分类图 classification chart
分模线 parting line
分区图 regional plan
分析图 analytical graph
分析图表 analytical chart
分析图学，图解分析 graphical analysis
分析线图 analytical diagram
分组视图 grouped view
风向徽（风玫瑰）[2] wind rose
封闭尺寸 closed dimension
否则 else
服务图 service chart
浮点数 float
符号 [3] symbol
符号表 list of symbols
符号表示法 symbolic representation
符号串 symbol string
符号大小 size of symbol
符号集 symbol set
符号区别 code type
符号图 symbolic drawing
符号细节 symbol detail
符号要素 symbol element
符号原图 symbol original
符号重要细节 symbol significant detail
符号族 symbol family
辐射算图 network chart

俯视图 top view
*俯视图 [4]，顶视图 *top view
‖例句: 完成截头棱锥的给定视图，并增加俯视图。Example sentence: Complete the given views of the truncated pyramid and add a top view.
辅助俯视图 auxiliary plan
辅助割面 auxiliary cutting plane, auxiliary cutting surface
辅助管道及仪表流程图 auxiliary piping and instrumentation flow diagram
辅助截平面法 auxiliary cutting plane method
辅助面 auxiliary surface
辅助平面 auxiliary plane
辅助平面法 auxiliary-plane method
辅助球面法 auxiliary-sphere method
辅助视图 [5] auxiliary view
‖例句①: 用辅助视图法确定两平面的交线，并判别可见性。Example sentence①: By the auxiliary-view method determine the intersection of the planes. Show visibility.
‖例句②: 辅助视图是投影在三个主要投影面（正立投影面、水平投影面、侧立投影面）以外的任一个平面上的视图。Example sentence②: An auxiliary view is a view projected on any plane other than one of the three principal planes of projection (frontal, horizontal, or profile).
辅助视图法 auxiliary view method
辅助素线法 auxiliary-element method
辅助投影 auxiliary projection

注：① *分角：又称"象限"。用水平和铅垂的两个投影面将空间分成的4个区域。前上方称第一分角，后上方称第二分角，后下方称第三分角，前下方称第四分角。
② 风向徽：又称风玫瑰。表示一年中各风向日数的标志。
③ 符号：表达一定事物或概念具有简化特征的视觉形象。
④ *俯视图：由上向下投射所得的视图。
⑤ 辅助视图：将物体向辅助投影面投影所得的视图。

辅助投影面 ① auxiliary projection plane, auxiliary plane, auxiliary plane of projection
辅助图 auxiliary figure
辅助圆法 auxiliary-circle method, auxiliary-curve method
辅助线 auxiliary line
复辅视图 double auxiliary view
复辅助视图 auxiliary-adjacent auxiliary view
复合符号 combined symbol
复合剖 ② compound section
复合条形图 compound bar chart
复合图形信号 composite picture signal
复接 multiple connection

复式比例尺 duplex scale
复式条形图 multiple bar chart
复式象形图 multiple pictorial chart
复位 reset /ˈriːˈset/
复线运行图 two-line train graph
复现 replicate（计）
*复制图 ③，复印图 *duplicate
复印纸 copy paper
复制 reproduction
复制方法 reproduction method
复制图表 reproduced chart
覆盖 overlay
覆盖图 covering graph

GKS 标准 graphical kernel system (GKS) standard
改善箭头 improve arrowheads
概念视图 conceptual view
干涉检验 interference checking
干线式接线图 trunk line type of connection diagram
甘特图 Gantt chart
感光纸 sensitized paper
钢 steel
钢笔杆 pen holder
钢结构 steel structure
钢结构图 steel structure drawing

钢筋编号 numbering of bars
钢筋表 ④ bending schedule, bar schedule
钢筋布置图 reinforcement drawing
钢筋成型图 bar schedule
钢筋混凝土单层厂房结构布置图 reinforced concrete single story factory framing plan
钢筋混凝土高、多层结构布置图 reinforced concrete structure framing plan for high-rise and multistory building
钢筋混凝土结构节点构造详图 reinforced concrete structure construc-tion details
钢筋混凝土结构图 reinforced concrete structure drawing

注：① 辅助投影面：绘制图样时，除基本投影面外，而选用的与一个基本投影面垂直、不与任何一个基本投影面平行的投影面。
② 复合剖：用组合的剖切平面剖开机件以获得剖视图的方法。
③ *复制图：由底图或原图复制成的图。
④ 钢筋表：为编造施工预算、统计用料，在配筋图上所附的图表。

钢筋混凝土桩详图 detail drawing of concrete pile
港口工程制图 harbor engineering drawing
高(度) height
高度线 height line
高级绘图扩展模块 advanced drafting extension
高级绘图系统 advanced cartographic system
格式化 formatting, formatted
格子线 grid (船)
格子纸 scale paper, section paper
隔断，间壁 partition wall
隔汽层 vapour barrier, vapour-proof course
隔热层 heat insulation course
个别零件图 individual detail drawing
个人计算机 personal computer
跟踪 tracking
跟踪符 tracking symbol
跟踪十字 tracking cross
更改 revision
更改表 revision table
更改栏，修正栏 change-record block
更新 updat
工厂布置图 plant layout
工厂配置图 industrial plats
工程地质图 engineering geological drawing
工程绘图机 engineering drafting machine
工程结构 engineering structure
工程建设标准 engineering construction standard
工程说明图 technical illustration
工程俗语 engineering parlance

工程图 technical drawing
工程图学[①] engineering graphics
工程图样 engineering drawing
工程用比例尺 engineer's scale
工程制图[②] engineering drawing
工程制图传输系统 engineering drawing transmission system
工地工程师 field engineer
工具模板 tooling template
工具图[③] instrument drawing
工序图[④] procedure drawing
工业场地平面图 mine arrangement plan, mine yard plan
工业电子图 industrial electronic drawing
工业徒手画 industrial freehand sketching
工业制图 industrial graphics
工艺标准 technique standard
工艺管道及仪表流程图 process piping and instrumentation flow diagram
工艺基准 technological datum
工艺建筑物联系图 technical flow be-tween work areas
工艺流程图 process flow diagram, process chart, technologic flow chart
工艺装备标准 standard for technical equipment
工装图[⑤] tooling drawing
工作草图 working sketch
*工作高度 *working height
工作面施工设计图 design of working face
工作圈数 number of working coils

注：[①] 工程图学：研究工程技术领域中有关图的理论及其应用的学科。
[②] 工程制图：又称"工程画"。它是研究绘制与阅读工程图样的学科。
[③] 工具图：指加工、检验、装配产品时使用的专用工具的图样。
[④] 工序图：又称工序草图、工序简图、工序示意图。在工序卡片或工艺卡片上，用简洁、形象的方式表示某道工序加工零件的详细过程的简图。
[⑤] 工装图：指各种工艺装备的图样，如工具图、量具图、夹具图等。

工作图

工作图① working drawing, shop drawing
工作位置原则 working position principle
工作站 working station
弓形板弹簧 leaf spring, semi-elliptic leaf-spring
公差 tolerance, variation
*公差单位② *standard tolerance unit
*公差等级③ *tolerance grade, grade of tolerance
公差代号 tolerance symbols
*公差带④ *tolerance zone
公差框格 tolerance frame
公差值 tolerance value
公称尺寸 nominal size
公称直径⑤ nominal diameter
公垂线 common perpendicular
公法线 common normal
公共边 common arm
公共信息图形符号 public information graphical symbol
公路布置图 highway alignment
公路工程图 highway engineering drawing
功能表图 function chart
功能键盘 function keyboard
功能设计 functional design
功能图⑥ function diagram
功能子程序 function subroutine
共轭直径 conjugate diameter
共轭轴 conjugate axis
共点算图 concurrent chart
共有线 common line

供电系统示意图 power system diagram
供电总平面图 general layout of electrical supply
勾缝 pointing
沟槽 slot
钩头斜键 gib-head taper key
构件代号 code letters for structural elements
构思草图 idea sketch
构想草图 diagrammatic arrangement, diagrammatic sketch
构造详图 detail of structure
估价图 estimate drawing
鼓形扫描数字化仪 rotating drum image scanning digitizer
固定叶片(鸭嘴笔) non-adjusting blade
故障，错误 bug, fault
刮线小刀 erasing knife
卦限 octant
挂图 wall diagram
挂图，展示图 demonstration
挂瓦条 batten
关键尺寸 critical dimension
关系图 graph of relation
管道布置图 piping layout drawing
管道材料表 piping material list
管道及仪表流程图 piping and instrumentation flow diagram
管道轴测图⑦ piping isometric drawing
管道综合图 general layout of pipe systems
管道纵断面图 transverse section of under-

注：① 工作图：在产品生产过程中使用的图样。
② *公差单位：计算标准公差的基本单位，它是基本尺寸的函数。
③ *公差等级：确定尺寸精确程度的等级。
④ *公差带：限制实际要素变动量的区域。
⑤ *公称直径：代表螺纹尺寸的直径，通常指螺纹大径的基本尺寸。
⑥ 功能图：表示理论的或理想的电路，不涉及实现方法的一种简图。在计算机绘图中指功能设计的框图。
⑦ 管道轴测图：按正等测投影绘制，表示管道及其所属阀门、仪表控制点和全部管件等布置安装情况的立体图。

ground drainage piping
管沟 trench, pipe trench
管架表 piping support list
管架图 pipe support detail
管件图 pipe piece detail
管口表 nozzle schedule
管口方位图 nozzle orientation
管理标准 administration standard
管螺纹 pipe thread
*管系图[①] *piping system drawing, pipe line drawing
管形图表 pipe organ chart
贯穿点 piercing point
‖ 例句: 线与面的贯穿点。*Example sentence:* piercing point of line and surface.
贯穿点法 piercing point method
灌区规划图(灌区平面布置图) irrigation project plan
光笔 light pen
光笔跟踪 light pen tracing
光笔中断 light pen attention
光标 cursor
光滑 smooth
光绘图机 photo plotter
光盘 CD (compact disk)
光盘只读存贮器 CD-ROM
光栅 raster
光栅单位 raster unit
光栅绘图仪 raster plotter
光栅扫描 raster scan
光栅图形 raster graphics
光栅显示器 raster display
光线 light ray, rays of light
光学方框图 block diagram of optics

光学胶合件图 gluey unit drawing of optics
光学零件图 detail drawing of optics
光学系统图 schematic diagram of optical system
光学制图[②] optics drawing
光学组件 optical module
光源 light source
规程 code
*规定画法[③] *specified representation
规范 specification
规范(格)化设备坐标 normalized device coordinate (NDC)
规格 specification
规格尺寸[④] characteristic dimension
规则波浪连续线 uniform wavy continuous line
规则锯齿连续线 uniform zigzag continuous line
规则螺旋连续线 uniform spiral continuous line
轨迹 locus
‖ 例句: 轨迹是点、直线或曲线按某一规定方式运动的路径。*Example sentence:* A *locus* is the path of a point, line, or curve moving in some specified manner.
滚动 scrolling (计)
滚动轴承 rolling bearing
滚筒式绘图机 drum plotter
滚珠轴承 ball bearing
滚子轴承 roller bearing
过程 procedure
过程图表,路线图表 route chart
过大 oversize

注:① *管系图:表示管道系统中介质的流向、流经的设备以及管件等连接、配置状况的图样。
② 光学制图:研究绘制与阅读光学图样的学科。
③ *规定画法:标准中对某些表达对象规定的特殊图示方法。
④ 规格尺寸:又称性能尺寸、特征尺寸。反映产品规格、性能和特征的尺寸。

*过渡配合 ^① *transition fit
过渡曲面 blend
过渡线 transition line
过梁 lintel
*过盈 ^② *interference
*过盈配合 ^③ *interference fit
过(包括)……作…… Represent...contains...
过……作……并…… Draw...contains..., is..., and...
过……作…… Drawing... from...
锅炉制图 boiler drawing
国家标准 national standard
国家标准化管理委员会 standardization administration of China (SAC)
国际标准 international standard
国际标准草案 draft international standard (DIS)
国际标准的补充草案 draft addendum to an international standard
国际标准的修正草案 draft amendment to an international standard
国际标准化组织 international organization standardization (ISO)
国际电子委员会 international electrotechnical commission (IEC)
国际公制标准 international metric standard
国际十进制分类法 universal decimal classification (UDC)

海图 bathymetric chart
函数标值(图尺) functional scale
罕用比例 odd proportion
汉卡 chinese character card
汉字 chinese character
汉字代码 chinese character code
汉字基本字符 chinese character basic character
汉字属性库 chinese character attribute library
汉字图形 chinese character image
汉字显示器 chinese character display
焊缝代号 welding symbol
*焊接螺柱 *welded stud
焊接图 ^④ welding drawing
行(式打)印机 line printer
航测地形图 aerial topographic map
航测图 aerial map
航海符号 marine symbol
航海图 marine chart, nautical chart, nautical map, navigation map
航空地形图 aeronautical planning chart
航(空)摄(影)地图 photo map
航(空)摄(影)相片镶嵌图 airphoto mosaic
航(空)摄(影)制图 aerial photomapping, aerial mapping work
号料草图 sketch for marking off (船)

注:① *过渡配合:可能具有间隙或过盈的配合。
② *过盈:孔的尺寸减去相配合的轴的尺寸所得的代数差,此差为负时是过盈。
③ *过盈配合:具有过盈(包括最小过盈等于零)的配合。
④ 焊接图:表示金属工件焊接方式及技术要求的图样。

合并 merge
合成图像 composite image
合成原图 composite artwork
合格标志 mark of conformity
合格认证 conformity certification
合格证书 certificate of conformity
合金钢 alloy steel
核心系统 core system
赫氏硬度 Herbert pendulum hardness
黑白复印 black-and-white reproduction
黑白复印机 white print machine
黑白图 black-and-white chart, black-and-white diagram, black-and-white graph
黑白图像 black and white picture
黑白图形 black and white pattern
黑白影印 white print
黑色笔 black pen
黑色发蓝 blackening
黑体字 blackface letter
黑线图 black-line print
恒等变换 identity transformation
横剖(断)面图 cross section, cross-sectional profile, transversal section
横剖面纸 cross-section paper
横剖线(船) body lines
横剖线图 body plan (船)
横坐标 abscissa
喉圆 throat circle, circle of gorge
后处理程序 postprocessor
*后视图① (背视) *rear view, back view, back elevation
厚(度) thickness
厚度规 thickness gage

弧 arc
弧成曲线 curve made up of successive arcs
弧度 radian
弧线的近似求长法 approximate rectification of an arc
弧有向图 arc-digraph
互换性 interchangeability, compatibility
互连接线表 interconnection table
互连接线图 interconnection diagram
*花键② *spline
花括号 brace
滑动配合, 适贴配合 slide fit, snug fit
化工工艺图 chemical technology diagram
化工设备图 chemical engineering equipment drawing
化工制图 chemical engineering drawing
划线器 scriber
画…… Drawing...
画法几何③ descriptive geometry
‖例句①: 画法几何是一门图示和图解空间问题的科学。*Example sentence*①: Descriptive geometry *is the science of graphic representation and solution of space problems.*
‖例句②: 画法几何是一种万能工具。*Example sentence*②: Descriptive geometry *is a versatile tool.*
画轮廓 delineate
画面④ picture plane
画面编辑 picture editing
画面文件 picture file
‖例句: 画平行线和垂直线。*Example sentence:* Drawing *parallel and perpendicular lines.*

注: ① *后视图: 由后向前投射所得的视图。
② *花键: 轴和轮毂上有多个凸起和凹槽构成的周向联接件。
③ 画法几何: 又称投影几何。它是研究在平面上图示空间几何形体和图解空间几何问题的理论和方法的学科。
④ 画面: 绘制透视图的投影平面。

画剖面线器 section liner
画颜色 draw color
环境保护标准 environmental protection standard
环面 torus, toroid
环绕 wraparound（计）
环头弹簧圆规 ring head bow
环形箭头（旋转符号）arc arrow, circular arrow
缓冲区 buffer
幻影图 phantom drawing
换面法 replacing projection plane method, method of substituting planes of projection
换算图 conversion chart
换页，进页 form advance
黄铜 brass
簧丝直径 diameter of spring wire
灰分等值线图 isogram of ash content
回火 tempering
回应，回显 echo
回转面 [1] surface of revolution
回转抛物面 paraboloid of revolution
回转体 revolving body, gyro-rotor
回转椭圆面 ellipsoid of revolution
汇编语言 assembly language
绘图板 [2] drawing board, plotting tablet
绘图笔 needle pen
绘图程序 graphical process
绘图方式 drawing mode
绘图辅助工具 drawing aid
绘图工具 drawing equipment, drafting equipment
绘图机 [3] plotter, drafting machine, drawing machine
绘图机步距 plotter step size
绘图模板 draft template, drawing template
绘图墨水 drawing ink
绘图铅笔 drawing pencil
绘图软件 graphics software
绘图头 plotting head
绘图仪 plotter
绘图仪步长 plotter step size
绘图仪器 drafting instrument, drawing instrument, plotter
绘图纸 drawing paper
绘图桌 drafting table
绘折线图 graph line
绘制位图 drawing bitmap
‖ 例句：绘制正投影图，一个重要步骤是正确判断构成视图的那些直线的可见性。
Example sentence: An essential step in drawing an orthographic view is the correct determination of the visibility of the lines that make up the view.
混凝土 concrete
活动百页窗 shutter
活动流图 action diagram
活动铅笔 mechanical pencil, removeable pencil
活动图像 active image
*活节螺栓 *eye bolt

注：① 回转面：由直母线或曲母线绕一轴线旋转形成的曲面。
② 绘图板（plottingtablet 图形输入板）；它是带有传感器的一块图形输入板，根据电磁感应、超声波等不同原理制作。由人操纵传感器的移动，从而可向计算机输入图形。绘图板（plottingboard 图形输出板）：绘图机输出部分的平板。
③ 绘图机（plotter）：一种自动化的绘图装置。它可将计算机的数据以图形的形式输出，从而在一个平面上绘出图形。

J

机车周转图 locomotive working diagram
机动铅笔刀 mechanical sharpener
机动示意图 kinematic diagram
机构原理示意图 mechanical schematic diagram
机械安装孔 mounting hole
机械工程图 mechanical engineering drawing
机械加工图 manufacturing drawing
机械制图① mechanical drawing
机器 machine
机制程序 machining process
机器语言 machine language
奇数 odd number
积聚 concentration
*基本尺寸 *basic size
基本符号 basic symbol
基本件 basic part
基本结构图 construction plan
基本模型 basic model
*基本偏差② *fundamental deviation, basic deviation
基本视图③ basic view
基本投影面④ basic projection plane
基本图型 basic pattern
基本形体 basic body
基本网格 basic grid
基本运行图 fundamental train graph
基本子程序 basic subroutine
基础标准 basic standard
基础平面图 foundation plan
基础图 foundation drawing
基础详图 base details, foundation details
基坑开挖线 line of excavation construction
*基孔制⑤ *hole-basic system of fits, basic-hole system
基数，基地址 base
基线式接线图 base line type of connection diagram
*基轴制⑥ *shaft-basic system of fits, basic-shaft system
基准 datum
基准尺寸 datum dimension
基准代号 datum symbols
基准点 datum point
基准符号 datum symbol
*基准孔⑦ *basic hole
基准面 datum plane, datum surface
基准面注尺寸法 parallel dimensioning

注：① 机械制图：研究绘制与阅读机械工程图样的学科。
② *基本偏差：标准规定的，用以确定公差带相对于零线位置的上偏差或下偏差。
③ 基本视图：将机件向基本投影面投影所得的视图。
④ 基本投影面：绘制视图时，作为投影面的正六面体的6个面称基本投影面。
⑤ *基孔制：基本偏差为一定的孔的公差带，与不同基本偏差的轴的公差带形成各种配合的一种制度。
⑥ *基轴制：基本偏差为一定的轴的公差带，与不同基本偏差的孔的公差带形成各种配合的一种制度。
⑦ *基准孔：基孔制中的孔为基准孔。标准规定的基准孔其下偏差为零。

基准目标　datum target
基准平面　datum plane, datum level
基准线　datum line
*基准直径　*gauge diameter
*基准轴①　*basic shaft
基线　molded base line（船）
基圆　base circle
基座图　foundation plan
激光打印机　laser printer
极三角形法　polar triangle method
极投影　polar projection
极线，极面　polar
*极限尺寸②　*limits of size
*极限高度　*height under ultimate load
*极限偏差③　*limit deviation, limits of deviation
极坐标　polar coordinate
极坐标算图　polar chart, polar diagram
极坐标的坐标网格法　grid method of polar coordinate system
集成电路　IC (integrated circuit)
集点　point of convergence
集合点　vanishing point
集中表示法　assembled representation
几何变换④　geometric transformation
几何公差　geometric tolerance
几何关系　geometric relationship
几何曲面　geometric curved surface
几何体　geometrics solid
几何图　geometric graph, geometrical drawing
几何图案　geometric pattern
几何图形　geometric figure
几何形体　geometric solid
几何形状　geometric shape, geometrical form
几何造型（建模）　geometric modelling
几何作图　geometrical construction
几何作图法　geometrograph
给水　watersupply
计曲线　contour curve
计数器　counter
计算机产生的全息图　computer generated hologram
计算机辅助工程　computer aided engineering
计算机辅助绘图　computer aided drafting
计算机辅助几何设计　computer aided geometric design (CAGD)
计算机辅助简图设计系统　computer aided schematic system
计算机辅助教学　computer assisted instruction (CAI)
计算机辅助设计⑤　computer aided design (CAD)
计算机辅助图形设计　computer aided graphic design
计算机辅助掩膜电路图设计　computer aided layout of masks
计算机辅助制造⑥　computer aided manufacturing (CAM)
计算机辅助学习系统　computer assisted learning (CAL)

注：①*基准轴：基轴制中的轴为基准轴。标准规定的基准轴其上偏差为零。
②*极限尺寸：允许尺寸变动的两个界限值，它以基本尺寸为基数来确定。
③*极限偏差：上偏差与下偏差的统称。
④几何变换：在显示技术中，指图形的放大、缩小、平移、旋转等，也包括三维图形的透视变换。
⑤计算机辅助设计：利用软件技术在具有图形显示功能的计算机系统上，对设计对象进行分析、计算、结构设计等工作，最后通过绘图机给出设计图。
⑥计算机辅助制造：在产品制造过程中，用计算机辅助处理制造过程的数据，控制机器的运行、材料的流动，以及对产品进行测试等的过程。

计算机绘图，计算机制图 computer graphics (CG)
计算机绘图系统 computer plotting system
计算机集成制造 computer intergrated manufacturing
计算机科学 computer science
计算机控制制图 computer directed drawing
计算机控制的绘图仪 computer directed drawing instrument
计算机逻辑图、计算机逻辑图示学 computer logic graphics
计算机生成的图像 computer generated image
计算机输出缩微图形 computer output micrographics
计算机缩微输出 computer output microfilming
计算机缩微图像技术 computer micrographics technology
计算机图示（形）系统 computer graphics system
计算机图像处理 computer image processing
计算机图像发生器 computer image generator
计算机图形 computer picture
计算机图形辅助三维交互作用 computer graphics aided three-dimensional interactive application
计算机图形接口 computer graphic interface (CGI)
计算机图形系统 computer graphic system
计算机图（形）学[1] computer graphics
计算机图形元文件 computer graphics metafile (CGM)
计算机微制图学 computer micrographics
计算机直接制图 computer direct drawing
计算机制图应用 computer graphic application
记号 marker
记录 record
记录，备忘录 memo
记名式机车周转图 locomotive working diagram with name
技术报告 technical report (TR)
技术标准 technical standard
技术产品文件用图形符号（tpd 符号）graphical symbols for use in technical product documentation
技术绘画[2] technical drawing
技术特性表 technical characteristic schedule
技术条件 technical provisions
技术要求[3] technical requirements
迹点（线）[4] trace
迹线平行线 spurparallel
既定处理 predefined process
加工符号 finish mark
加工面 finish surface
加工位置原则 manufacturing position principle
加工余量 extra metal amount
加浓（颜色）intensifying
加阴影，明暗法 shading
加阴影线 shadow lining
加注件号圆圈 ballooning
加注箭头 arrowed
夹具图[5] jig drawing
家俱制图 furniture drawing
甲板边线 deck line at side
甲板敷料图 deck covering plan

注：[1] 计算机图（形）学：用计算机技术和工程图学理论，研究显示和绘制图形的学科。
[2] 技术绘画：工程技术中，凭徒手、目测来画实物或设计对象的直观形象的绘图法。
[3] 技术要求：图样上对产品制造、检验、安装、调试、使用、修饰等规定的各种要求和指标。
[4] 迹线：平面与投影面的交线。
[5] 夹具图：指制造产品时使用的夹紧装置的图样。

甲板平面图 deck plan
甲板中线 deck line at center
假想画法 ① imaginary representation
假想剖面 phantom section
假想图 phantom drawing
假想线 phantom line, imaginary line
架设图 erection drawing
尖角 sharp corner
监视器 monitor
兼容性 compatibility
检错图 error check diagram
检验 inspection, approval
检验量规 inspection gage
减半投影 cabinet projection
剪取 clipping
简化 simplification
简化符号 simplified symbol
*简化画法 ② * simplified representation
简化图样 simplifying drawing
简化线 simplified line（船）
简化制图 simplified drafting
*简图 ③ * diagram, schematic diagram (drawing)
简图用符号 symbols for diagram
件号 parts number
件号圆圈 balloon
间壁, 隔断 partition board
间隔划线 dashed spaced line

*间距 * space
*间隙 ④ * clearance
*间隙配合 ⑤ * clearance fit
建筑比例尺 architect's scale
建筑材料 building materials, building materials, construction material
建筑草图 architectural sketch, architectural sketch, architectural freehand drawing
建筑符号 building symbols, architectural symbols
建筑符号和规定 building symbols and conventions
建筑规范 building codes
建筑规则 building regulations
建筑工程图 ⑥ architectural engineering drawing
建筑绘画 ⑦ building drawing
建筑平面图 ⑧ building plan, architectural plan
建筑施工图 ⑨ architetural working drawing
建筑示意图 architectural presentation drawing
建筑统一模数制 unified building modular system
建筑图 architectural drawing
建筑图字体 architect lettering
建筑物防雷接地平面图 lightning protection grounding layout
建筑学 Architectonics, architecture
建筑制图 architecture drawing

注：① 假想画法：在图样上，用规定的图线画出机件上某些部分的假想投影的图示方法。
② *简化画法：象规定画法、省略画法、示意画法等一类的图示方法。
③ *简图：由规定的符号、文字和图线组成示意性的图。
④ *间隙：孔的尺寸减去相配合的轴的尺寸所得的代数差，此差为正时是间隙。
⑤ *间隙配合：具有间隙（包括最小间隙等于零）的配合。
⑥ 建筑工程图：在建筑工程中绘制的总平面图以及建筑、结构、给排水、采暖通风、电气照明、动力设施等专业图样的总称。
⑦ 建筑绘画：简称建筑画。指表现建筑形象的种种绘画技法，如水墨渲染、水彩渲染、水粉画、铅笔画、钢笔画、工具线条画等。
⑧ 建筑平面图：假想用水平面将建筑物沿窗台以上部分剖切，将剖切面以下部分作水平投影所画的图。
⑨ 建筑施工图：主要表达建筑物内部情况、外部形状以及构造、装修、施工要求等的一套图样。

建筑坐标网 building orientation coordinate network
渐近线 asymptote
渐开线 involute
*渐开线花键① *involute spline
*键② *key
*键槽③ *key-way
键联接④ key joint
键盘 key board (KB)
箭头 arrowhead, arrow
交变剖面线 alternate crosshatching
交叉比 cross ratio
交叉线 skew lines
交错尺寸 staggered dimendions
交错角 stagger angle
交错排列数字 staggered numerals
交点 point of intersection
交点法 piercing point method
交互方式 interactive mode
交互式绘图 interactive graphics
交互式计算机绘图⑤ interactive computer graphics
交互式终端 interactive terminal
交互图形系统 interactive graphics system
交互图像处理 interactive image processing
交互作用 interaction
交线 intersecting lines, line of intersec-tion
‖例句：两平面的交线是两平面共有的一条直线。Example sentence: The intersection of two planes is a straight line common to the planes.

胶带 adhesive tape
胶合板 plywood, veneer board
胶面绘图板 veneer-faced board
胶木 vulcanite
角 angle
‖例句①：直线与平面之间的夹角位于包含直线并与给定平面垂直的平面内。Example sentence①: The angle between a line and a plane lies in a plane that is perpendicular to the given plane and contains the given line.
‖例句②：求直线AB与平面CDE之间夹角的真实大小。Example sentence②: Find the true size of the angle between AB and plane CDE.
角标 corner marking
角标志⑥ corner mark
角度标注 dimensioning angles
角度尺寸⑦ angular dimension
角度定位法 angular dimensioning
角度定位制 angular dimensioning system
角素变换，直射变换，共线变换 collineation transformation
角透视 angular perspective
铰接图 hinged connection diagram
脚手架图 false work plan
校对者 checker
校对 check, checking
校对记号 check mark
校对量规 checking gage
校对顺序 order of checking
校稿 proofreading
校核者，核准者 approving authority

注：① *渐开线花键：齿形是渐开线的花键。
② *键：置于轴和轴上零件的槽或座中，使二者周向固定以传递转矩的联接件。
③ *键槽：轴和轮毂孔表面上为安装键而制成的槽。
④ *键联接：用键将轴和轮毂联接成一体的联接方式。
⑤ 交互式计算机绘图：通过人机对话来进行绘图。用户对产生的图形可以修改和适当的处理，以获得要求的结果。
⑥ 角标志：用于基本图型（设计标志用图形符号的通用图形）最外沿四个拐角处的垂直相交线。
⑦ 角度尺寸：确定角度大小的度量数值。

校验图 check plot (计)
校阅 revised
阶梯剖 [1] echelon cutting, offset cutting
阶梯图表 staircase chart
阶梯线 staircase curve
接长杆 extension bar
接触点 point of contact
接触开关 contact switch
接触线 line of contact
接触网平面设计图 plane of contact system design
接触网平面图 plane of contact system
接地 connection to ground, earthing (电)
接地端子(电) earth terminal
接缝线 seam line
接合形式 type of joint
接入图,外线图(电) interconnection diagram
接线表 connection table
接线端点 terminal
接线夹 connecting clip
接线示意图 circuit-description diagram
*接线图 [2] * connection diagram, interconnecting diagram
接线箱 junction box
节点(结点) node
节点图 [3] joint detail
节点详图 details, details of joints
节径 pitch diameter
节径公差(螺纹) pitch diameter tolerance
节径公差符号(螺纹) pitch diameter tolerance symbol
节径上螺旋角 effective helix angle
*节距 * pitch

节距分布 pitch spacing
节流阀 throttling valve
节面 pitch surface
节线(齿轮,凸轮) pitch line
节圆 pitch circle
节圆喉径 pitch throat diameter
节圆直径 pitch diameter
节圆锥 pitch cone
节圆锥角 pitch cone angle
结构钢 structural steel
结构工程 structural works
结构框图 block diagrams
结构设计 structural design
结构施工图 [4] structural working drawing
结构施工图首页 specification on structaral working drawings
结构实体几何表示法 constructive solid geometry
结构图,构造图 structural drawing, construction drawing
结构详图 structural detail drawing
结构用型钢 structurall (steel) shapes
结构制图 structural drafting
结构制图员 structural draftsman
结束,完成 finish
截短半径尺寸线 interrupted dimension line of radius
截尖,截角 truncation
截交线 [5] line of section
截角柱,截棱柱 truncated prism
截角锥 truncated pyramid
截距 intercept

注:① 阶梯剖:用几个相互平行的部切平面剖开机件以得到剖视图的方法。
② *接线图:表示成套装置、设备或装置的电路连接关系的简图。
③ 节点图:用大于原图的比例,表示设备上某些局部结构形状及尺寸的一组视图。
④ 结构施工图:主要表示承重结构的布置情况,构件类型、大小及构造作法等的图样。
⑤ 截交线:截平面与立体表面的交线。

截面 ① section
截平面 cutting plane
截头棱锥 truncated pyramid
‖例句: 完成截头棱锥的给定视图,并增加俯视图。*Example sentence:* Complete the given views of the *truncated pyramid* and add a top view.
截圆柱 truncated cylinder
截圆锥 circular truncated cone, truncated cone
解剖图 exploded views
界面(接口) interface
界面需求 interface requirement
界限,上下限,约束 bound
金属 metal, metallic substance
金属材料剖面线的区别 code for metals
金属船体制图 drawing for metal hull of ships
金属化孔 plated through hole
*紧定螺钉 ② *set screw
紧固件 fastener
紧固级别 fastener series
紧固件类别 fastener group
紧滑配合 close sliding fit
紧密插口接头 close nipple
紧密度 tightness
紧配合 close fit, tight fit
紧转配合 close running fit
进位 carry
近景图,特景图 close up view
近似展开 approximate development
近似值 approximation
经度 longitude
经距 departures
经线,经度 meridian longitude
经验方程式 empirical equation
经验公式 empirical equation

经验规则 empirical rule
经验设计 empirical design
晶体管 transistor
晶体管符号 transistor symbol
晶体管线路 transistor circuit
精度 precision
精密公差 close tolerance
精密公差中心距 closely toleranced center distance
精密水平仪 precision level
精密直尺 precision straight edge
精密转动配合 precision running fit
精描图,写实图 rendering
井底车场平面图 shaft bottom plan
井上下对照图 above and underground comparative drawing, location map, site plan
井田区域地形图 topographic map of (underground) mine field
井下采掘关系图 a drawing showing the relation of underground mining method
井巷施工图 construction drawing of well and tunnel
颈 neck
警告 warning
径节 diametral pitch
径向尺寸 radial dimension
径向孔模式 radial hole pattern
径向偏转 radial runout
径向元线 radial element
径向轴承 radial bearing
净水力曲线图 diagram of hydrostatic curves
静电复印机 xerographic printer
静电绘图机(仪) electrostatic plotter, electrical plotter
静负荷 dead load

注: ①*截面: 假想用剖切平面将机件的某处切断,所得切断面形状的图形。
　　②*紧定螺钉: 用来固定两零件位置的螺钉。

静止图像 static image	矩阵连锁 matrix concatenation
镜像 mirror-image	句点 fullstop
镜像变换 mirroring	距离 distance
*镜像投影① *reflective projection	锯齿形螺纹 buttress thread
纠错图 error correction diagram	锯齿线，错纵线 zigzag line
九边形 nonagon	锯口 kerf
酒精复印机 ditto machine	聚氯乙烯 polyvinyl chloride (PVC)
救生设备布置图 arrangement plan of life saving appliances	聚乙烯 polyethelyne (PE)
	聚脂薄膜 mylar
*局部放大图② *drawing of partial enlargement, enlarged drawing	卷尺 flexible rule
	绝对标高 absolute elevation
局部辅助视图，部分辅助视图 partial auxiliary elevation	绝对对合 absolute involution
	绝对配极 absolute polarity
局部截面，部分截面 part section	绝对圆 absolute circle
局部剖视图③ partial section, broken-out section, local section	绝对锥 absolute conic
	绝对坐标 absolute coordinates
局部视图④ local view, partial view, broken view, part view	军用地图 ordnance map, military map
	军用透视 military perspective
矩形 rectangular form, rectangular quadrilateral	均方根 mean square root
	均方根值 root-mean-square value
*矩形花键⑤ *rectangle spline	均曲面 fair surface
矩形框 rectangular frame	均匀阴线 flat tint
矩形物 rectangular object	竣工图⑥ complation drawing
矩形锥 rectangular pyramid	
矩阵 matrix	

卡尺 caliper	卡规 snap gage
*卡箍螺栓 *clip bolt	卡华里尔投影 Cavalier projection

注：①*镜象投影：物体在平面镜中的反射图像的正投影。
②*局部放大图：将图样中所表示物体的部分结构，用大于原图形的比例所绘出的图形。
③ 局部剖视图：用剖切面局部剖开机件所得的剖视图。
④ 局部视图：将机件的某一部分向基本投影面投影所得的视图。
⑤*矩形花键：键齿两侧面为平行于通过轴线的径向平面的两平面的花键。
⑥ 竣工图：建筑物建造完毕后，按它实际建成的形式所绘的图样。

卡片，插件 card
卡片文件 file of cards
卡片组 card deck
开发阶段 development phase
开放系统互联参照模型 open system interconnection reference model
开关 switch
开关符号 switch symbol
开口螺栓 expansion bolt
开口皮带 open belt
开口曲线 open curve
*开口销 *cotter pin, split pin
开拓系统图 excavation system plan
开挖断面 excavated section
开挖线 line of excavation
开挖施工图 drawing of excavation construction
开尾式（箭头）open head
开尾式箭头画法 open-style arrowhead stroke
勘探工程分布图 layout sheet of exploratory engineering
勘探线地质剖面图 geological profile of exploratory line
可达尺寸 virtual size
可达情况 virtual condition
可定址的全点图形 all points addressable graphics
可互换性 interchangeability
可互换性组合 interchangeable assembly
可检测元素（图段）detectable element (segment)
可见部分 visible portion
可见点 visible point
可见外形线 visible outline
可见线 object line, visible line

可见性 visibility
‖例句①：判别可见性并完成视图。*Example sentence①: Determine the* visibility *and complete the views.*
‖例句②：完成棱锥的可见性并增加右视图。*Example sentence②: Complete the* visibility *of the pyramid and add a right-side view.*
可挠性材料 flexible material
可收缩图 contractible graph
可调三角板 adjustable triangle, geo-liner
可弯性 pliability
可用范围 available space
可展开者 developable
可展曲面 developable curved surface
刻度 graduation
刻度尺，分度标 graduated scale
空白范围 blank space
空白区域 blank area
*空白图① *blank drawing
空的 empty
空间 space
空间方向 space direction
空间曲线 skew curve
空间图形 space figure
空间锥面 space cone
空调机房平面图 airconditioning machine room layout
空调平面图 airconditioning layout
空图 empty graph
空心圆柱 hollow cylinder
*孔② *hole
孔排列模式 hole pattern, pattern of holes
孔位圆 hole circle, bolt circle
孔形模板 hole template
孔之中心 hole center

注：①*空白图：对结构相同的零件或部件，不按比例绘制且未标注尺寸的典型图样。
②*孔：主要指圆柱形内表面，也包括其他内表面中由单一尺寸确定的部分。

控制点 control point
控制符号 control symbol
控制图 control chart（统）
控制线路（电） control circuit
口令 password
扣环 snap ring
库 library（计）
夸大画法 overstate representation
块规，规矩规 gage block
宽度 breadth
宽（度） width
宽字体 extended letter
矿块采掘计划图 ore piece excavation plan
矿块地质图 ore piece geological drawing
矿块定型图 mine stopping set-on drawing
矿区地形图 mine area topographic map
矿区地形地质图 mine area topographic and geological map
矿区规划图 project plan of mine area
矿区水文地质图 mine area hydrogeological map
矿区图 plat of mineral claims
矿山测量图 mine survey drawing
矿山地质图 mine geological map
矿体几何学 geometry of ore deposits
矿体几何制图 geometrization of ore deposits
矿体等值线图 ore body isoline diagram
矿体顶底板等高线图 ore body top or bottom contours
矿体纵投影图 ore body vertical projection
矿田区域地形图 topographic map of mine field
矿图 map of a mine, mine maps
矿岩层对比图 ore and rock stratification correlatograph
框架图 framing diagram
*框图① *block diagram
括号 brackets, parenthesis, parenthesis mark

拉床 broaching machine
拉孔 broaching
拉伸弹簧 expansion spring
拉伸试验图 tensile test diagram
喇叭口接合 flare joint
蓝图 blue print, blue printing
蓝图分析 blueprint analysis
缆线 cable
累积公差 cumulative tolerance
累积误差 cumulative error
肋 rib
肋板边线 connecting lines for floor end on top
肋骨型线 frame lines
肋骨型线图 frame body plan
肋号 frame number
类似形 similar-shape
棱，棱边 edge
棱镜零件图 detail drawing of prism
棱柱 prism
棱锥 pyramid

注：①*框图：用框形符号或图形及连线和字符，表示某一系统工作原理的简图。

棱锥台 truncated pyramid, frustum of pyramid
冷冻机房平面图 chiller room layout
冷拉 cold drawn
冷缩配合，收缩配合 shrink fit
冷轧钢 cold-rolled steel (CRS)
理论曲线 theoretical curve
理论图（飞机）drawing of external shape
理论线图 theoretical lines plan
理论正确尺寸（基准尺寸）datum dimension
理论值 theoretical value
理想尺寸 basic dimension
理想几何形状 idea geometric form
理想余隙 basic clearance
立方体 cube
立面辅助视图 elevation auxiliary, elevation auxiliary view
*立面图① *elevation, elevational drawing
立面斜视图 elevation oblique drawing
立剖面图 sectional elevation
立体 solid
立体安装示意图 installation diagram
立体表示法 pictorial representation
立体草图 pictorial sketch
立体分解系统图 exploded assembly drawing, exploded pictorial drawing, illustrated working drawing
立体画 stereograph
立体几何 solid geometry
立体视图 pictorial view
立体说明图 pictorial illustration
立体投影 pictorial projection, stereographic projection
立体图 pictorial drawing, three dimensional view
立体图表 pictorial chart
立体图法 pictorial method
立体物 solid object
立体形 solid figure, three-dimensional shape
立体正投影图 axonometric drawing, axonometric projection
立体装配示意图 assembly diagram
连发运行图 continuous train graph
连接半径 connecting radius
连接点 connecting point
连接符，接插件 connector
连接弧 connecting circular arc
连接件 connector
连接螺纹 fastening thread
连接盘 land（电）
连接线 connecting line
连通超图 connected hypergraph
连通图 connected graph
连通无向图 connect-undirected graph
连续尺寸 continuous dimensions
连续点 consecutive points
连续断面 successive sections
连续二等分法 continued bisection process
连续辅助视图 successive auxiliary views
连续辅助投影 successive auxiliary projection
连续曲线 continuous curve
连续色调 continuous tone
连续色调描阴，深淡阴影法 continuous tone shading
连续元线 consecutive elements
连续注尺寸 successive dimensioning
连续注释 bulk annotation
联机，联成 on-line
联机系统 on-line system
联心线 line of centers
联珠熔接 bead weld
联轴节 coupling
链，链接 chain, link

注：①*立面图：建筑物、构筑物等在直立投影面上所得的图形。

链轮 sprocket wheel
链线 dot-and-dash line
链状尺寸 chain dimension
量表读数差 full indicator movement, full indicator reading, total indicator reading (TIR)
量度单位 unit of measurement
量度点 measuring point
量度点法（透视图）measuring point method
量度面 measuring plane
量度面法 measuring plane method
量度线 measuring line
量角器 protractor
量具 measuring device, measuring equipment
量具图 ① gauge drawing
梁板安装图 beam-slab installation diagram
梁拱线 camber curve (line)（船）
梁规 beam compass
两半合一 coupling of opposite half views
两端细分刻度尺 open divided scale
两对面 opposite sides
亮点 brilliant point
亮度 brightness
亮线 brilliant line
列表 tabulation
列表图 collective drawing
列表资料 tabular information
列车运行实绩图 train performance graph
列车运行图 train operation chart, train graph, train working diagram
列线图 alignment chart, alignment diagram, nomogram, nomograph
列线图术 nomography
邻边 adjacent side
邻角 adjacent angle
邻接地图 adjacent map

邻接件 adjacent part
邻接件线 adjacent parts line
邻接视图 adjacent view
邻接素线 consecative element
邻接组件 adjacent component
邻近区域 adjacent area
邻棱 adjacent edge
邻线 adjacent line
临地境界线 property line
临界公差 critical tolerance
临时支架 false work
临时支架图，塔架图（建筑）false work plan
檩条 purline
菱形 rhombus
菱形，斜方形 rhomboid, rhomboid quadrilateral, rhombus quadrilateral
零 cipher, naught, nil
零点 zero
零点偏置 zero offset
零公差 zero tolerance
零件 detail, part
零件表，明细表 parts list
零件编号法 parts designation system
零件测绘 ② survey and drawing of detail, measuring and sketching detail
零件草图 detail-sketching, detail sketch
零件分解图 parts explosion
零件蓝图 detail print
零件设计 detailing
零件图 ③ detail sheet, part drawing, shop detail drawing
零件图号 detail drawing number, item drawing number
零件要素 detail element
零件、装配联合图，集合图 combined detail

注：① 量具图：指制造产品时使用的专用测量工具的图样。
　　② 零件测绘：根据零件实物画其零件图的过程。
　　③ *零件图：表示零件结构、大小及技术要求的图样。

and assembly drawing
零件族 family of parts
零位坐标 zero coordinate
*零线① *zero line
零裕度 zero allowance
另一法 optional way
另一型 optional construction
流程图 flow chart, flow diagram, flow sheet
*流程图②，程序框图 *flow diagram, flow chart, flow sheet, flow block
流水作业图表 continuous production operation sheet
流线 flow line, stream line
流线形面 streamlined surface
‖例句: 流线型表面的画法。Example sentence: Representation of streamlined surfaces.
流向 direction of flow
流域规划图 valley project plan
柳叶法（柱面法）gore method (polycylindric method)
六边形 hexagon
六角承窝螺钉，内六角螺钉 hexagon socket screw
*六角法兰面螺母 *hexagon nut with flange
*六角螺母 *hexagon nut
六角头 hexagonal head
*六角头法兰面螺栓 *hexagon bolt with flange
*六角头盖形螺栓 *acorn hexagon head bolt
*六角头螺栓 *hexagon bolt
*六角头凸缘螺栓 *hexagon bolt with collar
*六角头自攻螺钉 *hexagon head tapping screw
*六角头自切螺钉 *hexagon head cutting screw

六角柱 hexagonal prism
六角锥 hexagonal pyramid
六面体 hexahedron
楼梯 stairs, stair case, stair way
炉体总图 general drawing of furnace body
炉型示意图 diagram of furnace
路径 path
路径图 route ordonnance map（电）
露天矿采掘进度计划图表 extavation progress plan and form of opencut
露天矿分层平面图 opencut division plan
露天矿基建终了平面图 opencut pre-production plan
露天矿开拓系统图 excavation system plan of opencut
露天矿年末综合平面图 year-end integrate plan of opencut
露天矿排水工程布置图 draining arrangement plan of opencut
露天矿终了平面图 final pit design of opencut
露天矿综合地质图 integrate geological map of opencast
铝 aluminum
铝合金 aluminum alloy
滤光镜零件图 detail drawing of filter
轮齿 gear tooth
轮辐 spoke, spoke of wheel
轮毂 hub
轮廓草图 outline sketch
轮廓图 profilagraph
轮廓元线 contour elemnet
轮廓，外形 profile
轮廓素线 contour element
轮廓线 outine, contour
轮廓组合图 outline assembly drawing

注：① *零线：公差与配合中确定偏差的一条基准直线。
② *流程图：表示生产过程中事物进行顺序的简图。

轮磨 grinding
轮缘 flange of wheel, rim
罗盘方位 rhumb
*逻辑图① *logic diagram
逻辑图形功能 logical graphic function
*螺钉② *screw
螺钉凸缘接头 screw flange joint
*螺距③ *pitch (of thread)
螺距规 screw pitch gage
螺孔 threaded hole, tapped hole
螺孔深度 depth of tapped hole
*螺母④ *nut
螺母的二面视图 two-face view of nut
螺母的三面视图 three-face view of nut
*螺栓⑤ *bolt
螺丝刀（改锥） screw driver
*螺尾⑥ *washout thread
螺尾长度 tail length
*螺纹⑦ *screw thread
螺纹半写实画法 semiconventional representaion of thread
螺纹部分 threaded part, threaded portion
螺纹长度 length of thread, thread length
螺纹大径 crest diameter, full diameter of thread
螺纹大径公差 crest diameter tolerance
螺纹大径公差符号 crest diameter tolerance symbol
螺纹大小径间隔 depth of thread spacing

螺纹峰线 crest line
*螺纹副⑧ *screw thread pair
螺纹规 thread gage
螺纹规格 thread specification
螺纹基本大径 basic major diameter
螺纹基本形式 basic form of thread, basic thread form
螺纹级别 thread series
螺纹简化符号（美式） simplified thread symbol
螺纹简化画法（美式） simplified representation of thread
螺纹件 threaded feature
螺纹角 thread angle
螺纹节径 effective diameter
螺纹理论深度 angular depth of thread
*螺纹联接⑨ *screwed joint
螺纹连接件 thread fasteners
螺纹轮廓 thread profile
螺纹配合等级 screw-thread class, thread classes
螺纹全深 total thread depth
螺纹伸出端，螺纹不配合部分 free end of the thread
螺纹深度 depth of thread, thread depth
*螺纹升角⑩ *lead angle
螺纹外形 thread contour
螺纹线 form line of thread, thread line
螺纹线数 number of threads

注：①*逻辑图：主要用二进制逻辑单元图形符号所绘制的电路简图。
②*螺钉：具有各种结构形状头部的螺纹紧固件。
③*螺距：相邻两牙在中径线上对应两点间的轴向距离。
④*螺母：(1)具有内螺纹并与螺栓配合使用的紧固件。(2)具有内螺纹并与螺杆配合使用，用以传递运动或动力的机械零件。
⑤*螺栓：配用螺母的圆柱形带螺纹的紧固件。
⑥*螺尾：向光滑表面过渡的牙底不完整的螺纹。
⑦*螺纹：牙型截面通过圆柱或圆锥的轴线并沿其表面的螺纹线运动所形成的连续凸起。
⑧*螺纹副：内、外螺纹相互旋合形成的联结。
⑨*螺纹联接：通过螺纹构成的联接，多为可拆卸联接。
⑩*螺纹升角：在中径圆柱或中径圆锥上，螺旋线的切线与垂直于螺纹轴线的平面间的夹角。

螺纹形状 form of thread, thread form
*螺纹旋合长度① *length of thread engagement
*螺纹牙型② *form of thread
螺纹真实表示法 true representation of screw thread
螺纹正规符号（美式）regular symbol, regular thread symbol
螺纹终止线 stop line of thread
*螺纹轴线 *axis of thread
螺隙 thread clearance

螺旋齿轮 spiral gear, helical gear
螺旋浆图 screw propeller drawing
螺旋角 angle of helix, helix angle
螺旋面 helicoid, helicoidal surface
螺旋盘旋面 helical convolute
*螺旋弹簧③ *helical spring, coil (spirel) spring
*螺旋线④ *helix, spiral
螺柱 stud
洛氏硬度 Rockwell hardness (HRC)
落锻 drop forging

麻花钻头 twist drill
马口铁皮 sheet tin
马尼拉硬纸板 manila paper
马赛克 mosaic
码 yard
埋头 countersunk head
埋头螺帽 stove nut
埋头螺栓 stove bolt
埋头铆钉 countersunk head rivet
埋头铆钉 countersunk head screw
漫游 panning（计）

毛刷 dusting brush
*毛坯图⑤ *model drawing, blank drawing
锚泊设备布置图 anchor arrangement plan
*铆钉⑥ *rivet
铆钉符号 rivet symbol
铆钉扣距 grip of rivets
*铆钉联接⑦ *riveted joint
*铆接图⑧ *rivet joint drawing
玫瑰线图 rose line chart
煤层等厚线图 isothickness map of coal seam
煤层等深线图 isobath map of coal seam

注：①*螺纹旋合长度：内外螺纹旋合时，螺旋面接触部分的轴向长度。
②*螺纹牙型：在通过螺纹轴线的剖面上，螺纹的轮廓形状。
③*螺旋弹簧：呈螺旋状的弹簧。
④*螺旋线：点沿圆柱或圆锥表面作螺旋运动的轨迹，该点的轴向位移与相应的角位移成正比。
⑤*毛坯图：零件制造过程中，为铸造、锻造等非切削加工方法制作坯料时提供详细资料的图样。
⑥*铆钉：一种金属制成的，一端有帽的杆状零件，穿入被联接的构件后，在杆的外端打、压出另一头，将构件压紧、固定。
⑦*铆钉联接：简称"铆接"。借助铆钉形成的不可拆卸联接。
⑧铆接图：表示多个构件用铆钉连接成零、部件的图样。

煤层底板等高线图 coal-seam floor contour map
煤层对比图 coal-seam correlation section
煤矿测量图 graphic documentations of coal mine survey
煤矿地质图 coal mine geological map
煤田地形地质图 coal topographic-geological map
每英尺锥度 taper per foot
每英寸螺纹数，每英寸牙数 thread per inch
美国 8 又 1/2″×14″ 标准纸 legal size
美国标准螺纹 American National screw thread
美国标准螺纹配合等级 American screw thread class
美国标准信纸大小 8 又 1/2″×11″ standard letter-sheet size
美国标准学会 American Standard Association
美国国家标准 American National Standard (ANS)
美国国家标准协会 American National Standard Institute (ANSI)
美国画法(第三角法) American method
美国机械工程学会 American Society of Mechanical Engineers (ASME)
美国机械工程学会锅炉法规 ASME boiler code
美国制投影(第三角投影) American projection
美学 aesthetics
门 door, hinged door (船)
门窗符号 door and window symbols
门窗五金 rough hardware

门框 doorframe
门楣 lintel
蒙日法 [1] Monge's methold, Mongean method
*密封圈 *seal
密封装置 sealing equipment (device)
面，表面 surface
*面板螺钉 *covers crew
面的分类 classification of surfaces
面的极限 surface limit
面的极限线 surface limit line
面的交线 intersection of surfaces
面积 area
面积计 planimeter
面积条图 area bar chart
面积图 area chart, area graph
面角 face angle, facial angle
面轮廓度 profile of any plane
面至面 F to F (face to face)
面坐标 areal coordinate
描绘 delineation, describe
描述图 discriptive graph
描图 tracing
描图布 tracing linen, translucent cloth
描图胶片 tracing film
描图顺序 order of tracing
描图纸 tracing paper
描写图 rendered drawing
描阴 shading
描阴片 shading screen
秒 second
灭点 [2] vanishing point
灭线 vanishing line

注：[1] 蒙日(Gaspard Monge, 1746—1818)，法国人，是 19 世纪著名的几何学家。他发展了 17 世纪的投影原理，创立了用平面图形准确表示空间立体的画法几何方法，成为现代工程制图的基础，著有画法几何学，被称为画法几何学之父。他推动了空间解析几何学的独立发展，奠定了空间微分几何学的基础，创立了偏微分方程的特征理论，引导了纯粹几何学在 19 世纪的复兴。
[2] 灭点：直线上无穷远点的透视。

名称线 appellative line（船）
名号牌 name plate
名数 denominate number
明暗画 light-and-shade drawing
明点 brilliant point
明细表（栏）① item list
*明细栏 *item block
明线 brilliant line
命令语言 command language
摹图，抄图 copying drawing
模板 template, stencil plate
模板图 template drawing, forming drawing, form work drawing
模糊图 fuzzy graph
模糊子图 fuzzy subgraph
模具图② die drawing
模拟 analog, analogy
模拟绘图方式 analog plot mode
模式发生 pattern generation（计）
模数 modulus
模线 lofting（飞）
模型 mould
模型打样 mock-up
模型图③ mould drawing

模型照像图 model photo drawing
模造纸 simili paper
模制 modeling
磨砂玻璃 sandblasted glass, rubbed glass ground glass
莫氏锥度 Moh's hardness, Morse taper
墨喷绘图机 ink jet plotter
墨水 ink
墨水描图 inked tracing
墨水瓶座 bottleholder
墨线 ink line
母点，动点 generating point
母点（线、面）④ generatrix, generator
母曲线 generating curve
母线，动线 generating line
母圆，动圆 generating circle
木材 woods
木结构 timber structure
木结构图 timber structure drawing
木模 wooden former
木模图 pattern drawing
目标符号 aiming symbol
目测法 ocular estimate, ocular estimation
目的程序 object program

N

耐火材料 refractory material
耐火砖 firebrick
南立面图 south elevation

挠四边形 skew quardilateral
挠性曲线尺 flexible curve (spline)
内摆线 hypocycloid

注：①*明细栏：由序号、代号、名称、数量、材料、质量、备注等内容组成的栏目。
②模具图：指模压成型工具的图样。如锻压模具图、冲压模具图、压铸模具图等。
③模型图：表示生产铸件的模型的图样。
④母线：形成规则曲面的动线。

内部构造图（飞机） inboard profile
内部形态 internal feature
内部形状 interior form, interior shape
内齿轮 internal gear
内错角 alternate-interior angle
内公切线 interior common tangent
内环面 inner-surface of toroid
内件 internal member, internal part
内角 interior angle
内接触 internal contact
内接多边形 inscribed polygon
内径分厘卡 inside micrometer
内卡 inside caliper
*内六角沉头螺钉 *hexagon socket countersunk flat cap head screw
*内六角圆柱头螺钉 *sexagon socket head cap screw
*内螺纹① *female thread, internal thread, nut thread
内螺纹简化符号 simplified internal-thread symbol
内螺纹件 internally threaded part
内螺纹让切 thread relief
内螺纹预钻孔 tap-drill hole
内螺纹正规符号（美式） regular internal-thread symbol

内面向上 inside face up
内切圆 incircle, inscribed circle
内心 incenter
内圆 inner circle
内圆角 filet
泥沼地 mud flat
拟合 fitting
拟圆柱曲面 pseudo cylindrical surface
拟圆锥曲面 pseudo conic surface
逆求法 reversible method
逆图 converse graph
逆有向图 converse digraph
鸟瞰图② bird's eye perspective
牛角面 horn surface
牛角体 cow's horn
牛皮纸 craft paper, kraft paper, vellum paper
牛头刨床 shaper
扭力弹簧 torsion spring
农场测量图 farm survey
农业机械 agricultural machinery
农业机械图形符号 graphical symbols for agricultural machines
女儿墙 parapet, parapet wall
暖色 warm color
诺漠图，列线图 nomogram, nomography, alignment chart

偶数 even number

偶数元 even digit

注：① *内螺纹：在圆柱或圆锥孔表面上形成的螺纹。
② 鸟瞰图：高视点的透视图。视点高出建筑群中最高建筑物，向下看所绘制的透视图。

P

排列图，帕累托图 Pareto chart, Pareto diagram
排水 sewerage
排水系统示意图 draining system diagram
排序，分类 sort
派生图、推导图、引导图 derivation graph
盘曲管 serpentine pipe
盘曲线 serpentine curve
盘头 pan head
盘形弹簧 disk spring
盘形图（圆形图）pie chart
盘旋变口体 convolute transition
盘旋面 convolute, convolute surface
判别 determine
判断 decision
抛光 glos finish, buff
抛物线 parabola
抛物线半余面 parabola spandrel
抛物线弓平面 parabola segment
抛物线宽 span of parabola, width of parabola
抛物线模板 parabola template
抛物线模板 parabola template
抛物线深度 depth of parabola
抛物线蜗线 parabolic spiral
抛物线形包络线 parabolic envelope
抛物线纵距 rise of parabola
配电盘 panel
配电盘 distribution pannel board

*配合 [1] *fit
配合长度 length of fit
配合尺寸 mating dimension
配合等级 fit quality, quality of fit
配合等级 classes of fits
*配合公差 [2] *variation of fit, *fit tolerance
*配合公差带 [3] *fit tolerance zone
配合件 mating part, mating piece
配合面 mating surface
配合制 mating surface, system of fits
配合制度 fit system
配合种类 type of fits
配筋图 [4] reinforcement drawing
配极变换 polarity transformation
配极对应 polarity
配极曲线 polar curve
配色法 scheme of colors
配置 config
配置计划 site planning
配置平面图 disposition plan
配置图 arrangement plan, block plan, plot plan
喷砂 sand blasting
批注 annotation
批注尺寸 note form size
批注方式 note form
批准 approval
皮带 belt
皮带轮的双弯幅 double curved arm of pulley

注：① *配合：基本尺寸相同的，相互配合的孔和轴公差带之间的关系。
② *配合公差：允许间隙或过盈的变动量。
③ *配合公差带：在公差带中，间隙或过盈的上下偏差所限定的区域。
④ 配筋图：又称钢筋布置图。表示钢筋混凝土构件内钢筋配置情况的图样。

偏差 deviation
偏差条形图 bar chart with plas and minus values
偏位尺寸 offset dimension
偏位角 offset angle
偏向角度 deflection angle
偏心度 degree of eccentricity, eccentricity
偏转公差 runout tolerance
频率搬移图 frequency translation diagram（电）
平板 surface plate
平板凸轮 plate cam
平板显示 panel display
平版印刷 lithoprinting, offset printing
平版印刷（石板）offset lithography (press work)
平版印刷（照相）offset lithography (camera work)
平版印刷法 lithographic process
平表面 plane face, plane surface
平承面 plain bearing surface
平尺 flat ruler
*平垫圈 *plain washer
平分 bisect
平分线 bisector
平衡超图 balanced hypergraph
*平键① *flat key
平角 straight angle
平距 horizontal distance
平均偏差 average deviation
平均曲率 average curvature
平均直径 average diameter
平面 plane, plane surface, flat surface

‖例句①：*平面是指这样一个面，该面上任意两点连成的直线都重合在该面内*。Example sentence①: *A* plane *is a surface such that a straight line connecting any two points in that surface lies wholly within the surface.*

‖例句②：*如果直线在平面内，直线上的点也都在平面内*。Example sentence②: *If a line is known to be in a plane, then any* point *on that line is in the plane.*

‖例句③：*平面内的两直线不是相交就是平行*。Example sentence③: *Any two lines in a plane must* either intersect or be parallel.

‖例句④：*在平面内过一点可作无数条直线*。Example sentence④: *An* infinite number of lines *that contain the point may be drawn in the plane.*

‖例句⑤：*过直线CD作一平面平行于直线AB*。Example sentence⑤: *Represent a plane that contains CD and is parallel to AB.*

平面的边视图（平面的重影视图）edge view of a plane
平面的真形视图 normal view of a plane
平面间的夹角 angle between planes
平面截口 plane section
平面的真形视图 normal view of a plane
平面度 flatness
平面方向 direction of plane
平面符号 plane symbol
平面规 surface gage
平面迹 plane trace
平面几何学 plane geometry
平面角 plane angle
平面立体 plane body
平面切削，平面车削 facing
平面球形图 planisphere
平面曲线 plane curve
平面体，多角体 plane solid
*平面图② *plan, plan view, plane graph
平面图（建筑）sectional plan
平面图形 plane figure

注：①*平键：矩形或方形截面而厚度、宽度不变的键。
　②*平面图：建筑物、构筑物等在水平投影面上所得的图形。

平面与球面反射镜和分光镜零件图 detail drawing of plane mirror spherical mirror and spectrometer

平面组成之几何形状 geometric plane figure

平面坐标 plane coordinates

平面坐标系 plane-coordinate system

平台式绘图机（平板绘图仪） flat bed plotter

平头螺钉 flat-headed screw

平斜尺，刀口平尺 flat bevel scale

平行 parallel

‖例句：两条水平线、两条正平线或两条侧平线，在两个主要视图中都平行，但两线在空间可能平行也可能不平行。Example sentence: Two horizontal, two frontal, or two profile lines that appear to be parallel in two principal views may or may not be actually parallel in space.

平行边 parallel edges

平行尺 parallel ruling straightedge

平行尺寸 parallel dimension

平行度 parallelism

平行光线 parallel light ray, parallel rays

平行六面体 parallelopiped

平行面 parallel planes

平行面法 paralled plane method

平行四边形 parallelogram

平行四边形法（椭圆） parallelogram method

平行投影 parallel projection

*平行投影法 [①] *parallel projection method

平行透视 parallel perspective

平行弦法 method of parallel chords

平行线 parallel lines

‖例句：平行直线在正投影中具有保持平行的性质。Example sentence: Parallelism of lines is a property that is preserved in orthographic projections.

平行斜线 parallel oblique lines

平行圆规 parallel compass

平行运行图 parallel train graph

平行直线 Parallelism of lines

平移 transiation, off set, translating

屏极线路（电） plate circuit

屏幕 screen

坡度 slope gradient, inclination

坡度线（结构） grade level

坡水板 water table

坡水板截面图 profile of water table

剖面区域 section area

*剖（断）面图 [②] *section

剖面线 section line, cross hatching, hatching, section lines, hatching lines

剖面详图 sectional detail drawing

剖面种类 type of sections

剖切 sectioning

剖切符号 [③] cutting symbol

剖切面 [④] cutting plane

剖切线 cutting line

剖视图 [⑤]（剖视） *sectional view

普通分规 ordinary dividers

普通螺纹（公制螺纹） metric thread, common thread

*普通平键 *general flat key

普通铅笔 wool-cased drawing pencil, writing pencil

注：① *平行投影法：投射线相互平行的投影法。
② *剖（断）面图（截面）：假想用剖切平面将机件的某处切断，所得切断面形状的图形。
③ 剖切符号：也称剖切位置线。绘制剖视图或截面时，在图样上表示剖切位置的符号。
④ 剖切面：绘制图样时，为了表达机件上沿视线方向被遮挡的部分，而假想剖开机件时所使用的平面或曲面。
⑤ *剖视图：假想用剖切面剖开机件，将处于观察者和剖切面之间的部分移去，将其余部分向投影面投影所得的图形。

普通识别法 general identification
普通图形符号 graph notation for ordinary graph
*普通楔键 *general taper key
普通圆规 common compass
*普通圆柱销 *general cylindrical pin
*普通圆锥销 *general taper pin

Q

七边形 heptagon
七面体 heptahedron
齐次坐标 homogeneous coordinate
齐头线（斜纹线）definite end lines
其余 rest, remains
企业标准 company standard
起槽焊接 groove weld
起槽件（焊接）grooved member
起动器 starter
起画面 starting plane
起止开关 start-stop button
气密斜管螺纹 dryseal taper pipe thread
气密直管螺纹 dryseal straight pipe thread
气体焊接 gas welding
汽车车身制图 automobile body drawing
契约图，发包图 contract drawing
砌砖图 drawing of brick works
千分之一英寸 mil
铅笔 pencil
铅笔刀 pencil sharpener, penknife
铅笔绘图 penciling
铅笔绘图顺序 order of penciling
铅笔记号 pencil mark
铅笔尖（圆规）pencil attachment, removed pencil
铅笔尖斜面 lead bevel
铅笔浓度，黑度 blackness
铅笔弹簧圆规 bow pencil
铅笔图 pencil drawing
铅笔图布 pencil cloth
铅笔线 pencil line
铅笔心 pencil lead
铅笔硬度 grade of pencil
铅笔圆规 pencil compass
铅垂面 vertical plane
铅垂线 vertical line, frontal profile line, plumb line
铅管 lead pipe
铅丝法 lead-wire method
铅芯 lead
签署的文件 signature document
前处理程序 preprocessor
前辅助视图 front auxiliary
前辅助视图，立面辅助视图 auxiliary elevation
前邻辅助视图 front-adjacent auxiliary
前景图像 foreground image
桥墩 bridge pier
桥面 bridge floor
翘曲面 warped surface
翘曲锥面 warped cone
切除高度（螺纹）height of truncation
切点 point of tangency, point of contact, tangent point

切割法 cutting method, by cutting
切弧 chamfer arc, chamfer curve, tangent arc
切平面 plane of tangency, tangent plane
切球面 tangent sphere
切线 tangent line, tangent
‖例句：过圆外一点作圆的切线。*Example sentence: Drawing a line tangent to a circle from a point outside the circle.*
切线键，吊楔键 lewis key
切削符号 machining mark, machining symbol, symbol for machining
切削品质 machining quality
切削裕度 machine finish allowance, machining allowance
切圆 tangent circle
青铜 bronze
轻打入配合 light drive fit
轻木 balsa wood
轻铅笔线 light pencil line
轻细铅笔线 faint pencil line
轻型 light duty
轻型槽形螺帽 light castle nut
轻型厚螺帽 light-thick nut
倾角 angle of dip, angle of slope
倾斜笔划 slanting stroke
倾斜度 angularity
倾斜透视 oblique perspective
倾斜位置 oblique position
求作线 required line
求…… find...
球半径 spherical radius
球阀 globe valve
球面 spherical surface
球面摆线 spherical cycloid
球面二角形 lune of a sphere
球面弓形 spherical segment
球面渐开线 spherical involute
球面经线展开法 gore method

球面内摆线 spherical hypocycloid
球面三角形 spherical triangles
球面扇形，球心角体 spherical sector
球面投影 spherical projection
球面外摆线 spherical epicycloid
球面纬线展开法 zone method
球切法 tangent-sphere method
球体 spheroid
球窝关节 socket and ball joint
球形 spherical shape
区，域 zone
区域地质图 regional geological map
区域分隔符 field separator
区域平面图 plot plan, block plan
区域图 plat
区域总平面图 general location plan
曲板 bend plate, bent plate
曲动线 curvilinear generatrix
曲合，折迭，卷缩 crimp
曲弧度 camber
曲架 bent
曲率 curvity, curvature
曲率半径 radius of curvature
曲面 curved surface
曲面加工 surface machining（计）
曲面轮廓 profile of any surface
曲面立体 body of curved surface, curved surface body, curved solid
曲面透视 perspective on curved surface
曲线 curve, curve line
曲线板 French curve, splines
曲线笔，等高线笔 contour pen
曲线笔划（字母） curved stroke
曲线长 curve length
曲线尺 curve ruler, curved rule
曲线发生器 curve generator
曲线回转面 double curved surface of revolution

曲线计 curvimeter
曲线轮廓 profile of any line
曲线条尺压铅 spline weight
曲元线 curvilinear element
曲折滑动接缝，肘管接缝 elbow slip joint
曲准线 curved diretrix, curvilinear directrix
取样长度（表面粗糙度）sampling length
取样和检查流程图 sampling and inspection flow chart
去角（倒角）chamfer
去角大小 chamfer angle
去角深度（焊接）depth of chamfering
去角线（倒角线）chamfer line
圈梁 girt, girth, peripheral beam
全波桥型整流器（电）full-wave bridge-type rectifer
全部光制 finish all over
全部刻度尺 chain scale (engineer fully divided scale)
全部切削 machine all over
全长 extreme length, overall length, total length, E to E (end to end)

全承面螺钉头 full bearing head
全辅助视图 complete auxiliary elevation
全高 extreme height, overall height, total height
全景透视 panoramic perspective
全刻度部分 full divided portion
全刻度尺 full divided scale
全宽 extreme breadth, overall spread, overall width, total width
*全螺纹螺柱 *stud bolt
全剖视图 ① full section, full section view, general section（建筑）
全深 total depth
全视图 fullview, complete view
全跳动 total runout
全圆 full circle
全值绘图机 absolute plotter
全周焊接 peripheral weld
全周焊接符号 weld all around symbol
全转 complete revolution
全字地位 full letter space
缺省，缺省值 default
缺省值选择 default selection

R

绕组图 winding diagram
热处理 heat treatment
热铁 WI (wrought iron), wrought iron
热铁板 wrought iron plate
热铁管，黑铁管 wrought iron pipe
热轧钢 hot-rolled steel
人工输入 manual input

人孔 manhole
人机学 human engineering, ergonomics
人字齿 double helical teeth
人字齿轮 herring bone gear
任务 task (计)
任意平面 general plane
任意曲面 surface of general form

注：① 全剖视图：用剖切面完全剖开机件所得的视图。

任意投影 arbitrary projection

‖例句：任意直线：是一条参考线，它的方向不一定符合基本方位。*Example sentence: Arbitrary line: A reference line, the direction of which does not necessarily coincide with cardinal direction.*（自*BS 3618-1:1969 Glossary of Mining terms —Term Definition, Confirmed January2011*）

任意直线 arbitrary line, general line
日本工业标准 Japanese Industrial Standard (JIS)
日光灯 fluorescent lamp
日期 date
容积图表 volume diagram
容量 capacity
容许变量 allowable variation
容许变异 permissible variation
容许承压 allowable bearing
容许误差 permissible error
熔接槽深度 depth of preparation
熔接尺寸标注 size specification of weld
熔接道表面 weld face
熔接道表面形状符号 contour symbol
熔接道符号 weld symbol
熔接道辅助符号 supplementary weld symbol
熔接道喉深 throat of weld
熔接道夹角 included angle of weld
熔接道节距 pitch of increments
熔接道每段长度 increment lenth of weld
熔接道形式 type of weld
熔接道有效长度 effective length of weld
熔接符号 welding symbol
熔接符号模板 selding symbol template, welding symbol template
熔接符号之尾义 tail of welding symbol
熔接基本符号 basic weld symbol
熔接接合 welded joint
熔接接头 welding fittings
熔接图 welding drawing
熔透量 root penetration
软，软磁盘 floppy
软件 software
软件包 software package
软拷贝 soft copy
软盘 floppy disk, diskette
锐角 acute-angle
润饰[①] polish

S

三点划线 dashed triplicate-dotted line
三点双划线 double-dashed, triplicate-dotted line
三点透视 three-point perspective
三点圆弧 circular arc three point
三度空间尺寸 three space dimensions
三交线法 three-intersecting line method
三角板 triangle（美国），set-square（英国）
三角形 triangle
三角形表图 trilinear chart
三角形法 triangulation method
三角形展开法 triangulation

注：① 润饰：使用阴、影、颜色等，使工程图样具有形象感的一种渲染性绘画手法。

三角学 trigonometry	晒蓝图器 blue print apparatus
三角折线图 trilinear chart	晒图 print, blue printing
三角柱，三棱柱 triangular prism	晒图机（复印机）printer, blue printing machine
三棱尺 triangular scale, three-edged rule, three-square rule, triangular rule	晒图架 printing frame
三面投影图 three-plane projection drawing	晒图纸 blue print paper
三视图 three-view drawing	晒图纸，影印纸 printing paper
三维空间 space of three dimension	山坡线 hachure
三维投影 three dimensional projection	删除 deletion
三维效果 pictorial effect	闪光熔接 flash weld
三线螺纹 triple-start thread	闪烁 blinking
三线坐标图 trilinear chart	扇形 sector
三相电动机，三相马达 three-phase motor	扇形公差域 wedge-shaped tolerance zone
三相图（电）three-phase diagram	商名 trade name
三轴等比正投影 isometric projection	商品限界（尺寸）commercial limits
三轴线图 triaxial diagram	商务图 business graphic
散布图 seatter diagram, dispersion pattern	商用哥德体字母 commercial Gothic letter
扫描 scan	商用零件（市售零件）commercial parts
扫描器 scanner	商用图 commercial drafting, commercial drawing
色彩 hue	上部结构（建筑）superstructure
色彩明度 color value	上层建筑及甲板室围壁图 trunk bulk-head plan of superstructure and deck house
色彩墨水 colored ink	上卷 scroll
色彩浓度 color intensity	上邻辅助视图 top-adjacent auxiliary
色调 intensity of color, tone	上模 upper die
色度图 chromaticity diagram	上墨 lining-in, lined in, inking
色片 colored sheet	上墨弹簧圆规 bow pen
色图 chromatic graph	上墨顺序 order of inking
色纸 colored paper	上墨图 ink drawing
沙黏土 sand clay	*上偏差[①] *upper deviation
沙丘 sand dunes	上下文 context
砂 sand	上限 high limit, superior limit, upper limit
砂浆 mortar	上游立面图 up-stream elevation
砂模铸造 sand casting	舍入 round off
砂纸 emery paper, sand paper, glass-paper block	设备 equipment
砂纸板 sandpaper pad, sandpaper block, sandpaper file	设备布置图 equipment placement drawing

注：①*上偏差：最大极限尺寸减其基本尺寸所得的代数值。

设备平台图 equipment plat form drawing
设备形象示意图 equipment illustration diagram
设备一览表 equipment list
设备元件图 parts list
设备用图形符号 graphical symbols for use on equipment
设备支架图 equipment support drawing
设备坐标系 device coordinate system
设计 design
设计标高 design elevation
设计草图 design layout, design draft, design sketch
设计尺寸 design size
设计工程师 design engineer
设计规格说明 design specification
设计规则检查 design requirement
设计阶段 design phase
设计基准 design datum
设计课 design staff
设计轮廓图 skeleton diagram
*设计图① *design drawing, design layout
设计图，草图 draft
设计图表 design chart
设计文件 design file
设计者 designer
设计字法(建筑) design lettering
射极(电) emitter section
射线 ray
射线锥 cone of ray
射影等价 projective equivalence
射影几何② projective geometry
射影平面 projective plane

射影群 projective group
射影坐标 projective coordinate
伸缩缝，伸缩接头 expansion joint
伸尾部分(字母) descender
深(度) depth
深度尺 depth rule
深度方向 depth direction
深度分厘卡 micrometer depth gage
深度缩小比例尺 foreshortened scale
深度轴线 depth axis
渗炭 cementation
渗碳硬化 pack hardening
升流图(电)，升位图(管路) riser diagram
升坡 rise of slope
生产图表 chart for production
生成式计算机图形 generative computer graphics
生石灰 calcium lime
省略尺寸 omitted dimension
*省略画法③ *omissive representation
施工测量 construction survey
施工机械 construction machine
施工计划 operations planning
施工建筑立面图 working drawing architecturalel evation
施工建筑平面图 working drawing architectural plan
施工建筑剖面图 working drawing architectural section
施工建筑图首页 specification on architectural working drawing
施工建筑详图 working drawing architectural details

注：①*设计图：在工程项目或产品进行构形和计算过程中所绘制的图样。
　②射影几何：研究图形射影性质和射影不变量的几何学。图形经过一切射影变换，保持不变的几何性质叫作射影性质；经过一切射影变换，保持不变的几何量叫作射影不变量。
　③*省略画法：通过省略重复要素的重复投影达到使图样简化的图示方法。

施工图

*施工图 [1] * production drawing, construction drawing, working drawing
施工详图 construction details
施工总平面布置图 general layout of construction (水)
施工总平面图 site plan working drawing
施力线 line of action
*十二角法兰面螺母 * 12 point flange nut
*十二角头法兰面螺栓 * 12 point flange screw, bihexagonal head screw
*十字槽半沉头螺钉 * cross recessed raised countersunk oval head screw
*十字槽半沉头木螺钉 * cross recessed raised countersunk oval head wood screw
*十字槽沉头螺钉 * cross recessed countersunk flat head screw
*十字槽沉头木螺钉 * cross recessed countersunk flat head wood screw
*十字槽盘头螺钉 * cross recessed pan head screw
*十字槽盘头木螺钉 * cross recessed pan head wood screw
*十字槽盘头自切螺钉 * cross re-cessed pan head thread cutting screw
十字形；十字接头，四通 cross
十字穴头 recessed head
石工结构图 masonry structure drawing
石工图，泥水工程图 masonry plan
石灰 lime, calcium oxide
石棉 asbesto
石棉浪板，石棉瓦 slate
识图，读图 visualization, blueprint reading
识图方向 reading direction

实测尺寸 measured size
实测图 surveyed drawing
实长图 true-length diagram
实长，真长 true length
实大，真大 true size
实际半径 true radius
*实际尺寸 [2] * actual size, real dimension
*实际偏差 [3] * actual deviation
*实际要素 [4] * real feature, *actual feature
实角 true angle
实体 entity
实体部分 material part of object
实(立)体镜仪器 stereoscopic instrument
实体模型 solid model
实体线路图 pictorial diagram
实线 continuous line, real-line
实形，真形 true shape
实验数据 experimental data
蚀刻指示图 etching indicator
矢量 vector
矢量发生器 vector generator
矢量投影 vector projection
矢量图 arrow diagram
矢量图形 line graphics (计)
示坡线 slope indication line
示出尺寸 show dimensions
示出待求数据 show required data
示出给定数据 show give data
示意草图 schematic sketch
*示意画法 [5] * schematic representation
示意图 schematic drawing, conventional diagram, ideograph, presentation drawing
示意图，略图 schematic diagram

注：① *施工图：表示施工对象的全部尺寸、用料、结构、构造及施工要求，以指导施工的图样。
② *实际尺寸：通过测量所得的尺寸。
③ *实际偏差：实际尺寸减其基本尺寸所得的代数值。
④ *实际要素：零件实际存在的要素。
⑤ *示意画法：用规定符号和 / 或较形象的图线绘制表意性图样的图示方法。

数据处理

世界坐标系（计） world coordinate system
市区图 city plat
视点 [1] observation point, point of sight, center of projection, point of view, station point, visual point
视见区 view port
视角 visual angle, angle of vision
视界 field of view
视空间 visual space
视口 view port
视口变换 view port (viewing) trans-formation
视面 view
视平面 horizon plane
视平线 horizon line, eye level, horizon, horizon line
*视图 [2] *view
视图的排列 arrangement of views, combination of views
视图换位 exchange of opposite views
视图上的交点 apparent intersection
视图选择 choice of views
视图之一部分 portion of view
视线 visual line, line of sight, visual ray
视线（界）角 aspec tangle
视向 view direction, viewing direction
视向平面线 viewing plane line
视向线 viewing-plane line
视心点 visual center, center of vision
视轴 axis of vision, visual axis
视锥 [3] vision cone
拭笔布 clpoth penwiper
拭笔器 penwiper
室内给排水平面图 indoor layout of plumbing system
室内给排水系统图 indoor plumbing system drawing
室内设计图 interior design graphics
室外给排水总平面图 general layout of outdoor water supply and drainage system
释放 release
手编程序 machine code programming
手柄 handle
手册 handbook, manual
手轮 hand wheel
首垂线 forward perpendicular（船）
首次衍生元线 element of first generation
首段结构图（首部结构图） structure plan of stem section
首曲线 stem curve
首柱图 stem post plan
受光面 illuminated face
书写模板 lettering guide
枢纽布置图 terminal layout
枢轴 pivot
输出设备 output device
输出原语 output primitive（计）
输入/输出 input / output (I/O)
输入设备 input device
熟石灰 calcium hydroxide
属性 attribute
鼠标器（鼠形器） mouse
鼠笼型电动机 squirrel-cage motor
术语标准 terminological standard
术语，命名 nomenclature
竖线 line up the board
竖向设计图 grading
数 number
数据 data
数据安全 data security
数据处理 data processing

注：① 视点：透视图中，观察者眼睛所在的位置，即投影中心。
② *视图：根据有关标准和规定，用正投影法将机件向投影面投影所得到的投影。
③ 视锥：以通过视点且垂直画面的视线为轴，视点为顶点，由视线形成的圆锥。

数据传送 data transfer
数据单 data sheet
数据的图示 graphical presentation of data
数据功能图 data functional diagram
数据绘图仪 data plotter
数据结构 data structure
数据块 block data
数据库 data base
数据库管理系统 data base management system
数据库集 data bank
数据流程图 data flow chart, data flow diagram
数据媒体 data medium
数据图形输入板 data tablet
数控绘图机 numerically controlled draughting
数控制图 numerical control graphics
数码 numerals
数学曲线 mathematical curve
数学图 mathematical graph
数元 digit
数值 numerical value
数字 numeral, digit
数字底线 bottom edge of digit
数字化 digitization（名词），digitize（动词）
数字化绘图 digitizing drawing
数字化仪 digitizer
数字交互式图形学 digital interactive graphics
数字控制 numerical control
数字象形图 numerical pictorial chart
数字之中高 midheight of digit
数字中高 midheight
数组，阵列 array
栓 cotter
栓槽 spline
栓槽轴 castellated shaft, keyed shaft, profile shaft, spline shaft
双倍比例尺 double size

双层比例尺作图 drawing with two scales
双点划线 dashed double-dotted line
双划点线 double-dashed dotted line
双划双点线 double-dashed double-dotted line
双极双投开关 double pole double throw switch
双连通图 biconnected graph
双连通子图 biconnected subgraph
双列对接 double strap butt joint
双面印制板零件图 two side printed board detail drawing
双曲率 double curvature
双曲面 hyperboloid
双曲抛物面 hyperbolic paraboloid
双曲线 hyperbola
双曲型对合 hyperbolic involution
双曲柱面 hyperbolic cylinder
双色图 biclourable graph
双三极管（电）twin triode
双凸面 biconvex
*双头螺柱 ① *stud
双弯形曲线 ogee curve
双线笔 railroad pen, double line pen, double line swivel pen
双线螺纹 double start screw thread
双线螺线 double thread
双向公差 bilateral tolerance
双向公差制 bilateral system of limits, bilateral tolerance system
双斜尺 opposite bevel scale
双心投影 dual central projection
双行数字（限界尺寸注法）numerals in two lines
双叶盘旋面 convolute of two nappes
双值图像 bi-level image
水彩 water color
水斗 rain-water head, leader head

注：①*双头螺柱：两端均有螺纹的圆柱形紧固件。

水力发电机层平面图 hydroelectric machine's floor plan
水利电力工程制图 hydraulic and hydroelectrical engineering drawing
水利工程鸟瞰图 hydraulic engineering bird's-eye view
水利枢纽布置图 hydro-junction's plan
水流测量图 hydrograph
水轮机层平面图 turbine's floor plan
水门 sluice
水磨石 terazzo
水泥船船体制图 drawing for concrete hull of ships
水平标度 horizontal scale
水平地质剖面图 horizontal geological section
水平短划（指线）short horizontal line
水平方向 horizontal direction
水平辅助投影面 horizontal reference plane
水平迹 horizontal trace, H-trace
水平距（斜度）run of slope
水平开拓掘进计划图 horizontal excavation plan
水平控制 horizontal control
水平剖面 horizontal section plane
水平面 horizontal plane
水平（面）投影 horizontal projection
水平投影 H-projection
水平投影面 horizontal projection plane, horizontal plane of projection, top plane
水平位置 horizontal position
水平线 horizontal line
水平斜轴测投影 horizontal oblique axonometric projection
水平轴线，横轴线 horizontal axis
水平主线 horizontal principal line
水平主要巷道平面图 main horizontal roadway plan
水刷石 granitic plaster, exposed aggregate finish
水位图 hydrographic map, water level map
水文地图 hydrographic map
水文地质图 hydrogeological map, hydrogeological drawing
水线 water lines
水准点 bench mark
顺理（木）with the grain
顺时针 clockwise
顺斜（量尺）reverse bevel
顺序编号 issue number
顺序字母 issue letter
顺游标 direct vernier
说明 designation, discription
说明表 explanatory table
说明图 explanatory drawing, illustrated drawing, illustration diagram
司蒂芬法（画椭圆）Stevens' methold
斯塔伯线号规 Stub's iron wire gage
四边形 quadrilateral, quad
四等分 quartile
四方屏幕 quadscreen
四分之一 quarter
四分之一圆 quarter circle
四氯化碳 carbon tetrachloride
四面体 tetrahedron
四坡屋顶 hip roof
四线螺杆 quadruple screw
四线螺纹 quadruple-start thread
四斜边尺 double bevel scale
四心近似法 four-center approximate method
四（圆）心椭圆法 four-center construction
松度 looseness
松节油 turpentine
松配合 loose fit
松转动配合 loose running fit
素线 element
素线平面法 element plane method

塑料 plastics
塑料材料 mylar, plastics material
塑料管 plastics pipe
塑料面制图板 plastics-faced drawing board
算法 algorithm
算术标度 arithmetic scale
算术分度水平坐标 arithmetical horizontal scale
算术分度图 arithmetic graph
算术内插法 arithmetical interpolation
算术平均（数） arithmetical average
*算图① *graph
随机数 random number
损益平衡图 break-even chart
榫，舌 tongue
榫头榫孔接合 mortise and tenon joint
缩尺 contraction scale, scale of foreshortening
缩尺分数 representative fraction (RF)
缩尺图 reduced size drawing

缩短 foreshortened
缩放 zooming
缩放图法 pantography
缩放仪 pantometer, pantograph
缩绘，缩图 contracted drawing
缩图器 omnigraph
缩微图像② microimage
缩微图形学③ micrographics
缩小比例 contracted scale
缩小视图 reduced view
缩小图 miniature drawing
缩小直径 reduced diameter
索引号码 reference number
索引图（概略原理图） key diagram, key drawing, key plan
索引字母 reference letter
锁紧装置 locking equipment (device)

T

T尺 T-square
T形槽 T-slot
*T形槽螺钉 *T-slot screw
T形件 T-piece
T形接合 Tee joint, T-joint
*T形螺栓 *T-head bolt, hammer head bolt
T形头螺栓 T-head bolt
T型钢 T-steel

tpd符号④ technical product documentation symbols (tpd symbols)
塌陷波及线图 cave-in affected drawing
塔位图 tower position drawing
胎架图 moulding bed plan
太粗 too heavy
弹出 pop up

注：① *算图：运用标有数值的几何图形或图线进行数字计算的图。
② 缩微图像：用摄影方法将图像缩小，并存放在底片上的图像。
③ 缩微图形学：一种研究把某种形式的图形信息转换为缩微印刷品，或把缩微印刷品转变为其他形式的信息的学科。
④ tpd符号：技术文件用图形符号。

*弹簧① *spring
弹簧笔舌鸭嘴笔 spring blade ruling pen
*弹簧垫圈② *helical spring lock-washer
弹簧分规 spring dividers, bow spacer, bow divider
弹簧内径 inside diameter of spring
弹簧外径 outside diameter of spring
弹簧圆规 spring bow
弹簧圆规 bow instrument
*弹簧中径③ *mean diameter of coil
*弹性挡圈④ *circlip, snap ring
*弹性垫圈⑤ *spring washer
碳化 carburizing
套节接合 socket joint
特定符号 specified symbol
特强管 extra strong pipe
特软橡皮 artgum
特殊比例，专用比例 special scale
特殊表达法 special representation
特殊工具 special tool
特殊螺纹 special thread
特殊视图 special view
特殊位置 special position
特殊位置视图 special-position view
特殊位置直线 special-position line
特殊阴影法 special shading method
特细螺纹 extra fine thread
特细螺纹级 extra fine thread series
特写画法 close up（建）

特性曲线 characteristic curve
特性形状 characteristic shape
特征图 characteristic graph
特种三角板 special triangle
特种图 special drawing
特重级（管）extra heavy series
梯形 trapezoid（美），trapezium（英）
梯形螺纹 acme thread, trapezoidal thread
提示 prompt
体对角线 body diagonal
体积 volume
体视对 stereopair
体视投影变换 stereographic projection, transformation
体视图⑥ stereogram
体系 system
天窗 skylight
天幕 awning plan
天气图 weather chart, synoptic weather chart, synoptical chart
添加 append
填角熔接 fillet weld
填角熔接宽度，填角熔接脚长 leg of fillet weld
填料 filler
填料函 gland box
填实式箭头 solid head
填实式箭头画法 solid-sytle arrowhead stroke
填心铅笔 mechanical pencil
挑檐 projected eaves

注：① *弹簧：利用材料的弹性和结构特点，使变形与载荷之间保持规定关系的一种弹性元件。
② *弹簧垫圈：弹簧丝断面为矩形的单圈螺旋形垫圈。
③ *弹簧中径：螺旋弹簧内径和外径的平均值。
④ *弹性挡圈：用弹簧制成的开口挡圈。
⑤ *弹性垫圈：具有弹性的可防止螺栓或螺母松动的垫圈。
⑥ 体视图：又称体视投影或体视图对。以人的两个眼睛分别为投影中心，将物体射影到纸面上，可得到一对投影，人们按互补色原理，将图对中左边的一个投影画成红色，右边的一个投影画成绿色，这样在用红、绿两种透明胶片做成的左红、右绿的滤色眼镜观看时，两个眼睛就会得出两个不同角度的红绿重合的深色影像，这两个不同角度的深色影像正好符合人的空间立体视觉构成的条件，于是在人的大脑里就形成鲜明而逼真的立体形象。所以这对投影就叫作体视图或体视图对。

条件图 conditional drawing
条形图 bar chart
条形图表 bar graph
调和共轭 harmonic conjugate
调试，排除故障 debug
调制解调器 modulator-demodulator (Mo-dem)
跳动 run-out
跳动公差① *run-out tolerance
跳过 skip /skip/
铁道工程图 railway engineering drawing
铁路 railroad, railway
铁路车站布置图 layout of railway stations
铁路曲线板 railway curves
铁路枢纽布置图 layout of railway junctions
铁砧 anvil
停机 halt /hɔ:lt/
通风、空调管道系统图 ventilation airconditioning duct system drawing
通风平面图② ventilation layout
通风网络图 ventilation network
通风系统示意图 ventilation system diagram
通风系统图 ventilation map
通风压力分布图 ventilation pressure map
通路（电）conducting pathway
通俗图表 popular chart
通用符号 common symbol
通用公差 general tolerance
通用公差说明 general tolerance note
通用化 generalization
通用计算机 general-purpose computer
通用剖面线 general hatching
通用图 current drawing, specified drawing
同平面 coplanar

同谱有向图 cospectral digraph
同上 ditto
同位角 corresponding angle
同心度 concentricity
同心度公差 concentricity tolerance
同心度公差 tolerance of concentricity
同心椭圆 concentric ellipses
同心圆 concentric circles
同心圆法 concentric circle method
同轴度 concentricity
同轴线 coaxial
同轴线度 coaxility
同轴圆柱体 coaxial cylinder
铜 copper
铜版纸 calendered pape
铜管 copper pipe
铜合金 copper composition
铜片图（电）foil drawing
统计地图 statistical map
统计图 statistical chart
统计图（表）statistical diagram
统计学 statistics
统一标准特细螺纹级 UNEF(unified extra-fine thread series)
统一螺纹 unified thread
统一螺纹配合等级 unified screw-thread class
头 head
头道螺丝攻 taper tap
头型 head style
投射角 angle of projection
投射面 plane projector
投射线 ray of projection, space projector
*投影③，射影 *projection

注：①*跳动公差：关联实际要素绕其基准线回转一周或连续回转时所允许的最大跳动量。
②通风平面图：表达通风管道和设备的平面布置图。
③*投影：根据投影法得到的图形。

投影变换① transformation of projections
*投影法② *projection method
投影规律 the law of projection
投影连线 connecting line of projection
*投影面③ *projection plane, plane of projection
投影面积 projected area
投影特性 characteristic of projection
投影图 projection drawing
投影线，投射线 projector, projection line
投影线法（透视图） direct-projection method
投影中心 center of projection
投影轴④ axis of projection
透镜零件图 detail drawing of lens
透明，透明度 transparence
透明边缘尺叶（丁字尺） transparent-edged blade
透明盒法 glass-box method
透明绘图胶片 transparent drawing film
透明胶纸 scotch tape
透明面 transparent surface
透明片 transparent-foil
透明平面 transparent plane
透明物 transparency
透视 perspective
透视平面图 perspective plan
透视平面图法 perspective plan method
透视同素对应，透视共线对应 perspective collineation
透视投影⑤ *perspective projection
透视图 perspective drawing
透视图表示法 perspective representation
透视性 perspectivity
透视中心 center of perspectivity
透射变换 homology transformation
透射对应，同调 homology
透射中心 center of homology
凸多边形 convex polygon
凸立体 convex solid
凸轮 cam
凸面 convex face
凸曲率 outward curvature
凸台 boss
凸缘 flange
*凸缘螺钉⑥ *collar screw
涂黑 blackened-in
涂黑描阴 smudge shading
*图⑦ *drawing, graphics
图案 chart-pattern, graphic pattern
图标 icon
图表 chart
图表符号 schematic symbol, diagram, graph
图表格式 chart format, chart layout
图表阅读器 chart reader
图的大小 drawing size
图的归档 filing drawings
图钉 thumb tack, tack, drawing pin, map pin
图段 associate segment, segment
图段动态属性 segment dynamic attributes

注：① 投影变换：又称投影改造。它是改变空间几何要素与投影面的相对位置，以便于图示或图解的方法。
② *投影法：投射线通过物体，向选定的面投射，在该面上得到图形的方法。
③ *投影面：投影法中得到投影的面。
④ 投影轴：投影法中，互相垂直的投影面的交线。
⑤ *透视投影：用中心投影法将物体投射在一个投影面上所得到的图形。
⑥ *凸缘螺钉：螺钉头下有直径较大的圆盘（代替垫圈）的螺钉。
⑦ *图：用点、线、符号、文字和数字等描绘事物几何特性、形态、位置及大小的一种形式。
图（diagram）：表示程序或设备的各部分之间在时间和空间上的相互关系的图形或符号。
（计）图（graph）：图论中，由若干节点和连续各节点的连线所组成的图。

图段属性

图段属性 segment attributes
图幅分区 grid reference system
图幅拼接缩引 sheet splieing index
图号 drawing number
图集 graphics set
图解代数 graphical algebra
图解法 graph method, graphic method, graphical method, graphic solutions
图解积分 graphical integration
图解计算 graphic calculation, graphic computation
图解力学 graphic statics
图解算术 graphical arithmetic
图解微分 graphical differentiation
图解微积分 graphical calculus
图库 graphics library
图块 block
*图框[1] *border
图框笔 border pen
图框线 border line
图例[2] legend
图论 graph theory
图面尺寸 actual measurement on the drawing
图面分区 zoning
图模型 graphical model
图目 list of drawings
图上表示法 drawing callout
图示法 graphic representation
图式 graph schema, schematism
图说线 legend line
图素[3] picture element
图文工艺 graphic arts
*图线[4] *line, alphabet of line
图线尺寸 line dimensions
图线构型 configuration of lines
图线画法 draughting of lines
图线宽度 line width
图像[5] image, picture
图像处理 picture processing
图像放大 blowback
图像扩张 dilatation
图像输出 exporting images
图像输出设备[6] image output device
图像输入设备[7] image input device
图像跳动 bouncing motion
图形[8] graphic, figure, picture
图形板 graphic panel
图形变比 zoom
图形标志[9] graphical sign
图形尺寸 dimension of picture
图形程序包 graphic package
图形插入 graphics insertion

注：[1] *图框：图纸上限定绘图区域的线框。
[2] 图例：用单线条表示材料、地质土壤、地形、建筑物或设施、设备等的一组示意图。
[3] 图素：又称图像元素、象素。构成数字图像的最小单位，通常是点、线，在低分辨度的图形显示器中，也可是类似字符、点阵等的图标。
[4] *图线：图中所采用各种型式的线。
[5] 图像(image)：绘制、摄制或印制的事物的形象。计算机制图中，图像是图形对象的模型在输出界面上的一种形象化表示。
图像(picture)：对文件的某一区域所作的显示。(计)
[6] 图像输出设备：一种从计算机取出与图像灰度或颜色相对应的数据，并能转换成为可见图像的设备，如显示器、硬拷贝机等。
[7] 图像输入设备：一种把图像的灰度或颜色变换成为计算机能够识别和处理的数据的设备，如摄像机、图像扫描仪、电传真扫描器等。
[8] 图形：描绘出来的物体的形状。
[9] 图形标志：以图形为主要特征，信息传递不依赖于语言的符号。

图形处理 graphic processing
图形窗口 graphics window
图形点阵打印机 graphic matrix printer
图形分析 graphic analysis
图形符号① graphical symbol
图形核心系统 graphical kernel system
图形监视器 graphic monitor
图形结构 graph structure
图形库 graphic library
图形码 graphic code
图形配置 graphic configuration
图形输出 graphic output
图形输出板② plotting board, graphic board
图形输入 graphic input (GIN)
图形输入板③ graphic tablet, plotting tablet
图形输入语言 graphic input language
图形设计系统 graphic design system
图形数据处理 graphic data processing
图形文件 picture file
图形显示 graphic display
图形显示装置 graphicz display unit
图形信号 figure signal
图形语言 graphic language
图形元 pattern primitive
图形阅读器 figure-reader
图形字符 graphic character
图学 graphics, graphic science
图学理论 graphical theory
*图样④ *drawing, draft
图语 graphic language
图元⑤ primitive
图元属性 primitive attributes
图之复制 reproduction of drawing
图之贮存 storing drawing
图纸 drafting paper, drawing paper, drawing sheet
图纸大小 drawing sheet size, size of drawing
*图纸幅面（图幅）⑥ *format, drawing sheet
图纸更改 drawing change
图纸目录 list of drawing
图纸外圆心 inaccessible center
图纸无光泽面 dull side of paper
图着色 graph coloring, tinting the drawing
图注 drawing statement
图组 set of drawing
徒手画（图）⑦ free-hand drawing, freehand
徒手画草图⑧ freehand sketch, freehand sketching
徒手线 irregular boundary line, freehand line
土地分区图 zoning map
土方工程 earthwork
土方工程图 earthwork drawing
土木工程师 civil engineer
土木工程师比例尺 civil engineer's scale
土木工程图 civil engineering drawing
土木制图 civil engineering graphics
推入配合 push fit
推力轴承 thrust bearing

注：① 图形符号：表示事物或概念的一种简单的图像、字符或标志。
② 图形输出板：绘图机输出部分的平板。
③ 图形输入板：它是带有传感器的一块图形输入图板，根据电磁感应、超声波等不同原理制作。由人操纵传感器的移动，从而可向计算机输入图形。
④ *图样：根据投影原理、标准或有关规定，表示工程对象，并有必要的技术说明的图。
⑤ 图元：计算机制图中，构成一个图形最基本的元素，如点、线段、折线、圆弧、圆、字符等。
⑥ *图纸幅面：图纸宽度与长度组成的图画。
⑦ 徒手画（图）：不使用绘图仪器和绘图工具，只凭手和笔，靠目测实物的大小来绘制图样的一种方法。徒手绘制的图样称为徒手图或草图。
⑧ 徒手画草图：计算机制图中，使用如光笔等图形输入设备，随手在屏幕上画草图的功能。

退出 quit
退刀槽 escape
退格，回退 backspace
退火 annealing
退隐面（垂直于主投影面之平面） receding plane
退隐线（垂直于主投影面之直线） receding line
拖动 dragging
拖放画图功能 drag and drop plot feature
脱机 off-line
脱机系统 off-line system

椭圆抛物面 elliptic paraboloid
椭圆 ellipse
椭圆板 ellipse, ellipse template
椭圆规 ellipsograph, trammel
椭圆面 ellipsoid
椭圆体坐标 ellipsoidal coordinate
椭圆形字母 ellipse letters
椭圆柱 elliptic cylinder
椭圆坐标 elliptic coordinate
拓扑变换 [1] topological transformation
拓印法 rubbing method, impression method, transfer by rubbing

W

挖与填 cut and fill
蛙式投影 worm's-eye projection
瓦斯预测图 gas emission forecast map
瓦特曼纸 [2] Whatman paper
歪倾 batter
外摆线 epicycloid
外板展开图 shell expansion and framings
外半径（肘管） heel radius
外部尺寸 external dimension
外部设备 peripheral device
外部形态 external feature
外错角 alternate-exterior angle
外割线 external secant
外公切线 exterior common tangent

外观（外貌）[3] appearance
外环面 outer-surface of toroid
外件 external member, external part
外角 exterior angle, external angle, outside angle
外径 outer diameter, outside diameter
外径系列 outside diameter series
外接圆 cicumscribed circle, circumcircle
外卡 outside caliper
外廓线 profile
‖例句：一个视图的外形轮廓线总是可见的。
Example sentence: The outline of a view will always be visibility.
外廓线（外形轮廓线） outline, profile
*外螺纹 [4] * external thread, male thread, bolt

注：① 拓扑变换：几何图形在正逆两方面都单值而又连续的变换。即将已知图形想象地拉伸、压缩、弯曲、扭转，而变成另一个图形时，不发生撕裂或黏结的变换。
② 瓦特曼纸：是一种高级绘画纸（板）。
③ 外观（外貌）：计算机制图中，输出图像的灰度、颜色、线型及是否闪烁等现象。
④ *外螺纹：在圆柱或圆锥外表面上形成的螺纹。

thread
外螺纹机件 externally threaded part
外螺纹简化符号 simplified external-thread symbol
外螺纹正规符号 regular external-thread symbol（美）
外切多边形 circumscribed polygon
外切矩形 enclosing rectangle
外切正方形 circumscribing square
外套件 female member
外推拔 external taper
外围尺寸 overall dimension
外围设备 peripheral equipment
外线图 external wirint diagram（电）
外心 circumcenter
*外形尺寸① *boundary dimension, outer dimension
外形轮廓线 outline
外形视图 external view, exterior view
*外形图② *figuration drawing, layout drawing, outline drawing
外形线 trim lines
外圆角 round
外缘线，边线 bounding line
弯度 degree of curvature
弯接 joggling
弯接头 joggle
弯曲半径 bend radius
弯曲离隙 bend relief
弯曲裕度 bend allowance
弯头 bend
弯折图 bending diagram
完成 done /dʌn/, complete
完工切削 finish machining
完美形状（几何公差） perfect form

完全多部图 complete n-partite graph
完全螺纹 perfect thread
完全螺纹，全牙 full [form] thread
完全图 complete graph
完全小数制 complete decimal system
完全有向图 complete digraph
完全子图 complete subgraph
完整螺纹 complete thread
万能绘图仪 universal drafting machine
万向鸭嘴笔 contour pen
网格 grid（计）
网格的网络 mesh network
网络图 network diagram
网格线空间 ruled space
网纹滚花 hatching knurling
微处理器 micro processor
微电路 micro circuit
微片 reduced microfilm
微调分规 hairspring divider
韦氏细螺纹 BSF thread (British Standard Fine thread)
维 dimension
维氏硬度 Vickers diamond hardness, diamond penetrator hardness
维氏硬度数 Vickers hardness number
尾垂线 after perpendicular
尾段结构图（尾部结构图） structure plan of stern section
尾水管层平面图 draught-tube plan
尾柱图 stern post plan
纬度 latitude
纬圆 parallel
未订标准之零件 non-standard parts
未对准 disalignment
未贯穿的孔，盲孔 blind hole

注：① 外形尺寸：部件或机器的总长、总宽、总高尺寸。
② *外形图：表示产品外形轮廓的图样。

未贯穿的螺孔，盲螺孔　blind tapped hole
未剖切之一半　unsectioned half
卫生标准　hygienic standard
卫生器材　sanitary ware
卫生设备　sanitary installation
卫生设备配件　sanitary fittings
位似变换 ①　homothetic transformation
位似对应　homothetic
位似中心　center of similarity
位图　bitmap（计）
位图编辑程序　bit editor
位图表示　bit map represeniation
位图大小　bit map dimension
位图函数　bit map functions
位图图像　bit map image
位图显示格式　bit map display format
位图显示器　bit map display
位图图形学　bit map graphics
位图字体　bit map font
位置尺寸　location dimension, position dimension, positional dimension, positioning dimension
位置度　true position
*位置公差 ②　*position, tolerance, tolerance of position, tolerance in position, positional tolerances
位置简图　location diagram
位置模式　location pattern
位置配合　locational fit
位置平面图　location plan
位置图　location drawing
位置误差　error of location
位置坐标　position coordinates
文件　file, document, documentation
文字符号　letter symbol
文件颁发表　document issuing list

文件内容安全性　document content security
文件（系统）安全性　documentation security
文字语言　word language
涡线弹簧　volute spring
*蜗杆　*worm
*蜗杆齿宽　*worm face width
蜗杆、蜗轮传动　worm and worm-gear transmission
*蜗轮　*worm wheel, worm-gear
蜗杆与蜗轮　worm and wheel
蜗线　spiral, spiral curve, spiral line
蜗线斜齿轮　spiral bevel gear
圬工结构　masonry structure
圬工结构图　masonry structure drawing
屋顶　roof
屋顶花园　roof garden
屋顶排水沟，屋顶排水管　roof gutter
屋顶平面图　roof plan
屋面材料　roofing material
屋瓦　roof tile
无槽圆头螺栓　unslotted round-head bolt
无缝金属软管　seamless flexible metal tubing
无格纸　plain paper
无刻度三角板　plain triangle
无效的　invalid /ˈɪnvəlɪd/
无线电罗盘站　radio compass station
无限长　indefinite length
无限平面　infinite plane
五边形　pentagon
*五角螺母　*pentagon nut
五边形的渐开线　involute of pentagon
物料平衡流程图　material balance diagram
物体　body object
物性计算　mass-properties calculation
误差　error

注：① 位似变换：在中心射影条件下，对应平面互相平行时的变换。
　　② *位置公差：关联实际要素的位置对基准所允许的变动全量。

西立面图 west elevation
舾装布置图 general arrangement of deck fittings
舾装结构图 structure plan of deck fittings
吸墨粉 french chalk, pounce
吸墨纸 pad
吸墨纸 blotter, blotting paper, pad
习惯画法 conventional drawing
习用表示法 conventional representation
习用方法 conventional method
习用符号 conventional sign, conventional symbol
习用光线 conventional lighting
习用画法 conventional practice
习用交线 conventional intersection
习用剖面线 hatching convention
习用剖切法 sectioning convention
习用位置 functioning position
细部剖面 detailed section
细点划线 chain thin line
细分刻度 close-divided
细节省略线 ditto line, repeat line
细螺纹 fine screw thread
细螺纹级 fine thread series
细铅笔线 sharp pencil line
细铅笔芯 fine line lead
细实线 continuous thin line, thin solid line
细线 fine line, thin line
细线阴影法（地图）hill shading by hachures
细牙 fine thread
细牙螺纹 fine thread
系泊布置图 anchor mooring and towing arrangement plan
系杆 sag ties（建筑）
系列化 serialization
*系统图 system diagram, system chart
狭体字 compressed letter
狭条 strip
狭窄部位 limited space
下部结构 substructure（建筑）
下角符号，下角数字，下角字母 subscript
下面加画横线 underlining
下模 lower die
*下偏差① lower deviation, under deviation
下倾角 downward inclination
下伸线（小楷字母）drop line
下限 lower limit
下限界，最小尺寸 interior limit
下陷 depression
下游立面图 down-stream elevation
掀开表示法 uncovered presentation
舷弧 sheer（船）
舷墙图 bulwark plan
弦 chord
弦长 chord length, chordal distance
弦角 angle of chord
显示 display, presentation
显示符号 display symbol
显示空间 display space
显示面 display surface
显示命令 display command
显示器 displayer

注：① *下偏差：最小极限尺寸减其基本尺寸所得的代数值。

显示软件 software for display
显示设备 display device
显示图像 display image
显示元素 display element
显示终端 video terminal
显著形体 outstanding feature
线，直线 line
‖例句①：除非另有规定，线这个名词通常用来表示直的线。Example sentence①: The term line is generally used to designate a straight line unless otherwise specified.
‖例句②：如果直线与平面不重合不平行则必定相交。Example sentence②: Unless a line is in or parallel to a plane, it must intersect the plane.
线的粗细 line thickness
线的规格 alphabet of lines
线的属性 character of line
线段 line segment, segment, segment of line
线段分析 segment analysis
线段之端点 extremity of a segment
线光顺 line smoothing（计）
线号规 wire gage
线及板金号规 wire & sheet-metal gage
线间距离 line spacing
线宽 line width, line weight
线框模型 wire frame model（计）
线路 wiring line
线路 circuit（电）
线路安装图 erection diagram, erection chart
线路断流板 circuit breaker panel（电）
线路符号 wiring symbol（电）
线路负荷 circuit load（电）
线路图 line diagram, layout, route map, road drawing, circuit drawing, electrical diagram
线轮廓度 profile of any line

线圈 coil
线束 line pencil
线素 line element
线条性质，线条种类 line quality
线条阴影 line shading
线条阴影法 line tone shading
线条优先级 precedence of lines
线条种类 type of line
线透视 linear perspective
线透视图 linear perspective drawing
线图 line drawing, line graph
线图图表 diagrammatic chart
线图形式 diagrammatic form
线性尺寸[①] linear dimension
线型 types of lines, line type, line kind, alphabet of lines, lines
线札图 harness drawing
线之粗细 thickness of line, weight of line, width of line
线坐标 line coordinate
现场草图 field sketch
现场测绘图 field sheet
现场浇制钢筋混凝土构件详图 cast-in situ reinforced concrete element details
现场全周熔接 field peripheral weld
限定符号 limit symbols
限界量规 limit gage
相错线 skew lines
相当值 equivalent
相等尺寸 equal dimensions
相对标高 relative elevation
相对方向 opposite direction
相对高度 relative elevation（地图）
相对位置 relative position
相对坐标 relative coordinates
相关尺寸标注 associative dimensioning（计）

注：① 线性尺寸：沿直线方向度量的尺寸。

相关位置 relative location（地图）
相关原则 correlation principle
相贯体 intersecting bodies, penetrated body
相贯线 [①] line of intersection
相交 intersect, intersecting
‖例句①：完成相交直线的主视图，并增加右视图。Example sentence①: Complete the front of the two intersecting lines and add a right-side view.
‖例句②：平面内的任何两直线不是相交就是平行。Example sentence②: Any two lines in a plane must either intersect or be parallel.
‖例句③：作直线AB平行于CD并与交叉直线EF、GH相交。Example sentence③: Pass a line AB parallel to CD and intersecting the skew lines EF and GH.
‖例句④：过点O作直线平行于平面ABC并与直线EF相交。Example sentence④: Draw a line that contains point O, is parallel to plane ABC, and intersects line EF.
‖例句⑤：任意两平面不平行必相交。Example sentence⑤: Any two planes either must be parallel or they must intersect.
‖例句⑥：平面和正圆锥体相交。Example sentence⑥: Intersection of plane and right circular cone.
相交线原则 intersecting-line principle
相交直线 intersecting lines
相邻视图 adjacent view
相邻素线 consecutive element
相切 tangency
相切元线 tangent element
相切圆 tangential circles
相似 similar
相似变换 similarity transformation

相似件，相对件 counterparts
相似三角形 similar triangle
相似投影法 the method resemble projection
相似投影图 projection resemble
相似形 similar figure
相似形态 similar feature
相似中心 center of similarity
相同方法 identical manner
相同面积 identical areas
*详图 [②] detail, detail drawing
详图标志 detail drawing mark
详图索引标志 detail index mark
详细草图 detail sketch
详细符号 detailed symbol
详细截面，个别截面，移动截面 separate section, shifted section
详细设计 detailed design
详细图 detail plan
向径 radius vector
向量 vector
向量图 vector chart
向上 upward
向视（图）法 direction view method
向心轴承 radial bearing
项目 item
项目代号 item designation
象限 quadrant
象素 pixel
象形表示法 pictorial descrition
象形图 pictorial chart（统计）
象元 image element
橡胶 rubber
橡皮 eraser, rubber
橡皮印章 rubber stamp
橡树 oak tree

注：① 相贯线：立体与立体相交表面的交线。
② *详图：表明生产过程中所需要的细部构造、尺寸及用料等全部资料的详细图样。

肖氏硬度 Shore hardness
削尖铅芯 sharpen lead
削铅笔机 mechanical sharpener, sharpening machine
消除隐藏线 removal of hidden line
消防水笼 fire hydrant
消去，除去 elimination
消失点 vanishing point
消失点法（透视图）combination method
消字膏 abrasive paste
*销 ① *pin
销键 pin key
*销联接 ② *pinned joint
小尺寸注法 decimal dimensioning
*小齿轮 *pinion
小刀 knife, pocketknife
*小径 ③ *minor diameter
小径（螺纹）root diameter
小量生产工作 small-production work
小螺钉 machine screw
小螺母 machine-screw nut
小螺栓 machine bolt
小数 decimal
小数尺寸 decimal size
小数当量 decimal equivalent
小数点 decimal marker, decimal point
小数分数并用注尺寸法 mixed fractional and decimal dimensioning
小数刻度 decimal scale
小数位数 decimal places, number of decimal, number of decimal places
小数制 decimal system
小图纸 demy paper

小写字母 lower-case letter, small letter
小圆规 drop compasses
楔 wedge
*楔键 ④ *taper key
楔形笔尖 chisel edge
楔形尖（铅笔）wedge point, chisel point
斜边（直角三角形），弦（直角三角形）hypotenuse
斜齿轮 helical gear, spiral gear
斜度 inclination, flat taper, taper on radius
斜度，坡度 slope
斜短线 oblique stroke
斜二等轴测投影（斜二测）oblique dimetric projection
斜二等轴测投影，半斜投影，减半投影 cabinet projection
斜方角柱 rhombic prism
斜方角锥 rhombic pyramid
斜方六面体 rhombohedron
斜高 slant height
斜拱（建筑）skew arch
斜管螺纹 taper pipe thread
斜脊，屋顶角 hip
斜交角，复斜角 skew angle
斜交接合 skew joint, splayed joint
斜角 oblique angle
斜角分划尺 diagonal scale
斜角面 bevel face
斜角柱 oblique prism, oblique pyramid
斜接合 miter joint
斜接头 oblique joint
斜截面 oblique section
斜矩形柱 oblique rectangular parallelopiped

注：① *销：俗称"销子"。贯穿于两个零件的孔中，主要用于定位，也可用于连接或作为安全装置中过载易剪断元件。

② *销联接：用销钉固定零件相对位置的连接。

③ *小径（螺纹小径）：与外螺纹牙底或内螺纹牙顶相切的假想圆柱或圆锥的直径。

④ *楔键：矩形或方形截面，宽度不变而厚度上有斜度的键。

斜矩形柱 oblique rectangular prism
斜孔 inclined hole
斜棱 beveled edge
斜棱（角锥）slant edge
斜六角形模板 tilt-hex drafting template
斜螺纹 taper thread
斜面，一般位置平面 oblique plane
斜面（角锥）slant face
斜面倾斜 declivity
斜内螺纹 taper internal thread
斜劈锥体 oblique conoid
斜平面 sloping plane
斜剖① oblique cutting
斜剖线 diagonal（船）
斜视图② oblique view, oblique drawing
斜体大写字母 inclined capital letter
斜体小写字母 Reinhardt alphabet, Reinhardt letter
斜体字 inclined letter, italics, slope charcter
斜体字母 italic letter, slant letter
*斜投影③ *oblique projection
*斜投影法④ *oblique projection method
斜椭圆柱 oblique elliptical cylinder
斜透视 oblique perspective
斜外螺纹 taper external thread
斜五角柱 oblique pentagonal prism
斜线 oblique line, inclined line, slant line, sloping line
‖例句①：斜线是与任何主要投影面（正立投影面、水平投影面、侧立投影面）都不平行的直线。*Example sentence①: An* oblique line *is one that is not parallel to any of the principal planes: frontal, horizontal, or profile.*

‖例句②：斜线在两个或两个以上的主要视图中平行，他们在空间也平行。*Example sentence②:* Oblique lines *that appear parallel in two or more principal views are parallel in space.*
斜线螺旋面 oblique helicoid
斜销 taper pin
斜圆柱 oblique cylinder
斜圆柱体 oblique cylinder
斜照明法 oblique illumination system
斜直线 oblique straight line
斜轴 oblique axis
斜轴测投影 oblique axonometric projection
斜锥体 oblique cone
斜坐标 oblique coordinates
写保护 write protect
写生地图 pictorial map
写入板 tablet（计）
写字笔尖 speedball pen, writing nib
写字辅助器 lettering aids
写字用具 lettering device
心型，砂心 core
心型箱，砂心箱 core box
心型座 core print
心轴 arbor
芯片 chip
锌白 Chinese white (zinc white)
锌板 sheet zinc
锌板布置图 zinc plate arrangement plan
信号联锁图表 signal interlocking chart
信号设计图 signal design drawing
信号总布置图 general layout of signal
信息 information, info, message
信息处理 information processing

注：① 斜剖：用不平行于任何基本投影面的剖切平面剖开机件以获得剖视图的方法。
② 斜视图：将机件向不平行任何基本投影面的平面投影所得的视图。
③ *斜投影：根据斜投影法所得到的图形。
④ *斜投影法：投射法与投影面相倾斜的平行投影法。

星形布置图（建筑）star-shaped plan
形式 type
形象、形态、状态 configuration
形状 shape
*形状公差① * form tolerances, tolerance in form, tolerance of form
形状误差 error of form
形状中心 center of figure
型材图② formed material drawing
型钢 section iron
型钢标注 designation of structural steel sections
型线 molded lines
*型线图③ * lines plan
型砧 swage
型值 offsets
型值表 table of offsets
醒目 highlighting
性能尺寸 characteristic dimension
修护图表 repair chart
修改 amend
修改阶段 revision phase
修圆 round out
修正尺寸 common trimmed size
袖珍图（矿）pocket drawing
锈 rust
虚拟空间 virtual space
虚拟视图 false view
虚线 dashed line, dot line, hidden line
虚线笔 dotted line pen
虚线剖面线 dotted crosshatching, dotted seetion line
虚线圆 dashed circle

序号 serial number
旋入长度 screw-in length
旋入长度（螺纹）length of engagement
旋入深度（螺纹）entrance length
旋向 direction of turning
旋压成形 spinning
旋转 revolve, rotate, rotating（计），rotation, revolution
‖例句：当直线绕轴旋转时，直线上所有的点都旋转了同样角度。Example sentence: As a line revolves about an axis, all-points of the line revolve through the same angle.
旋转变换 rotation transformation
旋转单曲面 single curved surface of revolution
旋转电钮 rotary switch
旋转断面 revolved section
旋转法 revolution method, method of revolution, method of rotation
旋转复曲面 double curved surface of revolution
旋转面 surface of revolution
旋转剖④ aligned cutting
旋转剖面，转直剖面 aligned section
旋转剖视图 revolved section
旋转视图⑤ revolved view, aligned view
旋转轴线 axis of revolution, axis of rotation
选矿工艺管路图 pipe drawing of ore selecting
选煤工艺原则流程图 basic flow sheet of coal preparation
选煤设备流程图 equipment flow sheet of coal preparation
选线法（交线求法）selected-line method

注：① *形状公差：单一实际要素的形状所允许的变动全量。
② 型材图：指成型材料的图样。
③ *型线图：用成组曲线表示物体特征曲面（如船体、汽车车身、飞机机身等型表面）的图样。
④ 旋转剖：用两个相交的剖切平面剖开机件，假想将一截面旋转以得到剖视图的方法。
⑤ 旋转视图：假想将机件的倾斜部分旋转到与某一选定的基本投影面平行后投影所得的视图。

选择装配 selective assembling
渲染图① rendering

循环图 cyclic graph

Y

压光 burnishing
压痕接缝 button punched standing seam
压块 ducks
压力角 angle of pressure
压力弹簧 compression spring
压铅 lead weight, draftsmen duck
压入配合 press fit
压入收缩配合类 force or shrink fit
压塑 compression molding
压缩弹簧 compression spring
压印 coin
压铸件 D cast (die casting)
鸭嘴笔 blade ruling pen
鸭嘴笔，直线笔 ruling pen
鸭嘴笔杆 pen handle
鸭嘴笔尖 nib, pen point
鸭嘴笔尖（圆规） pen attachment
鸭嘴笔叶片 blade of pen
*牙侧② flank
*牙底③，根部 *root
*牙顶，顶部 *crest

牙型 form of thread
*牙型角④ *thread angle
亚历山大图纸⑤ Alexandria paper
延迟 delay /dɪˈleɪ/
延伸杆 extension bar, lengthening bar
延伸公差带⑥ projection tolerance zone
延时继电器 timedelay relay
延性铸铁 ductile /cast/ iron
延展性 ductility
沿边剖面线 outline sectioning
沿子午线之剖面 meridian section
研光 lapping
研心器 lead pointer, pencil pointer, pencil sharpener
颜色冲淡 diluting of color
颜色对比 contrast of color
颜色加深，线条加粗 darken
颜色铅笔 colored pencil
颜色索引 colour index
檐口 eaves
眼图 eye pattern

注：① 渲染图：对建筑物、建筑物内部、建筑物细部等，采用阴影或各种不同颜色绘制出的立体图、透视图等。用以形象而鲜明地表达建筑物的立体形式和建筑物各垂直平面间的关系，环境气氛，甚至室内的光影、装饰等。
② *牙侧：连接牙顶与牙底的螺纹侧表面。
③ *牙底：螺纹沟槽的底部，连接相邻两个牙侧的螺纹表面。
④ *牙型角：螺纹牙型上，两相邻牙侧间的夹角。
⑤ 亚历山大图纸：是一种中等绘图纸，以埃及亚历山大港命名。
⑥ *延伸公差带：根据零件的功能要求，位置度和对称度公差带需延伸到被测要素的长度界限之外时的公差带。

燕尾槽 dovetail slot, dovetail groove
燕尾榫 dovetail tongue; dovetail
阳面 illuminated part of a surface
阳台 balcony
仰角 elevation angle
*仰视图 ① *bottom view, inverted plan
氧化 oxidation
氧乙炔熔接 oxyacetylene welding
样板图（飞）template drawing, drawing of sample
样品 sample
样条函数 spline function
野外测图 field mapping
业务报告图 besiness report diagram
页边 margin
一般符号 general symbol
一般工作 ordinary work
一般级别（爱克姆螺纹）general-purpose class (ACME thread)
一般配合类别 general fit class
一般批注 general note, general remarks
一般图文学 general graphic arts
一般位置平面 oblique plane
一般位置线 oblique line, general position line
一层平面图（首层平面图，底层平面图）first floor plan [美]; ground floor plan [英]
一点透视 one-point perspective
一览表 synoptic table
一元式机车周转图 single locomotive working diagram
一字尺 parallel ruler

移出断面 ② removed section
移位，移数，换挡 shift
仪器 instruments
仪器图 ③ instrument drawing
遗漏，缺失 missing
以图表示 show construction
以直线分隔 ruled off
已裁标准大小 standard trimmed size
已开发地 cleared land
已切削表面 machined surface
已切削部分 machined part
已知尺寸，给定尺寸 given dimension
已知大小，给定大小 given size
已知线，给定线 given line
艺术字 freakish printing
*翼形螺钉 ④ *wing screw
*翼形螺母 *wing nut
异步程序图 asynchronous graph
异常结束 abort /əˈbɔːt/
异素变换，对射变 correlation transformation
因变数 dependent variable
因果分析图（鱼骨图，树枝图）cause and effect diagram
阴 ⑤ shade
阴暗面 dark face
阴暗效果 shaded effect
阴极（电）cathode
阴线 ⑥ line of shade
阴影 shade and shadow
阴影法 hill shading
阴影画 shadow graph

注：① *仰视图：由下向上投射所得的视图。
② 移出断面：将切断面图形画在相关视图之外的断面。
③ 仪器图：用绘图仪器和绘图工具，按一定的绘图原理和方法绘制图样的一种方法。所绘制的图样称为仪器图。
④ *翼形螺钉：又称"蝶形螺钉"。螺钉头有伸出两翼的螺钉。
⑤ 阴：又称阴面或暗面。在光线照射下，物体背光的表面。
⑥ 阴线：在光线照射下，物体上受光面与背光面的分界线。

引出线 ① （电） leader, leader line
引入线（电） service entrance
引线孔 terminal hole
隐蔽部分 invisible portion
隐蔽点 hidden point
隐蔽外形线 hidden outline
隐蔽细部 hidden detail
隐蔽线 invisible line
隐蔽形体 hidden feature
隐藏轮廓线 hidden outline
隐藏面 hidden surface
隐（藏）线 hidden line
隐含 imply /im'plai/
印刷体 black letter
印制板 printed board
印制板零件图 printed board detail drawing
印制板制图 printed board drawing
印制板装配图 printed board assembly drawing
印制板组装件 printed board assembly
印制电路 printed circuit
印制线路 printed wiring
英镑符号 pound mark
英寸尺寸标注 inch dimensioning
英国铜匠螺纹 brass thread
英寻线 fathom-line
英制比例尺 imperial scale
英制单位 imperial unit
英制线号规 Imperial Wire Gage
影 ② shadow
影线 ③ line of shadow
影像 image
影印 photo printing, photocopying, reprographic process
影印法 diazo process

影印件 photocopy print, photostat print
影域 umbra
应答（计） echo (in computer graphics)
应力图 stress diagram
应用几何 graphic geometry
应用程序 application program
应用程序图标 application icon
应用软件 application software
映射 mapping
映象生成 map generalization
映像管 video tube
硬度 grade of hardness
硬度（铅笔） degree of hardness
硬化 hardening
硬件 hardware
硬拷贝 hard copy
硬拉 hard drawn
硬盘 fixed disk
硬橡皮 vulcanite
用户 subscriber, user
用户坐标 user coordinate
用器画 instrument drawing
用铅笔描图 pencil tracing
用……确定……并…… By... determine... and
用仪器画的线 ruled line
用引线 use guide lines
优化 optimization
优化设计 optimization design
优弓形 major segment
优弧 superior arc
优角 reflex angle, superior angle
油槽 oil groove
油断路器（电） oil circuit breaker (OCB)
油封 oil seal
油光锉 lance-tooth steel file, smooth (cut) file

注：① 引出线：又称指引线。用细实线从图中某处引出到图纸的空白处，以加注说明该处情况的线。
② 影：光线照射下，被物体遮掩而不能受光的部分在其表面上形成的暗象。
③ 影线：影的轮廓线。

油路线 oiling diagram
油石 oilstone
油毡 asphalt felt
油纸 asphalt paper, oilpaper
游标 vernier
游标高度规 vernier height gage
游标卡尺 vernier caliper
游标深度规 vernier depth gage
有槽螺帽 slotted nut
有格纸 preprinted grid sheet, print paper
有机表面处理 organic finish
有肩针（圆规）shouldered needle
有头螺钉 cap screw
有头螺母 cap nut
有限逻辑视图 finite logical view
有限平面 limited plane
有限图 finite graph
有限元分析 finite element analysis
有向图 digraph（计）
有效长度 effective length
有效螺纹长度 useful length of screw thread
*有效圈数 *effective coil number
右辅助视图 right auxiliary, right auxiliary view
右立面图 right elevation
右上角 upper right corner
右上象限 upper right-hand quadrant
*右视图① *right view, right side view, right-hand view
右手件 right-hand part
右下角 lower right-hand corner
*右旋螺纹② *right-hand thread
隅撑 knee brace
余角 complementary angle
余角法 complementary-line method

余面角 lap angle
余面圆 lap circle
鱼眼坑 spot face
鱼眼孔 spot-faced hole
语句 statement
语句标号 statement label
与X对准 in line with X
与/或树 AND/OR tree
与/或图 AND/OR graph
与/或形 AND/OR form
雨水管 downspout, rainwater pipe, down pipe
裕量 allowance
预处理器 preprocessor
预制钢筋混凝土构件详图 precast reinforced concrete element details
预钻孔深（攻螺纹用）depth of tap drill
原尺寸（足尺）full dimension
原点 origin
*原理图③ *schematic diagram, elementary diagram
原色 basic color
原始备份 original backup
原始粗糙铸形 original rough-cast form
原始介质 original medium
原始数据 original data
原始图纸 original drawing
原始文件 original document
原素 primitive
*原 图④ *original drawing, original artwork, artwork
原图表 primary chart
原图处理系统 artwork process system
原则性流程图 principled flow chart

注：①*右视图：由右向左投射所得的视图。
②*右旋螺纹：顺时针旋转时旋入的螺纹。
③*原理图：表示系统、设备的工作原理及其组成部分的相互关系的简图。
④*原图：经审核，认可作为原稿的图。

源图与终图[①] source figure and destination figure
元件面 component side
元素 element
元线 element line
圆 circle
圆(形)螺母 round nut
圆的尺寸标注法 dimensioning of circle
圆的渐开线 involute of circle
圆点 dot
圆钉 thumb pin
圆顶螺纹 knuckle thread
圆拱 circular arch
圆度 roundness
圆规 compasses
圆规的可换笔头 interchangeable point of compass
圆规附件 compass attachment
圆规架 compass frame
圆规接长杆 lengthening bar
圆规用起子 compass key
圆规用铅笔尖 pencil leg of compass, pencil point of compass
圆规用上墨笔头 inking point of compass
圆规用鸭嘴笔尖 pen leg of compass
圆弧 arc, circle arc, circle curve, circular arc
圆弧规 arcograph
圆弧模板 circle curve template
圆弧切线模板 radius tangent template
圆弧曲线 circular curve
圆弧三角形 circular triangle
圆弧插补 cicular interpolation
圆弧样条 arc splines
圆滑曲线 smooth curve
圆环 ring, annular torus, annulus
圆环形字母 loop letter
圆及共轭径法(画椭圆) circle method for conjugate diameter
圆角模板 radius template
*圆螺母 *round nut
圆模板 circle template
圆内接四边形 cyclic quadrilateral
圆盘图表 pie chart
圆球 sphere
圆球面 spheric surface
圆式条形图 radial bar chart
圆束 pencil of circles
圆跳动 cycle runout
圆头 rounded end
圆头短方颈螺栓 round head short square-neck bolt
圆头方颈螺栓 round head square-neck bolt
圆头螺钉 round head cap screw
圆头螺栓 round head bolt
圆头平键 Pratt-Whitney key
*圆头带榫螺栓 *cup nib bolt
圆心片 bone center
圆心圆弧 circular arc centre
圆形 circular
圆形部分 circular portion
圆形剖面 circular section
圆形视图 circular view
圆形双曲面体 circular hyperboloid
圆形图 pie chart
圆形元线 circular element
圆形中心线 circular center line
圆周角 angle of cirumference
圆周偏转 circular runout
圆柱 cylinder, circular cylinder
*圆柱齿轮 *cylindrical gear

注：① 源图与终图：在几何图形模拟中，如果给定图A使其变换至图B，则前者称为源图后者称为终图。

圆柱齿轮传动 cylindrical gear transmission
圆柱底 cylinder bottom
圆柱度 cylindricity
*圆柱螺纹① *straight thread, parallel screw thread
*圆柱螺旋弹簧② *cylindrical helical spring
圆柱面 cylindrical surface, circular cylindrical surface
圆柱体配合 cylindrical fit
圆柱投影 cylindrical projection
*圆柱销 *cylindrical pin, *straight pin, round pin
圆柱形公差带 cylindrical tolerance zone
圆柱性面 cylindroid
圆柱轴线 cylinder axis
圆柱坐标 circular cylinder coordinates
圆周 circumference
圆锥 cone, circular cone
圆锥齿轮 bevel gear, cone gear
圆锥齿轮传动 bevel gear transmission*
圆锥角 coning angle
圆锥螺旋线 conic helix
圆锥螺纹③ *taper screw thread, conical thread
圆锥面 circular conical surface, conic surface
圆锥（二次）曲线④ conical curves
圆锥曲线，二次曲线，割锥线 conic sections
圆锥曲线线束，二次曲线线束 pencil of conics
圆锥台 truncated cone, frustum of cone, cone frustum
圆锥投影 conical projection
*圆锥销 *conical pin, taper pin
圆锥形螺旋线 conical helix
圆锥形盘旋面 conical convolute
圆锥形体 conic shape
圆锥形斜螺旋体 conical oblique helicoid
圆锥形直螺旋体 conical right helicoid
圆锥锥度 conic taper
圆准线 circular directrix
缘 rim
允差 allowance
运动轨迹图 drawing of motion trace
运输系统示意图 transportation system diagram

Z

Z 形算图 Z chart
Z 型钢 zee
再次衍生元线 element of second generation
暂存 plasing out
暂停 pause

增加 add, increase
增量 increment
增量绘图机 incremental plotter
增量向量 incremental vector
增量坐标 incremental coordinate

注：①*圆柱螺纹：在圆柱表面上形成的螺纹。
②*圆柱螺旋弹簧：外廓呈圆柱形的螺旋弹簧。
③*圆锥螺纹：在圆锥表面上形成的螺纹。
④圆锥曲线：又称二次曲线、割锥线。平面截割圆锥所形成的种种曲线。

轧制钢 rolled steel
轧制型材 rolled shape
闸阀 gate valve
闸流晶体管 thumbtack
窄截面 thin sections
展开 development
展开画法① (铁) representation of develop, development representation, aligned view
展开视图 expansion view*
展开图② (矿)* developing drawing, developed view, unfolding drawing
展开图基线 girth line, stretchout line
展览图表 chart for display
展性铸件 malleable castings, malleable cast iron
栈桥 (土木) trestle
站 station
站缝 standing seam
站线 station ordinates
照比例绘制 drawing to scale
照明法 method of illumination
照明配电盘 lighting panel (L.P)
照明平面图③ lighting layout
照明系统图④ lighting system drawing
照相测量地图 photogrammetric map
照相底图 artwork master
照相平版 (印刷) photo-lithography
照相缩小尺寸 photogrphic reduction dimension
照相图 photo-drawing
折边 edging
折尺 folding rule, folding scale

折叠 fold
折叠记号 fold mark
折断画法⑤ interrupted view, conventional breaks representation, conventional break
折断式条形图 bar chart
折断线 break line
折痕 cease
折角管 angular tube
折角线 knuckle lines
折线 fold line, folding line
针笔 technical drawing pen, technical fountain pen, technical pen, needle-in-tube pen
针笔笔尖 technical pen point
针笔写字接头 lettering joint
针尖 (圆规) needle-point
针尖脚 (圆规) needle-point leg
针盘比测器 dial compartor
针盘指示器 dial indicator
真分数 proper fraction
真平度 flatness
真平度公差 flatness tolerance
真实尺寸，真实大小 true size
真实投影 true projection
真实椭圆 true ellipse
真实中心 true-center
真形视图 normal view
真直度 straightness
整数 integer
整数部分 (带分数) whole number portion, whole units
正八面体 regular octahedron
正齿轮 spur gear

注: ① 展开画法：在图样上，将某些曲面或不处于同一平面而互相衔接的一些图形，依次连续摊平画出的图示方法。
② *展开图：空间形体的表面摊平后的图形。
③ 照明平面图：表达建筑物内各种照明设施和照明线路布置的图样。
④ 照明系统图：用单线轴测图的形式表达建筑物内照明系统的图样。
⑤ 折断画法：对机件沿长度方向的截面形状相同或按一定规律变化时，将其断开去掉一段缩短绘图的图示方法。

正垂面 normal surface
正垂视图 normal view
正垂线 horizontal-profile line
正等轴测格纸 isometric paper
正等轴测投影（正等测） isometric projection
正多边形 regular polygon
正多面体 regular polyhedron, regular solid
正二等轴测投影（正二测） dimetric projection
正二十面体 regular dodecahedron, regular icosahedron, regular hexahedron
正方棱柱 square prism
正方形 square
正负复合符号 combined plus and minus sign
正规级 regular series
正规模式 regular pattern
正极（电） positive section
正交，直交 orthogonal
正交轴线 orthogonal axes
正角锥体 regular pyramid
正截口 right section
正楷字 hand printing
正理（木） straight-grain
正立体 right solid
正立面图 front elevation
正立投影面 frontal projection plane
正螺旋面 right helicoid
正面投影 frontal projection
正面斜轴测投影 frontal oblique axonometric projection
正平面 frontal plane
正平线 frontal line
正四面体 regular tetrahedron
正视图 frontview, elevation
*正投影 ① *orthogonal projection, orthographic projection

‖例句：正投影是在与投影线垂直的投影面上画出线条来表示物体的方法。*Example sentence:* Orthographic projection *is a method of representing an object by a line drawing on a projection plane that is perpendicular to parallel projectors.*

正投影草图 orthographic sketch
*正投影法 ② *orthogonal projection method
正投影图 orthograph, orthographic view, orthographic drawing
正图纸（晒图） autopositive paper
正位度 true position
正位度公差 true position tolerance
正位平面 true position plane
正位透视 normal perspective
正位轴线 true position axis
正弦曲线 sine curve
正裕度（最小余隙） positive allowance
正圆度 roundness
正则曲线 regular curve
正值 positive value
正轴测投影 rectangular axonometric projection
*支承圈数 *number of end coils
支距标注（曲线）尺寸法 dimensioning by offsets
支距法 offset method
支距法作图 offset construction
支柱 strut
直尺 linear scale, straight blade, straight rule
直齿轮 spur gear
直方图 bar graph
直管螺纹 straight pipe thread
直角 right angle
直角边 leg of a right triangle

注：① *正投影：根据正投影法所得到的图形。
② *正投影法：投射线与投影面相垂直的平行投影法。

直角回转 quarter turn
直角三角形 right triangle
直角三角形法 right triangle method
直角双曲线 rectangular hyperbola
直角投影 orthogonal projection
直角弯管 quarter bend
直角柱 right prism
直角锥 right pyramid
直角坐标 rectangular coordinate, Cartesian coordinates
直角坐标点绘 rectangular plot
直角坐标图表 rectilinear chart
直角坐标系的坐标网格法 grid method of rectangular corrdinate system
直接注入公差 direct tolerancing
直径 diameter
直立边（熔接）perpendicular leg
直立方向 vertical direction
直立割面 vertical section plane
直立股（直角三角形）vertical leg
直立迹 vertical trace, V-trace
直立剖面，竖直剖面 vertical section
直立投影，垂直投影 vertical projection, V-projection
直立投影面 vertical projection plane
直立圆柱体 vertical cylinderical piece
直立主线 vertical principal line
直流电 direct current
直流电动机控制器 DC motor controller
直六角柱 right hexagonal prism
直螺纹 parallel thread, straight thread
直内螺纹 straight internal thread
直劈锥面 right conoid
直三角柱 right triangular prism
直式大写字母 vertical capital letter
直式小写字母 vertical lower-case letter

直式字母 upright style of letters
直体字 vertical letter
直体字母 vertical letter, vertical character
直外螺纹 straight external thread
直纹滚花 straight knurling
直纹面 ruled surface
直线 line, straight line
‖例句: 直线在平面上的投影，是直线两端点在平面上投影的连线。Example sentence: The projection of a line on a plane is the line connecting the projection on the plane of the endpoints of the line.
直线的点视图（直线的端视图）point view of a line
‖例句: 直线段的位置由它的两端点的位置确定。Example sentence: A straight-line segment is established by locating its endpoints.
直线笔（鸭嘴笔）ruling pen
直线部分（字法）straight-line portion
直线插补 linear interpolation
直线度 straightness
直线段 A straight line segment
直线方程式，一次方程式 linear equation
直线距离 linear distance
直线面① ruled surface
直线描阴 line shading
直线式接线图 straight line type of connection diagram
直线束 pencil of lines
直线与平面的贯穿点 piercing point of line and surface
直线字母 straight-line letters
直线坐标 rectilinear coordinates
直演生线 rectilinear generatrix
直元线 rectilinear element
直圆柱 right cylinder, right circular cylinder

注：① 直线面：由直母线运动形成的曲面。

直圆锥 cone of revolution, right circular cone, right cone
直准线 rectilinear directrix, straight directrix
执行草图 executive sketch
纸 paper
纸板 paste board
纸带 paper-tape
纸面亮度 paper texture
纸莎草纸① papyrus
纸上走线 route location on paper
纸张大小 paper size
只读存储器 read only memory
指北标志 cardinal points of the compass
指北针 north point
指定长度 specified length
指定事项 specific note
指定转折（熔接）definite break
指令 instruction
指捻螺钉，翼形螺钉 thumb screw
指捻螺母，翼形螺母 thumb nut
指线 leader, leader line
指向 pointing
质量管理图 control chart
制程中途图 drawing of an intermediate step
制表 tab
制模 model making
制模工 template maker
制图 drafting
制图板 drafting board, drawing board
制图比例 scale of drawing
制图比例尺 drawing scale
制图笔尖 drawing pennib
制图布 drawing cloth
制图材料 drawing material
制图尺 plotting scale
制图法 drafting method
制图符号 drawing symbol
制图刮刀 drawing knife
制图规则 graphical law
制图机 draft machine
制图技术 drawing technique
制图胶带 drafting tape
制图胶片 drafting film
制图墨水 drawing ink
制图铅笔 draftsman's pencil, drawing pencil
制图室 drafting room, drawing office, drawing room
制图室标准 drafting room standard
制图室记录 drafting room record
制图术 draftsmanship
制图顺序 order of drawing
制图仪，绘图仪 drafting machine
制图仪器 drawing instruments, graphic instruments
制图用尺寸比例 drawing proportion
制图圆规 drawing compass
制图员 draftsman
制图桌 drafting table, drawing table, drawing desk
制造程序 manufacturing process
制造程序，生产程序 production process
制造方法 manufacturing method, method of manufacturing
制造公差 manufacturing limits, manufacturing tolerance
制造精度 manufacturing precision
制造说明书 manufacturing illustration
制造条件 manufacturing requirement
制造用图 production drawing
终端 terminal

注：① 纸莎草纸：又名纸草纸。纸莎草是一种禾状水生植物，盛产于尼罗河一带，基茎为木质，古埃及人用以造纸叫纸莎草纸，此纸可以书写、绘画。

终止 terminating
终止，界限 termination
中点 mid-point
中垂线 perpendicular bisector
中段复合地质平面图 midsection complex geological plan
中断 interrupt
中断视图 interrupted view
中断线式接线图 interrupted line type of connection diagram
中国标准化协会 Chinese Association for Standardization
中国工程图学会 China Engineering Graphics Society (CEGS)
中横剖面图 midship section plan
中级打入配合 medium drive fit
中级压入配合 medium force fit
中级转动配合 medium running fit
中间螺钉弹簧圆规 center-wheel bow
中间面 middle plane, middle surface
中间色 medium color
*中径① *pitch diameter
*中径线② *pitch line
中位节圆（蜗轮） median pitch circle
中位直径 median diameter
中线面（船） middle-line plane
中线（线条粗细） medium line
中心 center, centre
中心冲 center punch

中心角 angle at the center, central angle
中心角等分线 radial division line
中心距 center distance, center to center, between center
中心距量规 center distance gage
中心孔 central hole
中心孔钻 center drill
中心面 central plane
中心平面 median plane
中心投影 central projection
*中心投影法③ *central projection method
中心线 center line
中心重合 coinciding center
中央处理单元 central processing unit
中央轴线 central axis
中站面（船） midstation plane
重打入配合 heavy drive fit
重级 heavy duty, heavy series
重力线（结构） gravity line
重压缩配合 heavy force and shrink fit
周长 perimeter
周期 period
周围，周边 periphery
周围长度 girth
*轴④ *shaft, axle
轴测分解图⑤ axonometric xeploded drawing
*轴测投影⑥ *axonometric projection
轴测投影面 axonometric projection plane
轴测投影轴 axonometric axis

注：①*中径：又称螺纹中径，假想圆柱或圆锥的直径，该圆柱或圆锥的母线通过螺纹牙型上的沟槽和牙厚宽度相等。该假想圆柱或圆锥称为中径圆柱或中径圆锥。
②*中径线：中径圆柱或中径圆锥的母线。
③*中心投影法：投射线汇交一点的投影法。
④*轴：(1)支承转动件，传递运动或动力的机械零件。(2)主要指圆柱形外表面，也包括其他外表面中由单一尺寸确定的部分。
⑤轴测分解图：按一组装配零件的装、卸次序，将各零件排列在一条轴线上绘制出各零件的轴测图。
⑥*轴测投影：将物体连同其参考直角坐标系，沿不平行于任一坐标面的方向，用平行投影法将其投射在一个投影面上所得到的图形。

*轴测图 [1] *axonometric drawing
*轴承 *bearing
*轴承内径 *bearing bore diameter
轴肩 shoulder
轴间角 [2] axis angle, angle between axes
轴颈 shaft neck
轴线 axis
轴线的平面 axial plane
轴向伸缩系数（轴向缩短率）[3] coefficient of axial deformation, ratio of foreshortening for any axis
轴向剖面 axial section
逐次断面 separte section, seccssive section
逐点比较法 point by point comparative method
主程序 mast program, main program
主点 [4] main point, principal point
主机 mainframe（计）
主基线 primary ground line
主基准 primary datum
主接线图 main wiring diagram
主控钥匙系统 turnkey system
主剖面 principal section
*主视图 [5] front view, principal view
‖例句: 完成平面图形的主视图。*Example sentence:* Complete *the front view of the plane figure.*
主投影面 principal plane (of projection)
主要曲线 principal curve
主要特征视图 master view
主要细节 dominant detail
主要制造方位 main position of manufacture
主坐标平面 principal coordinate planes
注尺寸顺序 order of dimensioning
注尺寸用模板 dimensioning template
注释 annotation, descriptive note
注限界尺寸法 limit dimensioning method
注销 logout
驻场工程师 resident engineer
驻点 standpoint
柱 column
柱脚，柱基，础 plinth
柱坑 counter bore
柱坑孔 counterbored hole
柱面 [6] cylindrical surface
柱面螺旋体 cylindrical helix
柱条图表 column chart
柱形螺旋体 cylindrical helicoid
柱形斜螺旋体 cylindrical oblique helicoid
柱形直螺旋体 cylindrical right helicoid
柱状面 cylindroid
柱状图 [7] columnar section
铸件 casting, cast part
铸件图 casting drawing
铸孔 cored hole
铸铁 cast iron
铸铁管 cast iron pipe
铸铁接头 cast iron fitting
铸造公差 casting tolerance
铸造图 production drawing for casting

注：① *轴测图：用平行投影法将空间形体和确定其位置的空间直角坐标系，投影到投影面上所得到的图形。
② 轴间角：轴测投影中，两根轴测轴间的夹角。
③ 轴向伸缩系数：轴测轴上的单位长度与相应投影轴上的单位长度的比值。
④ 主点：视点在画面上的正投影。
⑤ *主视图：由前向后投射所得的视图。
⑥ 柱面：直母线沿着某一曲导线移动，并始终平行某一直导线所形成的曲面。
⑦ 柱状图：又称地质柱状剖面图。根据钻探所得地质资料，由下到上按地层时代的先后，以序排列绘制的柱状图样。

专利图 ① patent drawing, patent office drawing
专门名词，术语 terminology
专业标准 specialized standard
专用公差 special tolerance
专用计算机 special-purpose computer
专用字符 special symbol
砖型图 sketch of brick types
转储，卸下 unload /ˌʌnˈləud/
转动配合 running fit
转角接合 corner joint
转接顺序图（开关的）sequence chart
转量线 mitre line
转向点 point of change in direction, turning point
转移 jump
转移，分支，支线 branch /brɑ:ntʃ/
转印膜 transfer film
转圆线（船）rounded line
转折割面 offset (cutting) plane
装配程序图 assembling process chart
装配尺寸 ② assembly dimension
装配尺寸链 dimensional chain for assembly
装配工艺 assembling technic
装配螺栓 assembling bolt
装配示意图 ③ assembly diagrammatic drawing
*装配图 ④ *assembly drawing, assembly print, assembly sheet
装配图号 assembly drawing number
装饰线（字法）serif
装饰性笔画 appendge
装妥零件 assembled parts

桩位布置图 piling plan
桩位图 piling foundation plan
追踪运行图 following train graph
锥柄 taper shank
*锥齿轮 *level gear
锥底 base of cone
锥顶点 conic apex
锥度 taper, taper on diameter
锥度方向 direction of taper
锥度公差 taper tolerance
锥度公差域 taper tolerance zone
锥坑 counter sink
锥坑孔 countersunk hole
锥孔 taper hole, taper socket
锥面 ⑤ conical surface
锥曲面 conic curved surface
锥台 frustum
锥体尺寸注法 form dimensioning of tapers
锥形(铅)笔尖 round point, plain point of pencil
锥形投影 conical projection
锥状面 conoid
锥轴 axis of cone
准点 point directrix
准面，导面 director
准平面 directing plane, plane director, plane directing surface
准线 base line, directrix
准锥面 cone director
着色 toning
着色硬化 color-harden
资料图纸一览表 data drawing list
子程序 subroutine, subprogram

注：① 专利图：申请发明专利权时，表明其发明特征的图样。
② 装配尺寸：保证机器或部件的正确装配关系，满足其工作精度和性能要求的尺寸。
③ 装配示意图：用规定符号和线条简单表示机器或部件的结构特点、零件间的相对位置和装配连接关系的简图。
④ *装配图：表示产品及其组成部分的连接、装配关系的图样。
⑤ 锥面：直母线沿着某一曲导线移动，并始终通过某一定点所形成的曲面。

子图 child diagram
子午面 meridian plane
子午线 meridian, meridian line, principal meridian
字 word
字段（计）fiele
字法 lettering
字符 character
字符串 character string
字符发生器 character generator
字符行 character row
字符大小 character size
字符绘图 character plotting
字符集 character set
字符间隔 character spacing
字符图形 character graphic
字符图形算法 character graphic algorithm
字符显示设备 character display device
字符旋转 character rotation
字高 character height
字格板 Ames instrument, Ames lettering instrument
字格三角板 Braddock-Rowe triangle
字规 lettering guide, lettering plate, lettering stencil, lettering template
字宽 width of letter
字母 alphabet
字母符号 letter symbol
字母间的空白 background area
字母间空白 white space
字母宽度 letterspace
字母顺序 alphabetic order
字母之尾 tail of letter
字母组合 composition in lettering
*字体① *lettering, character style
字图点式传输、字图照相法显示 alphaphotographic mode
字行 row of letters
字之翘角（字法）spur
字之中高 midheight of letter
自变数 independent variable
自动化 automatic
自动标注尺寸 automatic dimensioning
自动打印控制语言 autotype control language
自动工程设计程序 automatic engineering design program
自动画线 autoscore
自动绘图 autodraft, automatic draft, automatic plotting, automation plotting
自动绘图机 drafting, plotter, autoplotter, autographic machine, automatic drafting machine, automated drafting
自动绘图设备 automatic drawing device, automatic drawing equipment
自动绘图数字化 automatic drawing digitizing
自动绘图系统 automatic drafting system
自动绘图仪 automatic plotter, autoplotter
自动铅笔按钮 push button top
自动图像恢复系统 automatic image retrieval system
自动图像传输 automatic picture transmission
自动图像传输系统 automatic picture transmission system
自动制图 automated drafting
*自攻螺钉 *tapping screw
自举 boot
自来水直线笔（绘图笔）fountain ruling pen
*自切螺钉 *thread cutting screw
自然空间坐标系 world coordinate system
自位轴承 self-aligning bearing
*自由高度 *free-height
自由转动配合 free running fit

注：①*字体：图中文字、字母、数字的书写形式。

综合材料表 comprehensive material list
综合分号运行图 compositive sectional train graph
综合平面图 composite plan
综合折线图 multiline chart
棕图 brown print
总保险丝 main fuse
*总布置图① *general plan
总电路(电) main circuit
总段结构图 structure plan of block
总高 height overall
总工艺流程图 general diagram
总和 total
总平面图 general plan, site plan, plot plan, block plan
*总圈数 *total number of coils
总图② general arrangement, general arrangement drawing, general drawing, general plan, general view, master plan
总线，母线 bus
总线图 master diagram
纵断面图(地) longitudinal section, longitudinal section profile
纵缝 longitudinal seam
纵剖面纸 profile paper
纵剖木纹 along grain of wood
纵剖线 buttock (line)
纵剖线图(船) sheer plan
纵向尺寸 longitudinal dimension
纵向视图 longitudinal view
纵轴线 longitudinal axis
纵坐标 ordinate

纵坐标轴 axis of ordinate
足尺 full size scale
足尺(比例) full scale, full size
组合、装配 assembling
组合草图、装配草图 assembly sketch
组合符号 compound symbols
组合工作图 assembly working drawing
组合角尺 combination set
组合剖面 assembly section
组合图 association graph, composition diagram
组件 assembly, component
组件装配图 subassembly drawing
组织图 organization chart
钻 drilling
钻尖 drill point
钻孔 drilled hole, drill
钻孔导向点 center spot
钻孔深度 depth of drill
钻孔柱状图 bore hole columnar section
钻鱼眼孔 spot-facing
最初构想 original idea
最粗线 widest line
最大尺寸 maximum size
最大公差 maximum tolerance
*最大过盈③ *maximum interference
最大喉径 outer throat diameter
*最大极限尺寸④ *maximum limit size
*最大间隙⑤ *maximum clearance
最大倾斜线 line of greatest inclination, line of maximum inclinatin
最大容许尺寸 largest permissible dimension
最大容许间隙 maximum allowable clearance

注：①*总布置图：表示特定区域的地形和所有建筑物布局及邻近情况的平面图。
②总图：表示产品总体结构及基本性能的图样。
③*最大过盈：对过盈配合或过渡配合，孔的最小极限尺寸减轴的最大极限尺寸所得的代数差。
④*最大极限尺寸：允许尺寸变动的两个界限值中较大的极限尺寸。
⑤*最大间隙：对间隙配合或过渡配合，孔的最大极限尺寸减轴的最小极限尺寸所得的代数差。

*最大实体尺寸① *maximum material size
*最大实体状态② *maximum material condition
最大限界 maximum limit
最大斜度线 spurnomal
最大直径 overall diameter
最短距离 shortest distance
最多留料极限 maximum material limit
最多留料情况 maximum material condition
最后模型 finish model
最后图样 final drawing
最少留料极限 minimum material limit
最少材料情况 least material condition (LMC), minimum material condition
最外尺寸 outer most dimension
最小尺寸 minimum size
最小公差 minimum tolerance
*最小过盈③ *minimum interference
*最小极限尺寸④ *minimum limit size
*最小间隙⑤ *minimum clearance
最小螺纹根弧半径 minimum root radius
最小容许间隙 minimum allowable clearance
*最小实体尺寸⑥ *least material size
*最小实体状态⑦ *least material condition
最小限界 minimum limit
左辅助视图 left auxiliary
左辅助图，左立面图 left elevation
*左视图⑧ *left view, left side view, left-hand view
‖ 例句：给定直线 AB 的主视图和俯视图，要求增加左视图。Example sentence: Add a left view *of a line AB, given the* front and top views.
左手件 left-hand part
左下角 lower left-hand corner
*左旋螺纹⑨ *left-hand thread
作废 superseded
作图法 method of construction
作图函数 drawing function
作图线 witness line
作业 job
作……并…… pass... and...
坐标 coordinate
坐标变换 coordinate transformation
坐标点法 coordinate method
坐标方格网（矿）mining coordinate paper
坐标面 coordinate plane
坐标图形 coordinate graphics
坐标网格 grid（电），coordinate division
坐标系 system of coordinates
坐标原点 origin of coordinates
坐标纸，方格纸 plotting paper
坐标注尺寸法 dimensioning by coordinates, dimensioning by ordinates, ordinate dimensioning
坐标轴 coordinate axis

注：①*最大实体尺寸：最大实体状态下的极限尺寸。
②*最大实体状态：孔或轴具有允许的材料量为最多时的状态。
③*最小过盈：对过盈配合，孔的最大极限尺寸减轴的最小极限尺寸所得的代数差。
④*最小极限尺寸：允许尺寸变动的两个界限值中较小的极限尺寸。
⑤*最小间隙：对间隙配合，孔的最小极限尺寸减轴的最大极限尺寸所得的代数差。
⑥*最小实体尺寸：最小实体状态下的极限尺寸。
⑦*最小实体状态：孔或轴具有允许的材料量为最少时的状态。
⑧*左视图：由左向右投射所得的视图。
⑨*左旋螺纹：逆时针旋转时旋入的螺纹。

附 录

APPENDIX I（附录一）

工程制图英文常用缩写词一览

缩写或符号	英文名词	中文含义
A/F	Across flats	对边
ASSY	Assembly	装配
CRS	Centres	中心
CL	Centres line	中心线
CHAM	Chamfered	倒角，去角
CH HD	Cheese head	圆头
CSK	Countersunk	埋头孔
CSK HD	Countersunk head	埋头
C'BORE	Counterbore	平底锪孔
CYL	Cylinder or cylindrical	圆柱，柱面
DIA	Diameter	直径
ø	Diameter (preceding a dimension)	直径（注在尺寸前）
DRG	Drawing	图样
EV	Edge view	重影视图
EXT	External	外部
FIG	Figure	图形
GL	Use guide lines	用引线
HEX	Hexagon	六角形
HEX HD	Hexagon head	六角头
I/D	Internal diameter	内直径

续表

缩写或符号	英文名词	中文含义
INT	Internal	内部
LH	Left hand	左边
LI	Line of intersection	交线
MATL	Material	材料
MAX	Maximum	最大
M/C	Machine	机械
M/CD	Machined	加工
M/CY	Machinery	机构
MIN	Minimum	最小
NTS	Not to scale	不按比例
NO.	Number	号数
O/D	Outside diameter	外径
PCD	Pitch circle diameter	节圆直径，中心线圆直径
R	Radius (preceding a dimension, capital letter only)	半径（大写，注在尺寸前）
RADR	Radius	半径
RH	Right hand	右边
RD HD	Round head	半圆形
SCR	Screwed	螺丝，螺纹
SØ	Spherical diameter (preceding a dimension)	球形直径（注在尺寸前）
SR	Spherical radius (preceding a dimension)	球形半径（注在尺寸前）
S'FACE	Spotface	锪孔
SQ	Square (in a note)	方形（用于注释）
□	Square (preceding a dimension)	方形（注在尺寸前）

续表

缩写或符号	英文名词	中文含义
STD	Standard	标准
SWG	Standard wire gauge	标准线规
TL	True length	实长
TS	True size	真实尺寸
U'CUT	Under cut	切槽，凹割
VOL	Volume	体积
WT	Weight	重量

APPENDIX Ⅱ（附录二）

中国内地与港澳台地区工程制图部分术语对照

表 1 以中国内地术语首字汉语拼音首字母为序

中国内地术语	港澳台地区术语
安装尺寸	裝置尺寸
安装图	裝置圖
半剖视	半剖面
部件	分裝配，次總成，次組合
槽形螺帽	堡形螺帽
侧垂线	前橫線
尺寸界线	延長線，引伸線
尺寸线	寸法線
齿顶	齒冠
赤平极射投影	立體投影
带状法	球面緯線展開法
倒角	去角（倒角）
等高线笔、曲线笔、轮廓笔	萬向鴨嘴筆
等轴测立方体	等角立方體
等轴测视图	等角視圖
等轴测图	等角圖
等轴测坐标	等角坐標
底图	描圖
点划线，点划相间线	鏈線

续表

中国内地术语	港澳台地区术语
电力平面图	電源圖
丁字尺	T尺
定位孔	參考洞
二等轴测图	二等角圖
分辨率	解像度
分格转绘法	分格法
分规	兩腳規
分角	象限
俯视图	頂視圖，上視圖，平面圖
辅助截面	輔助割面
辅助视图	輔視圖
辅助线	輔線
辅助正视图	前輔助視圖，立面輔助視圖
辅助锥（面）	輔錐
复制图	摹圖，抄圖
工程图学	工程圖學，工程畫
工程图样	工程畫，工程圖
工艺流程图	製造流程圖
功能图	機能線圖
规格尺寸	基準尺寸
过盈配合	干涉配合
焊接图	熔接圖
喉圆	隘圓
后视图	背視圖

续表

中国内地术语	港澳台地区术语
画法几何	投影幾何
回转面	旋轉成面
绘图笔	針筆
绘图机	描線器，製圖機
机械安装孔	固定孔
基本尺寸	基本大小
基座图	基礎平面圖
夹具图	工模圖
间隙配合	餘隙配合
减半投影	半斜投影
简图	示意圖，線圖
角透视	成角透視，斜透視
阶梯剖	轉折的剖面
截平面，切平面	割面
紧滑配合	緊密滑動配合
紧配合	緊密配合
紧转配合	緊密轉動配合
局部视图	部分圖，中斷視圖
可展曲面	展開可能曲面
框图	方塊圖
连接盘	焊襯墊
量角器	量角規，角度規，量角尺
描阴	明暗法
灭点	消失點

续表

中国内地术语	港澳台地区术语
灭点法	消失點法（透視圖）
明细栏	零件表
母曲线	演生曲線
母线	演生線，動線
母圆	演生圆，动圆
挠性曲线尺	可繞曲線規
平距	水準距離
剖切符号	割面線
剖切面	割面
切平面法	割面法（求交線）
切球面法	割球法（求交線）
倾斜直线	斜傾直線
曲线笔	萬向鴨嘴筆
曲线尺	蛇尺
曲折线	鋸齒線，錯縱線
全剖视	全剖面
全剖视图	全剖面
上偏差	上尺寸差
视点	照準點
双头螺纹	雙紋螺紋
双头螺柱，螺柱	螺椿
双斜面比例尺	四斜邊尺
体视图	立體畫
投影	投射

续表

中国内地术语	港澳台地区术语
图例	圖說
图纸幅面	版式
徒手图	草圖
外形线	外廓線
下偏差	下尺寸差
舷墙图	甲板欄柵圖
详细剖面（多个剖面）	細部剖面
小圆规	彈簧圓規
斜等轴测图	等斜圖
斜二等轴测投影	斜兩等角投影
斜二等轴测图	半斜圖
斜剖	斜斷面
斜轴测投影	斜不等角投影
旋转剖	轉正剖面
仰视图	底視圖
一字尺	平行尺
移出断面	移轉剖面圖
引出线	引線
印制板制图	印刷電路製圖
右视图	右側視圖
圆度	真圓度
照相底图	原版工作圖
折断画法	中斷視圖
正垂线	側橫線

续表

中国内地术语	港澳台地区术语
正等测投影	三軸等比正投影
正等轴测投影	等角投影
正二等轴测投影	二等角投影，兩等角投影
正截口	正斷面
正平面	前平面
正平线	前平線
正投影	正（交）投影
直观图	寫生圖，立體圖
中心投影法	集中投影法
轴测草图	立體正投影草圖
轴测投影	不等角投影
轴测图	立體正投影圖
主视图	前視圖
装配工作图	組合工作圖
装配图	組合圖
锥度	推拔
锥面	錐形面
锥状面	劈錐曲面
字体	字法
纵剖线图	側視圖
纵向视图	縱視圖
足尺	足比例尺
最大实体原则	最多留料原理
最大实体状态	最多留料情況

续表

中国内地术语	港澳台地区术语
左视图	左侧视图
坐标网格	栅格，方格

表2 以港澳台地区术语首字汉语拼音首字母为序

港澳台地区术语	中国内地术语
T尺	丁字尺
腽圆	喉圆
版式	图纸幅面
半剖面	半剖视
半斜投影	减半投影
半斜图	斜二等轴测图
堡形螺帽	槽形螺帽
背视图	后视图
不等角投影	轴测投影
部分图，中断视图	局部视图
参考洞	定位孔
草图	徒手图
侧横线	正垂线
侧视图	纵剖线图
成角透视，斜透视	角透视
齿冠	齿顶
寸法线	尺寸线
弹簧圆规	小圆规
等角立方体	等轴测立方体

续表

港澳台地区术语	中国内地术语
等角視圖	等轴测视图
等角投影	正等轴测投影
等角圖	等轴测图
等角坐標	等轴测坐标
等斜圖	斜等轴测图
底視圖	仰视图
電源圖	电力平面图
頂視圖，上視圖，平面圖	俯视图
二等角投影	正二等轴测投影
二等角圖	二等轴测图
分格法	分格转绘法
分裝配，次總成，次組合	部件
輔線	辅助线
輔助割面	辅助截面
輔助視圖	辅视图
輔錐	辅助锥（面）
干涉配合	过盈配合
割面	截平面，切平面
割面	剖切面
割面法（求交線）	切平面法
割面線	剖切符号
割球法（求交線）	切球面法
工程畫，工程圖	工程图样
工程圖學，工程畫	工程图学

续表

港澳台地区术语	中国内地术语
工模圖	夹具图
固定孔	机械安装孔
海拔，高度	标高
基本大小	基本尺寸
基礎平面圖	基座图
基準尺寸	规格尺寸
集中投影法	中心投影法
甲板欄柵圖	舷墙图
緊密滑動配合	紧滑配合
緊密配合	紧配合
緊密轉動配合	紧转配合
可撓曲線規	挠性曲线带，挠性曲线尺
立體畫	体视图
立體投影	赤平极射投影
立體正投影草圖	轴测草图
立體正投影圖	轴测图
鏈線	点划线，点划相间线
量角規，角度規，量角尺	量角器
兩等角投影	正二等轴测投影
兩腳規	分规
零件表	明细栏
螺椿	双头螺柱，螺柱
描圖	底图
明暗法	描阴

续表

港澳台地区术语	中国内地术语
摹圖，抄圖	复制图
劈錐曲面	锥状面
平行尺	一字尺
前輔助視圖，立面輔助視圖	辅助正视图
前橫綫	侧垂线
前平面	正平面
前平綫	正平线
前視圖	主视图
傾斜度	斜度
去角（倒角）	倒角
全剖面	全剖视图
熔接圖	焊接图
三軸等比正投影	正等测投影
上尺寸差	上偏差
蛇尺	曲线尺
示意圖，綫圖	简图
雙紋螺紋	双头螺纹
水準距離	平距
投射	投影
投影幾何	画法几何
圖表	表图
圖說	图例
推拔	锥度
萬向鴨嘴筆	曲线笔

续表

港澳台地区术语	中国内地术语
習用畫法	习惯(规定)画法
細部剖面	详细剖面(多个剖面)
下尺寸差	下偏差
線的粗細	线宽
線束圖	线扎图
象限	分角
消失點	灭点
消失點法(透視圖)	灭点法
斜不等角投影	斜轴测投影
斜斷面	斜剖
斜兩等角投影	斜二等轴测投影
寫生圖，立體圖	直观图
延長線，引伸線	尺寸界线
演生曲線	母曲线
演生線，動線	母线
演生圓，動圓	母圆
移轉剖面圖	移出断面
引線	引出线
印刷電路製圖	印制板制图
右側視圖	右视图
餘隙配合	间隙配合
原版工作圖	照相底图
圓柱形投影	圆柱投影
造船用曲線規	船体曲线板

续表

港澳台地区术语	中国内地术语
柵格，方格	坐标网格
展開可能曲面	可展曲面
照準點	视点
折斷習用畫法	习惯断裂画法
真圓度	圆度
針筆	绘图笔
正（交）投影	正投影
正斷面	正截口
中斷視圖	折断画法
轉折的剖面	阶梯剖
轉正的視圖	旋转视图
轉正剖面	旋转剖
裝置尺寸	安装尺寸
裝置圖	安装图
字法	字体
足比例尺	足尺
組合工作圖	装配工作图
組合圖	装配图
最多留料情況	最大实体状态
最多留料原理	最大实体原则
左側視圖	左视图

APPENDIX III（附录三）

常见国际、区域及主要国家标准代号

1. 国际标准代号

国际航空运输协会标准（ATA）
国际人造纤维标准化局标准（BISFA）
国际计量局（BIPM）
食品法典委员会标准（CAC）
关税合作理事会标准（CCC）
国际照明委员会标准（CIE）
国际无线电干扰特别委员会标准（CISPR）
国际原子能机构标准（IAEA）
国际航空运输协会（IATA）
国际民航组织标准（ICAO）
国际辐射防护委员会标准（ICRP）
国际辐射单位和测量委员会标准（ICRU）
国际乳制品联合会标准（IDF）
国际电工委员会标准（IEC）
国际签书馆协会和学会联合会标准（IFLA）

国际制冷学会标准（IIR）
国际劳工组织标准（ILO）
国际海事组织标准（IMO）
国际橄榄油理事会标准（IOOC）
国际标准化组织标准（ISO）
国际电信联盟标准（ITU）
国际兽疾局标准（OIE）
国际法制计量组织标准（OIML）
国际葡萄与葡萄酒局标准（OIV）
国际铁路联盟标准（UIC）
联合国教科文组织标准（UNESCO）
世界卫生组织标准（WHO）
世界知识产权组织标准（WIPO）
世界贸易组织（WTO）

2. 区域标准代号

非洲地区标准（ARS）
阿拉伯标准（ASMO）
欧洲标准化委员会（CEN）
欧洲电工标准化委员会（CENELEC）

经互会标准化常设委员会（CMEA）
欧洲标准（EN）
欧洲电信标准（ETS）
泛美标准（PAS）

3. 主要国家标准代号

美国国家标准（ANSI）
美国石油学会标准（API）
澳大利亚标准（AS）
美国机械工程师协会标准（ASME）
美国试验与材料协会标准（ASTM）

英国国家标准（BS）
加拿大标准（CSA）
捷克标准（CSN）
德国国家标准（DIN）
丹麦标准（DS）

美国食品与药物管理局标准（FDA）
中国国家标准（GB）
俄罗斯国家标准（ΓOCT）
印度标准（IS）
日本工业标准（JIS）
美国军用标准（MIL）
巴西标准（NB）
比利时标准（NBN）
美国电气制造商协会标准（NEMA）
法国国家标准（NF）

挪威标准（NS）
奥地利标准（ŌNORM）
波兰标准（PN）
美国机动车工程师协会标准（SAE）
瑞典标准（SIS）
瑞士标准协会标准（SNV）
美国电信工业协会标准（TIA）
意大利标准（UNI）
德国电气工程师协会标准（VDE）

APPENDIX Ⅳ（附录四）

国际标准化组织（ISO）成员团体代号

Algeria 阿尔及利亚 (IANOR)
Argentina 阿根廷 (IRAM)
Armenia 亚美尼亚 (SARM)
Australia 澳大利亚 (SA)
Austria 奥地利 (ON)
Bahrain 巴林 (BSMD)
Bangladesh 孟加拉国 (BSTI)
Barbados 巴巴多斯 (BNSI)
Belgium 比利时 (NBN)
Botswana 博茨瓦纳 (BOBS)
Brazil 巴西 (ABNT)
Bulgaria 保加利亚 (BDS)
Canada 加拿大 (SCC)
Chile 智利 (INN)
China 中国 (SAC)
Colombia 哥伦比亚 (ICONTEC)
CostaRica 哥斯达黎加 (INTECO)
Cuba 古巴 (NC)
Cyprus 塞浦路斯 (CYS)
Czech Republic 捷克共和国 (UNMZ)
Denmark 丹麦 (DS)
Ecuador 厄瓜多尔 (INEN)
Egypt 埃及 (EOS)
Ethiopia 埃塞俄比亚 (QSAE)
Finland 芬兰 (SFS)
France 法国 (AFNOR)
Germany 德国 (DIN)
Ghana 加纳 (GSB)
Grece 希腊 (ELOT)
Hong Kong, China 中国香港 (ITCHKSAR)（通讯成员）
Hungary 匈牙利 (MSZT)
Iceland 冰岛 (IST)
India 印度 (BIS)
Indonesia 印度尼西亚 (BSN)

Iran, Islamic Republic of Iran 伊朗 (ISIRI)
Iraq 伊拉克 (COSQC)
Ireland 爱尔兰 (NSAI)
Israel 以色列 (SI)
Italy 意大利 (UNI)
Jamaica 牙买加 (JBS)
Japan 日本 (JISC)
Jordan 约旦 (JISM)
Kazakhstan 哈萨克斯坦 (KAZMEMST)
Kenya 肯尼亚 (KEBS)
Korea, Democratic people's Republic of Korea 朝鲜民主主义人民共和国 (CSK)
Korea, Republic of Korea 韩国 (KATS)
Kuwait 科威特 (KOWSMD)
Luxembourg 卢森堡 (ILNAS)
Macau, China 中国澳门 (CPTTM)（通讯成员）
Malaysia 马来西亚 (DSM)
Malta 马耳他 (MSA)
Mauritius 毛里求斯 (MSB)
Mexico 墨西哥 (DGN)
Mongolia 蒙古 (MASM)
Moroco 摩洛哥 (SNIMA)
Netherlands 荷兰 (NEN)
New Zealand 新西兰 (SNZ)
Nigeria 尼日利亚 (SON)
Norway 挪威 (SN)
Oman 阿曼 (DGSM)
Pakistan 巴基斯坦 (PSQCA)
Panama 巴拿马 (COPANIT)
Philippines 菲律宾 (BPS)
Poland 波兰 (RKN)
Portugal 葡萄牙 (IPQ)
Romania 罗马尼亚 (ASRO)
Rusian Federation 俄罗斯联邦 (GOSTR)
Saudi Arabia 沙特阿拉伯 (SASO)

Singapore 新加坡 (SPRINGSG)
Slovakia 斯洛伐克 (SUTN)
Slovenia 斯洛文尼亚 (SIST)
South Africa 南非 (SABS)
Spain 西班牙 (AENOR)
Sri Lanka 斯里兰卡 (SLSI)
Sweden 瑞典 (SIS)
Switzerland 瑞士 (SNV)
Syrian Arab Republic 叙利亚 (SASMO)
Tanzania, United Republic of Tanzania 坦桑尼亚 (TBS)
Thailand 泰国 (TISI)
The Republic of Trinidad and Tobago 特立尼达和多巴哥 (TTBS)
Tunisia 突尼斯 (INNORPI)
Turkey 土耳其 (TSE)
USA 美国 (ANSI)
Ukraine 乌克兰 (DSSU)
United Arab Emirates 阿拉伯联合酋长国 (ESMA)
United Kingdom 英国 (BSI)
Uruguay 乌拉圭 (UNIT)
Uzbekistan 乌兹别克斯坦 (UZSTANDARD)
Venezuela 委内瑞拉（FONDONORMA）
Viet Nam 越南（STAMEQ）
Zimbabwe 津巴布韦（SAZ）

APPENDIX V（附录五）

ISO及有关TC、SC简介

　　ISO是国际标准化组织的英语简称，其全称是International Organization for Standardization或International Standard Organized。"ISO"源于希腊语"ISOS"，即"EQUAL"——平等的意思。国际标准化组织（ISO）是由各国标准化团体（ISO成员团体）组成的世界性的联合组织。制定国际标准工作通常由ISO的技术委员会（TC），或TC下设的分技术委员会（SC）完成。各成员团体若对某技术委员会确立的项目感兴趣，均有权参加该委员会的工作，与ISO保持联系的各国际组织（官方的和非官方的）也可参加有关工作。中国是ISO的常任理事国及积极成员国（Participating-member, P-member），代表中国的组织为中国国家标准化管理委员会（Standardization Administration of China, SAC）。

　　按照国际合作部的记载，截至2011年1月26日，ISO有727个技术委员会。ISO下和工程制图密切相关的技术委员会主要有TC10和TC145。

　　ISO/TC10是国际标准化组织第十技术委员会，它的名称是"技术产品文件"（Technical product documentation），成立于1946年，担负着对全世界各类工程技术文件进行标准化的工作。每1~2年就要召开一次ISO/TC10全会，基本上每年都有分委会会议或其下设的各种工作组会议，来研究有关技术文件的标准化。ISO/TC10有10个分技术委员会（SC），各分技术委员会按工作需要可设若干个工作组（WG）。分技术委员会及工作组按制、修订标准的需要而设立或撤销。

　　中国是1978年参加ISO/TC10的，是积极成员国（P-member）。与ISO/TC10对口的中国的技术委员会是国家标准化管理委员会第146分会（SAC/TC146），名称为"产品技术文件"，负责全国制造业有关的技术产品文件标准化工作。秘书处设在机械科学研究院。

　　ISO/TC145是国际标准化组织第145技术委员会，成立于1970年，它的名称是"图形符号"（Graphical symbols），主要任务是全面负责图形符号国际标准化工作，负责图形符号的协调工作和基础性与公用图形符号国际标准的制、修订工作。其分技术委员会，SC1负责公共信息图形符号；SC2负责安全识别、标志、形状、符号和颜色；SC3负责设备用图形符号。

　　中国是ISO/TC145的积极成员国（P-member）。与ISO/TC145对口的中国的技术委员会是国家标准化管理委员会第59分会（SAC/TC59），名称为"图形符号"，负责全国公用性图形符号和图形符号的基础性综合标准化工作。秘书处设在中国标准研究院。

APPENDIX Ⅵ（附录六）

IEC及有关TC、SC简介

　　IEC 是国际电工委员会的英语简称，其全称是 International Electrotechnical Commission。它成立于 1906 年，至今已有 100 多年历史，是世界上最早的国际性电工标准化机构，负责有关电气工程和电子工程领域中的国际标准化工作。中国于 1957 年加入，是积极成员国（P-member）。按照国际合作部的记载，截至 2011 年 1 月 26 日，IEC 有 174 个技术委员会。

　　国际电工委员会（IEC）让其第 3 技术委员会（IEC/TC3，英文名称：Information Structures, Documentation and Graphical Symbols，中文名称：信息结构，文件编制和图形符号）全面负责电气技术和相关领域中图形符号的标准化工作。IEC 与 ISO 在标准化方面保持密切合作的关系，这些工作往往是成立联合技术委员会（JTC）完成。

　　IEC/TC3 几个有关的分技术委员会担负着如下的工作：

（1）SC3A 负责简图用图形符号的标准化工作。

（2）SC3B 负责制定用于技术文件的总则：包括简图、表图和图表，说明书和项目号。

（3）SC3C 负责设备用图形符号的标准化工作。

（4）SC3D 负责数据库用数据系。

　　特定的工作可委托分技术委员会下设工作组。

　　中国是 IEC 的常任理事国及积极成员国（P-member）。与 IEC/TC3 对口的中国的技术委员会是国家标准化管理委员会第 27 分会（SAC/TC27），名称为"电气信息结构，文件编制和图形符号"，负责全国电气信息结构、电气图形符号、电气文件编制和电气制图等专业领域标准化工作。秘书处设在机械科学研究院。

APPENDIX Ⅶ（附录七）

中华人民共和国标准代号

标准名称	代号	标准名称	代号
国家标准	GB	粮食	LS
推荐性国家标准	GB/T	林业	LY
国家标准指导性技术文件	GB/Z	民用航空	MH
		煤炭	MT
强制性地方标准	DB + *（+* 为行政区划代码前两位数）	民政	MZ
		农业	NY
推荐性地方标准	DB + */T	轻工	QB
安全生产	AQ	汽车	QC
包装	BB	航天	QJ
船舶	CB	气象	QX
测绘	CH	国内贸易	SB
城镇建设	CJ	水产	SC
新闻出版	CY	石油化工	SH
档案	DA	电子	SJ
地震	DB	水利	SL
电力	DL	商检	SN
地质矿产	DZ	石油天然气	SY
核工业	EJ	海洋石油天然气	SY（10000 号以后）
纺织	FZ	铁道	TB
公共安全	GA	土地管理	TD
供销	GH	体育	TY
广播电影电视	GY	物资管理	WB
航空工业	HB	文化	WH
化工	HG	兵工民品	WJ
环境保护	HJ	卫生	WS
海关	HS	文物保护	WW
海洋	HY	稀土	XB
机械	JB	黑色冶金	YB
建材	JC	烟草	YC
建筑工业	JG	通信	YD
金融	JR	有色冶金（YS）	
交通	JT	医药	YY
教育	JY	邮政	YZ
旅游	LB	中医药	ZY
劳动和劳动安全	LD		

资料来源：http://www.sac.gov.cn/.

APPENDIX VIII（附录八）

工程制图ISO国际标准号及中国相对应的GB国家标准号

ISO 标准编号	ISO 标准名称	GB 标准号
128-20:1996	技术制图　图线	GB/T 17450—1998
128-21:1997	技术制图　CAD系统用图线的表示	GB/T 18686—2002
128-22:1999	技术制图　画法的一般原则　第22部分：引线的规范及应用	
128-23:1999	技术制图　画法的一般原则　第23部分：建筑制图的图线	GB/T 50104—2001
128-25:1999	技术制图　画法的一般原则　第25部分：造船制图的图线	
128-40:2001	技术制图　图样画法　剖视图和断面图	GB/T 17452—1998 GB/T 4458.6—2002
128-50:2001	技术制图　图样画法　剖面区域的表示方法	GB/T 17453—2005 GB/T 4458.6—2002
1101:2004	产品几何技术规范(GPS)　几何公差　形状、方向、位置和跳动公差标注	GB/T 1182—2018
1302:2002	产品几何技术规范(GPS)　技术产品文件中表面结构的表示法	GB/T 131—2006
2594:1972	建筑工程制图　投影法	GB/T 50001—2017
2595:1973	建筑工程制图　施工图的尺寸注法　制造尺寸和使用尺寸的表示	GB/T 50001—2001
3040:1990	技术制图　圆锥的尺寸和公差注法	GB/T 15754—1995
3098-0:1997	技术产品文件　字体　第0部分一般要求	GB/T 14691—1993
3098-2:2000	技术产品文件　字体　第2部分：拉丁字母、数字和符号	GB/T 14691—1993

续表

ISO标准编号	ISO标准名称	GB标准号
3098-3:2000	技术产品文件 字体 第3部分：希腊字母	GB/T 146914—1993
3098-4:1997	技术产品文件 字体 第4部分：拉丁字母的表示区别与特殊标识	GB/T 14691.4—2005
3098-5:1997	技术产品文件 字体 第5部分：拉丁字母、数字和符号的CAD字体	GB/T 14691—1993
3098-6:2000	技术产品文件 字体 第6部分：西里尔字母	GB/T 14691.6—2005
3511-1:1977	过程测量控制功能及仪表位置 符号表示法 第1部分：基本要求	GB/T 2625—1981
3511-2:1984	过程测量控制功能及仪表位置 符号表示法 第2部分：基本要求的扩充	GB/T 2625—1981
3511-3:1984	过程测量控制功能及仪表位置 符号表示法 第3部分：仪表接线图用详细符号	GB/T 2625—1981
3511-4:1985	过程测量控制功能及仪表位置 符号表示法 第4部分：过程计算机和接口及共同显示/控制功能的基本符号	GB/T 2625—1981
3753:1977	真空技术 图形符号	GB/T 3164—2007
3766:2003	建筑制图 钢筋混凝土的简化表示法	GB/T 50105—2010
3952-1:1981/Amd1:2002	机构运动简图 图形符号 第1部分：机构构件运动	GB/T 4460—2013
3952-2:1981	机构运动简图 图形符号 第2部分：运动副	GB/T 4460—2013
3952-3:1979	机构运动简图 图形符号 第3部分：机构及其组成部分的连接	GB/T 4460—2013
3952-4:1985	机构运动简图 图形符号 第4部分：多杆机构及其组成部分	GB/T 4460—2013
4067-2:1980	房屋建筑和土木工程制图 设备 第2部分：卫生器具的简化表示法	GB/T 50106—2010
4067-6:1985	技术制图 设备 第6部分：地下供水与排水系统的图形符号	GB/T 50106—2001

续表

ISO标准编号	ISO标准名称	GB标准号
4069:1977	房屋建筑和土木工程制图 视图与剖面图上区域的表示法 一般原则	GB/T 50001—2017
4157-1:1998	建筑制图 符号表示系统 第1部分：建筑物和建筑构件	GB/T 50105—2010
4172:1991	技术制图 建筑制图 预制结构的安装图	GB/T 50105—2010
5455:1979	技术制图 比例	GB/T 14690—1993
5456-1:1996	技术制图 投影法 第1部分：提要	GB/T 19692—2008
5456-2:1996	技术制图 投影法 第2部分：正投影表示法	GB/T 14692—2008
5457:1999	技术产品文件 图纸幅面和格式	GB/T 14689—2008
5459:1981	技术制图 几何公差 几何公差的基准和基准体系	GB/T 1958—2017
TR5460:1985	技术制图 几何公差 形状、方向、位置和跳动公差 检验原则和方法指南	GB 1958—2017
6414:1982	技术制图 玻璃器具表示法	GB/T 12213—1990
7200-1984	技术制图 标题栏	GB/T 10609.1—2008
7518:1983	技术制图 建筑制图 拆除和重建的简化表示法	GB/T 50104—2010
7573:1983	技术制图 明细栏	GB/T 10609.2—2009
8015:1985	公差原则	GB/T 4249—2018
8048:1984	技术制图 建筑制图 视图、剖面图与断面图的表示法	GB/T 50001—2017
8560:1986	技术制图 建筑制图 模数大小、图线和网格的表示法	GB/T 50001—2017
10578:1992	形状和位置公差 延伸公差带及其表示方法	GB/T 17773—1999

APPENDIX IX（附录九）

常见简化汉字、汉字简化偏旁及其繁体字对照

（右上角打 * 的简化字也用作简化偏旁）

A

爱* = 愛
碍 = 礙
肮 = 骯
袄 = 襖

B

坝 = 壩
罢* = 罷
板 = 闆
办 = 辦
帮 = 幫
宝 = 寶
报 = 報
贝* = 貝
备* = 備
笔* = 筆
币 = 幣
毕* = 畢
毙 = 斃
边* = 邊
标 = 標
表 = 錶
别 = 彆
宾* = 賓
卜 = 蔔
补 = 補

C

才 = 纔
参* = 參
蚕 = 蠶
仓* = 倉
层 = 層
搀 = 攙
谗 = 讒
馋 = 饞
缠 = 纏
产* = 產
忏 = 懺
长* = 長
尝* = 嘗
偿 = 償
厂 = 廠
车* = 車
彻 = 徹
尘 = 塵
衬 = 襯
称 = 稱
惩 = 懲
迟 = 遲
齿* = 齒
冲 = 衝
虫* = 蟲
丑 = 醜

出 = 齣
刍* = 芻
础 = 礎
处 = 處
触 = 觸
辞 = 辭
从* = 從
聪 = 聰
丛 = 叢
窜* = 竄

D

达* = 達
带 = 帶
单* = 單
担 = 擔
胆 = 膽
当 = 當
党 = 黨
导 = 導
灯 = 燈
邓 = 鄧
籴 = 糴
敌 = 敵
递 = 遞
点 = 點
电 = 電
淀 = 澱

迭 = 叠
东* = 東
冬 = 鼕
动* = 動
斗 = 鬥
独 = 獨
断* = 斷
队* = 隊
对* = 對
吨 = 噸
夺 = 奪
堕 = 墮

E

儿 = 兒
尔* = 爾

F

发* = 發
矾 = 礬
范 = 範
飞 = 飛
坟 = 墳
奋 = 奮
粪 = 糞
丰* = 豐
风* = 風
凤 = 鳳

肤 = 膚
妇 = 婦
复 = 複

G

盖 = 蓋
干 = 幹
赶 = 趕
冈* = 岡
个 = 個
巩 = 鞏
沟 = 溝
构 = 構
购 = 購
谷 = 穀
顾 = 顧
刮 = 颳
关 = 關
观 = 觀
广* = 廣
归* = 歸
龟* = 龜
柜 = 櫃
国* = 國
过* = 過

H

汉 = 漢
号 = 號
合 = 閤
轰 = 轟
后 = 後
胡 = 鬍
壶 = 壺
护 = 護
沪 = 滬

华* = 華
划 = 劃
画* = 畫
怀 = 懷
坏 = 壞
欢 = 歡
还 = 還
环 = 環
回 = 迴
汇* = 彙、匯
会* = 會
伙 = 夥
获 = 獲、穫

J

击 = 擊
鸡 = 雞
积 = 積
极 = 極
几* = 幾
际 = 際
继 = 繼
夹* = 夾
家 = 傢
价 = 價
戋* = 戔
歼 = 殲
艰 = 艱
监* = 監
拣 = 揀
茧 = 繭
硷 = 鹼
见* = 見
荐 = 薦
舰 = 艦
姜 = 薑
将* = 將

浆 = 漿
讲 = 講
奖 = 獎
桨 = 槳
酱 = 醬
胶 = 膠
阶 = 階
疖 = 癤
节* = 節
洁 = 潔
借 = 藉
仅 = 僅
尽* = 盡
进* = 進
惊 = 驚
竟 = 競
旧 = 舊
举* = 舉
剧 = 劇
惧 = 懼
据 = 據
卷 = 捲

K

开 = 開
壳* = 殼
克 = 剋
垦 = 墾
恳 = 懇
夸 = 誇
块 = 塊
亏 = 虧
困 = 睏

L

腊 = 臘

蜡 = 蠟
来* = 來
兰 = 蘭
拦 = 攔
栏 = 欄
烂 = 爛
乐* = 樂
垒 = 壘
类 = 類
累 = 纍
离* = 離
礼 = 禮
里 = 裏
历* = 曆、歷
丽* = 麗
隶 = 隸
帘 = 簾
怜 = 憐
联 = 聯
练 = 練
炼 = 煉
粮 = 糧
两* = 兩
辽 = 遼
疗 = 療
了 = 瞭
猎 = 獵
邻 = 鄰
临 = 臨
灵* = 靈
岭 = 嶺
刘 = 劉
龙 = 龍
娄* = 婁
卢 = 盧
庐 = 廬
芦 = 蘆
炉 = 爐

卤＊＝鹵、滷
虏＊＝虜
陆＝陸
录＊＝錄
驴＝驢
虑＊＝慮
乱＝亂
仑＊＝侖
罗＊＝羅

M

马＊＝馬
买＊＝買
麦＊＝麥
卖＊＝賣
么＝麼
霉＝黴
门＊＝門
蒙＝矇、濛
梦＝夢
面＝麵
庙＝廟
灭＝滅
蔑＝衊
亩＝畝

N

难＊＝難
恼＝惱
脑＝腦
拟＝擬
酿＝釀
鸟＊＝鳥
聂＊＝聶
宁＊＝寧
农＊＝農

疟＝瘧

P

盘＝盤
辟＝闢
凭＝憑
苹＝蘋
仆＝僕
扑＝撲
朴＝樸

Q

齐＊＝齊
岂＊＝豈
启＝啓
气＊＝氣
千＝韆
迁＊＝遷
牵＝牽
纤＝纖、縴
佥＊＝僉
签＝簽
乔＊＝喬
窍＝竅
窃＝竊
亲＊＝親
寝＝寢
庆＝慶
穷＊＝窮
琼＝瓊
秋＝鞦
区＊＝區
曲＝麯
权＝權
劝＝勸
确＝確

R

让＝讓
扰＝擾
热＝熱
认＝認

S

洒＝灑
伞＝傘
丧＝喪
扫＝掃
涩＝澀
啬＊＝嗇
杀＝殺
晒＝曬
伤＝傷
舍＝捨
沈＝瀋
审＊＝審
声＝聲
胜＝勝
圣＊＝聖
师＊＝師
湿＝濕
时＊＝時
实＝實
势＝勢
适＝適
寿＊＝壽
兽＝獸
书＝書
属＝屬
术＝術
树＝樹
帅＝帥
双＊＝雙

松＝鬆
苏＝蘇
肃＊＝肅
虽＝雖
随＝隨
岁＊＝歲
孙＊＝孫

T

台＝臺、檯
态＝態
坛＝壇、罎
叹＝歎
誊＝謄
体＝體
条＊＝條
粜＝糶
铁＝鐵
厅＝廳
听＝聽
头＝頭
图＝圖
涂＝塗
团＝團
椭＝橢

W

洼＝窪
袜＝襪
万＊＝萬
网＝網
为＊＝爲
韦＊＝韋
卫＝衛
稳＝穩
乌＊＝烏

无* = 無	严* = 嚴	园 = 園	烛 = 燭
务 = 務	盐 = 鹽	远 = 遠	筑 = 築
雾 = 霧	厌* = 厭	愿 = 願	专* = 專
	阳 = 陽	钥 = 鑰	妆 = 妝
X	养 = 養	跃 = 躍	庄 = 莊
	痒 = 癢	云* = 雲	桩 = 樁
牺 = 犧	样 = 樣	运 = 運	装 = 裝
习 = 習	尧* = 堯	酝 = 醞	壮 = 壯
戏 = 戲	钥 = 鑰		状 = 狀
系 = 係、繫	药 = 藥	**Z**	准 = 準
虾 = 蝦	爷 = 爺		浊 = 濁
吓 = 嚇	业* = 業	杂 = 雜	总 = 總
纤 = 纖	叶 = 葉	赃 = 髒	钻 = 鑽
咸 = 鹹	页* = 頁	脏 = 臟	
显 = 顯	医 = 醫	凿 = 鑿	**简化偏旁及其**
县 = 縣	义* = 義	枣 = 棗	**繁体**
宪 = 憲	亿 = 億	灶 = 竈	
献* = 獻	忆 = 憶	斋 = 齋	讠[言]
乡* = 鄉	艺 = 藝	毡 = 氈	饣[食]
响 = 響	阴* = 陰	战 = 戰	多 [昜]
向 = 嚮	隐* = 隱	赵 = 趙	纟 [糸]
象 = 像	应 = 應	折 = 摺	収 [取]
协 = 協	佣 = 傭	这 = 這	艹 [艸]
胁 = 脅	拥 = 擁	征 = 徵	収 [臨]
写* = 寫	痈 = 癰	证 = 證	只 [戠]
衷 = 褻	踊 = 踴	郑* = 鄭	钅 [金]
衅 = 釁	优 = 優	症 = 癥	𰀉 [學]
兴 = 興	忧 = 憂	执* = 執	𥁕 [睪]
须 = 鬚	犹* = 猶	只 = 只、隻、祇	圣 [𦔮]
悬 = 懸	邮 = 郵	质* = 質	亦 [繼]
旋 = 鏇	余 = 餘	制 = 製	呙 [咼]
选 = 選	鱼 = 魚	致 = 緻	
寻* = 尋	与* = 與	钟 = 鍾、鐘	
	吁 = 籲	肿 = 腫	
Y	郁 = 鬱	种 = 種	
	御 = 禦	众 = 眾	
压 = 壓	誉 = 譽	昼 = 晝	
亚* = 亞	渊 = 淵	朱 = 硃	

APPENDIX X（附录十）

希腊字母

希 腊 字 母	英 文 读 音
A α ∝*	alpha /ˈælfə/
B β	beta /ˈbiːtə, ˈbeitə/
Γ γ	gamma /ˈgæmə/
Δ δ ∂*	delta /ˈdeltə/
E ε	epsilon /ˈepsilən, epˈsailən/
Z ζ	zeta /ˈziːtə/
H η	eta /ˈiːtə, ˈeitə/
Θ θ ϑ*	theta /ˈθiːtə/
I ι	iota /aiˈəutə/
K κ	kappa /ˈkæpə/
Λ λ	lambda /ˈlæmdə/
M μ	mu /mjuː/
N ν	nu /njuː/
Ξ ξ	xi /ksai, sai, zai/
O o	omicron /əuˈmaikrɒn/
Π π	pi /pai/
P ρ	rho /rəu/
Σ σ s**	sigma /ˈsigmə/
T τ	tau /tau/
γ υ	upsilon /ˈjuːpsilən, juːpˈsailən/
Φ ϕ φ*	phi /fai/
X χ	chi /kai/
Ψ ψ	psi /psai/
Ω ω	omega /ˈəumigə/

注：*旧体字母，**后缀字母。